STATISTICAL METHODS

HERBERT ARKIN/RAYMOND R. COLTON

FIFTH EDITION

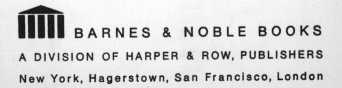

BARNES & NOBLE BOOKS

A DIVISION OF HARPER & ROW, PUBLISHERS

New York, Hagerstown, San Francisco, London

Preface

The increasingly widespread use of statistical methods in a broadening spectrum of disciplines has necessitated a broader acquaintance with these techniques.

The number of courses in this field as well as the literature of the subject has mushroomed. However, the basic books are usually lengthy and complex, frequently treating the subject as a branch of mathematics rather than an applied tool.

It is the object of this book to provide a concise and simple treatment of the basic statistical methods required in a wide variety of fields. The treatment has been simplified and does not assume a knowledge of higher mathematics.

On the other hand, derivations for most important formulas are provided and are included in the Technical Appendices for the perusal of those who desire to study them.

Although the emphasis is on application, the underlying theory, where appropriate, is briefly explained.

This edition has been generally revised and updated with additional material and new chapters on "Chi Square and Tests of Hypotheses" and "Analysis of Variance." In addition the numerous changes and additions since the first edition resulted in a complete resetting of the entire volume, greatly improving the legibility of the text.

About the Authors

Herbert Arkin received the degrees of B.S.S. and M.B.A. from the City University of New York and the degree of Ph.D. from Columbia University. He holds the rank of Professor at the Bernard M. Baruch College of the City University of New York where he is chairman of the Statistics Department. He has been on the teaching staff of the City University since 1931, and has also taught in the Graduate School of Economics at New York University. He has been a statistical and quality control consultant to a wide variety of business concerns; is co-author with R. R. Colton of *Tables for Statisticians* in the Barnes & Noble Outline Series; and is the author of several other books including *Handbook of Sampling for Auditing and Accounting*.

Raymond R. Colton received the degree of B.S.S. from the City University of New York and the degrees of M.A. and Ph.D. from Columbia University. He has taught at the School of Business of the City University since 1931 and is now Professor of Management. He has contributed articles to leading professional magazines and has served as a consultant to many firms. He is the author of *Industrial Purchasing: Principles and Practices* and *Savings Bank and Trust Investments in Gas and Electrical Securities*, and co-author of *Graphs, Tables for Statisticians, Modern Economics*, and *Research and Report Writing for Business and Economics*.

Table of Contents

Chapter I

Introduction

DEFINITION OF STATISTICAL METHOD
A statistical method is any technique used to obtain, analyze, and present numerical data.

ELEMENTS OF STATISTICAL TECHNIQUE
The elements of statistical technique include the
1. collection and assembling of data
2. classification and summarization of data
3. presentation of data in
   a. textual form
   b. tabular form
   c. graphic form
4. analysis of data

CHARACTERISTICS AND LIMITATIONS OF STATISTICAL METHODS
1. Statistical methods are the only means for handling large masses of numerical data.
2. Statistical techniques apply only to data which are reducible to quantitative form.
3. Statistical techniques are *objective*. The results, however, are affected by the necessarily *subjective* interpretation.
4. Statistical techniques are the same for the social as for the physical sciences; e.g., for economics, education, sociology, and psychology as contrasted with biology, chemistry, and astronomy. Method and theory apply alike to these divergent fields.

1

STATISTICAL DISTRIBUTIONS (SERIES)

Before analyzing numerical data, it is necessary to arrange them systematically. The data may be arranged in a number of different ways. Technically such an arrangement is called a **distribution** or a **series**. Different types of distributions are mentioned below.

When data are grouped according to	The resulting series is called a
1. magnitude	frequency distribution
2. time of occurrence	time series
3. geographic location	spatial distribution

In addition there are a number of special types of distribution in which the data may be arranged by *kind* or by *degree*.

THE FREQUENCY DISTRIBUTION
Definition

The frequency distribution is an arrangement of numerical data according to size or magnitude. For each value-grouping (class interval), the number of observations of that magnitude are indicated.

Construction

A frequency distribution is constructed in the following manner.

1. Using the **range** of the data (the interval between the highest and the lowest figure) as a guide, the data are divided into a number

Table 1.1-Score Sheet
City Tax Rate of "True" Valuation in 261 Cities in the United States

Rate Per Thousand Dollars class intervals	tally	Total Number of Cities frequency
4– 7.99	Ⅱ/	5
8–11.99	Ⅱ/ Ⅱ/ Ⅱ/	15
12–15.99	Ⅱ/ Ⅱ/ Ⅱ/ Ⅱ/ Ⅱ/ Ⅱ/ Ⅱ/ Ⅱ/ Ⅱ/ /	46
16–19.99	Ⅱ/ Ⅱ/ Ⅱ/ Ⅱ/ Ⅱ/ Ⅱ/ Ⅱ/ Ⅱ/ Ⅱ/ Ⅱ/ Ⅱ/ Ⅱ/ Ⅱ/ ///	68
20–23.99	Ⅱ/ Ⅱ/ Ⅱ/ Ⅱ/ Ⅱ/ Ⅱ/ Ⅱ/ Ⅱ/ Ⅱ/ Ⅱ/ Ⅱ/ ///	58
24–27.99	Ⅱ/ Ⅱ/ Ⅱ/ Ⅱ/ Ⅱ/ Ⅱ/ //	32
28–31.99	Ⅱ/ Ⅱ/ Ⅱ/ Ⅱ///	22
32–35.99	Ⅱ/ Ⅱ/	10
36–39.99	//	2
40–43.99	//	2
44–47.99		0
48–51.99	/	1
		261

Source: *United States Department of Commerce, Financial Statistics of Cities.*

of conveniently sized groups. The groups are called **class intervals**. (See Table 1.1, Column 1).[1]

The size of the class interval is dependent upon the number of values to be included in the distribution. The range of the values is determined by finding the difference between the highest and lowest values; and this difference is divided by the number of class intervals desired. The resulting size is rounded off.

Few class intervals are used when a limited number of values are included, and a large number of class intervals are used when the distribution is to be compiled from many values. The most efficient number of class intervals usually lies between ten and twenty.

Other requirements for the determination of the class interval are[2]

a. The class intervals should not overlap; 0–4.9, 5–9.9 etc., should be used in preference to 0–5, 5–10, etc.

b. When the values tabulated coincide with the integers or with selected values, these values, ordinarily, should constitute the midpoints of the groups.

c. When possible the class intervals should be of uniform size.

2. The groups are then placed in a column with the lowest class interval at the top and the rest of the class intervals following according to size.

3. The data are then scored. Each item is checked once next to the class interval into which it falls. (See tally, Table 1.1.)[3]

---

[1] As a preliminary step the raw data may be arranged according to size. The series is then called an **array**.

[2] A formula sometimes used to determine the size of the class interval is

$$C = \frac{\text{range}}{1 + (3.322 \log N)}$$

where

C = class interval size or width,

range = difference between highest and lowest value in data to be tabulated,

N = total number of observations to be tabulated.

This is sometimes referred to as the Sturges formula.

[3] In scoring data an efficient procedure is to connect the first and fourth score by the fifth. Using this procedure totals are obtained merely by adding the resulting units, multiplying by five, and adding the odd scores. (See Table 1.1.) A tally of this type saves time in counting frequencies and also eliminates possible inaccuracies.

Graphic Presentation

If two lines are drawn perpendicular to one another and are divided according to a scale of values, given data may be represented by reference to the scale. The horizontal line is known as the **X-axis** and the vertical line as the **Y-axis**. If the values for a point are given, the point may be located on the graph. For example, in Figure 1.1 the point $x = 2$, $y = 3$ may be located at the point marked a.

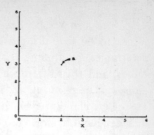

Fig. 1.1—Location of plotted point $x = 2$, $y = 3$.

If the two axes are marked with a scale in terms of the units of the given data, the frequency distribution may be presented graphically.

1. The class-interval grouping, which will be termed the **independent variable**, is placed on the X-axis (horizontal) and the frequency or dependent variable is placed on the Y-axis (vertical).[1]

2. The number of cases falling within a class interval is plotted at the midpoint of that class interval at the level indicated for that number of cases by the scale on the Y-axis.[2]

3. When connected, the plotted points form a **frequency polygon**. (See Figure 1.2.)

4. Rectangles may be constructed by using as the width the size of the class interval and as the height the frequency in each class interval. When the class intervals are equal, these rectangles form a **histogram** (also known as a **rectangular frequency polygon**).[3]

Types

The more usual types of frequency distributions are given below. In addition to these there are a number of unusual types such as the bimodal curve (two peaks), and the "j" and inverted "j" curves (curves shaped to resemble a "j").

1. **Symmetrical distribution.** The "normal" curve is the best known example of a symmetrical distribution. (See Figure 1.3.) This type of distribution is discussed in detail in Chapter XII.

---

[1] The distance along the X-axis is called the **abscissa** while that along the Y-axis is called the **ordinate**.

[2] A cross sectional background or ruling may be used to aid in this work.

[3] This instruction for drawing a histogram needs modification when class intervals vary in size.

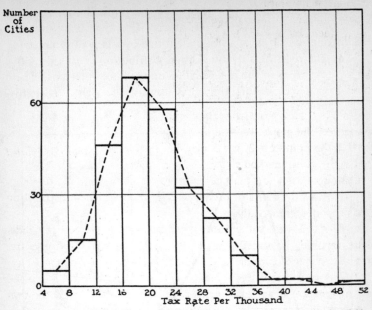

Fig. 1.2–Distribution of tax rates on "true" valuation for 261 cities in the United States.

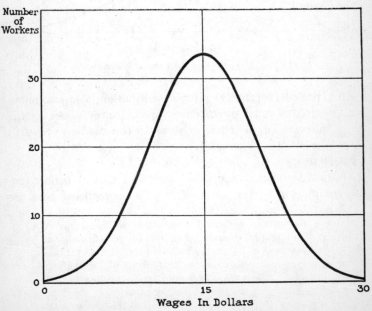

Fig. 1.3–Hypothetical 'normal" distribution.

2. **Skewed distribution.** Most frequency distributions extend
farther in one direction than in the other. This type is known as a
skewed distribution. It is, of course, identified by a lack of symmetry.

> a. The **right (positively) skewed distribution** is caused by the
> extremes in the higher values on the horizontal axis. These
> extremes distort the curve toward the right. (See Fig-
> ure 1.4.)

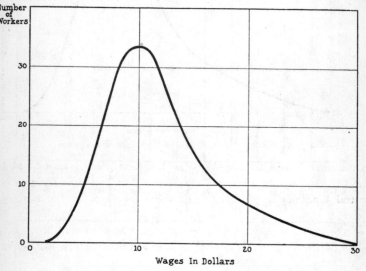

Fig. 1.4—Hypothetical right skewed distribution.

> b. The **left (negatively) skewed distribution**, a less common
> type, is caused by extremes in the lower values on the
> horizontal axis. These extremes distort the curve towards
> the left. (See Figure 1.5.)

Characteristics[1]

1. Natural, economic, and sociological data show a distinct tendency to group about a given point. This grouping tendency gives rise

---

[1]Data in the form of numerical observations are commonly divided into two types, **continuous** and **discrete** (non-continuous). In
the continuous series it is possible to have every size of the variables between given limits, i.e., in the distribution of weights, or of
ages of individuals, where any size is possible. In the discrete series
only limited gradations are possible, as for instance, in the distribution of numbers of pupils in public schools in the United States.
Fractions of a unit cannot appear in this series.

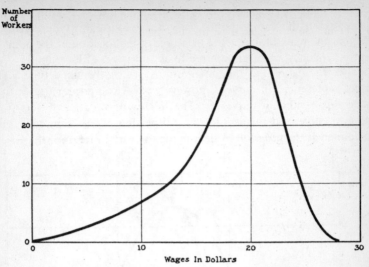

Fig. 1.5—Hypothetical left skewed distribution.

to the peak which always occurs in frequency distributions. In Figure 1.6, the two distributions are identical except that the **points of central tendency** are located at different positions on the horizontal scale.

Since the location of the peak (or, the point of central tendency) is one of the frequency-distribution characteristics which may be mea-

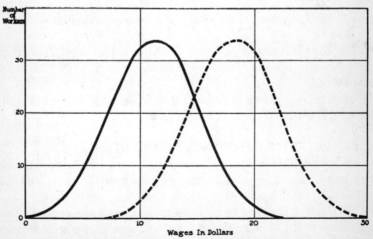

Fig. 1.6—Hypothetical distributions showing different points of central tendency.

sured (the **average** is its measure), and since the group of values tends to cluster about this central point, it is possible to use this typical value to describe the mass of data. The measures of the location of the point of central tendency are often referred to as "measures of location."

2. In Figure 1.7 the distributions are of the same type. The values represented in curve *a,* however, vary to a greater degree than those represented in curve *b.* The *degree* of variation differs from curve to curve, and this is known as **dispersion**. Dispersion, then, may be defined as the *variation* in size among the various items of the series.

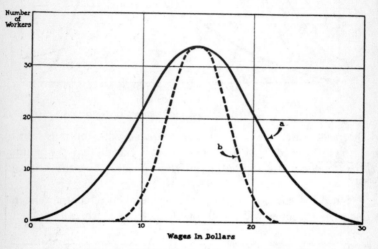

Fig. 1.7–Hypothetical distributions showing different dispersions.

3. The distributions in Figure 1.8 differ in that curve *a* is symmetric while curve *b* is not. Lack of symmetry (as in curve *b*) is known as **skewness**.

4. The two curves in Figure 1.9 differ in their degree of "peakedness." This characteristic is known as **kurtosis**.

CUMULATIVE FREQUENCY DISTRIBUTION

A distribution in which the frequencies are cumulated is known as an **ogive**. Examples of the ogive are shown in Tables 1.2 and 1.3, p. 10.

In Table 1.2 a **"less than"** ogive is used. In some cases, an "and over" ogive might be more useful. An **"and over"** ogive is constructed by cumulating the items on an "and over" basis as in Table 1.3.

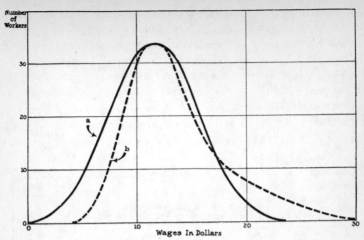

Fig. 1.8—Hypothetical distribution showing skewness (curve b).

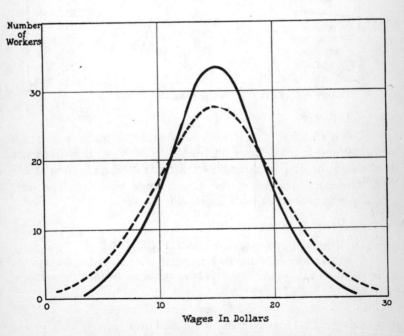

Fig. 1.9—Hypothetical distributions showing difference in kurtosis.

Table 1.2–"Less Than" Ogive
Distribution of Wholesale Sales by Size of Firm
for the United States.

Size of Firm (Thousands of Dollars of Sales)	Number of Firms
less than 100	51,563
less than 200	90,108
less than 300	113,669
less than 500	140,309
less than 1,000	168,921
less than 2,000	186,537
less than 5,000	196,750
less than 10,000	199,404
less than 100,000	200,777

Source: United States Department of Commerce.

Table 1.3–"And Over" Ogive
Distribution of Farms in United States by Size, 1959

Size in Acres	Number of Farms
0 and over	3,703,894
10 and over	3,463,161
50 and over	2,651,959
100 and over	1,994,274
180 and over	1,222,033
260 and over	807,660
500 and over	336,264
1000 and over	136,299

Source: United States Department of Commerce.

ANALYSIS

Unless a mass of data is grouped in some systematic manner, it is unwieldly and in many cases impossible to analyze. In a frequency distribution the data are arranged so that, with the application of furthur techniques, analysis of the data is made possible. The mere grouping of data does not itself accomplish an analysis.

ADDITIONAL BIBLIOGRAPHY

DUNCAN, A. J., *Quality Control and Industrial Statistics*, pp. 35-44, Richard D. Irwin, Inc. Homewood, Ill., 3rd ed., 1965.

YULE, G. U., & KENDALL, M. G., *An Introduction to the Theory of Statistics*, Chapter IV, Hafner Publishing Co., New York, 1958.

Chapter II

Analysis of the Frequency Distribution—Central Tendency—Arithmetic Mean

MEASURES OF CENTRAL TENDENCY—AVERAGES
An **average** is a typical value which is intended to sum up or describe the mass of data. It also serves as a basis for measuring or evaluating extreme or unusual values. The average is a measure of the location of the point of central tendency.

KINDS OF AVERAGES
The most important averages are
1. the arithmetic mean   (discussed in Chapter II)
2. the median
3. the mode
4. the geometric mean   (discussed in Chapter III)
5. the quadratic mean
6. the harmonic mean

THE ARITHMETIC MEAN
Definition
Due to ease of computation and long usage, the arithmetic mean is the best known and most commonly used of all the averages. When the word *mean* is used without qualification, the reference is to the arithmetic mean.

11

Calculation—Ungrouped Data

Long Method

The arithmetic mean of a small group of individual values may be obtained by dividing the sum of the values by the number of items used.

The computation of the arithmetic mean is expressed in formula form as[1]

$$\overline{X} = \frac{\Sigma(X)}{N}$$

where

$\overline{X}$ = arithmetic mean
Σ = symbol meaning "sum of"[2]
X = data expressed as individual values
N = number of items

Short Method

A simpler method may be devised by an examination of the characteristics of the arithmetic mean. If the mean is computed for a number of individual items (see Table 2.1) and the deviation (distance) of each of the items from the mean is obtained, these deviations will total up to zero.[3]

If, however, some point other than the true arithmetic mean is selected, the sum of the deviations from this point will not be zero. In the same series, for instance, 90% may be selected as an arbitrary starting point (known technically as the **guessed mean**, and identified by the symbol Z).[4]

If the *average* deviation of each item from the guessed mean is

---

[1] This formula is also given in various textbooks, using different symbols, as

$$M = \frac{1}{N}\Sigma(X) \text{ or } M = \frac{\Sigma(m)}{N}$$

[2] The symbol Σ is a Greek capital letter sigma.
[3] See Technical Appendix I for mathematical proof.
[4] This symbol is also given in various textbooks as

$$\overline{X}_d, a, \text{ or } A$$

Table 2.1–Deviations from the Mean
Grades of Ten Students on an Examination in Arithmetic

Student number	Grade (Per cent)	Deviation from mean[1] (x)
1	95%	15%
2	92	12
3	90	10
4	86	6
5	86	6
6	80	0
7	75	−5
8	72	−8
9	64	−16
10	60	−20
	800%	0

$$\text{Mean } (\overline{X}) = \frac{800\%}{10} = 80\%$$

obtained, and if this value is added to the arbitrary starting point $(\overline{Z})$, the result will be the arithmetic mean.[2]

$$\frac{\Sigma(d)}{N} = \frac{-100\%}{10} = -10\% \text{ (average deviation)}$$

Table 2.2–Deviations from the Guessed Mean

Student number	Grade (Per cent)	Deviation from guessed mean (d)
1	95%	+5%
2	92	+2
3	90	0
4	86	−4
5	86	−4
6	80	−10
7	75	−15
8	72	−18
9	64	−26
10	60	−30
	800%	−100%

---

[1] The letter x is assigned as the symbol for the deviation from the arithmetic mean $(X - \overline{X})$.

[2] See Technical Appendix II for mathematical proof.

$$\overline{X} = \overline{Z} + \frac{\Sigma(d)}{N} = 90\% + \frac{(-100\%)}{10} = 80\%$$

where

$\overline{Z}$ = guessed mean
d = deviation of each value from guessed mean
N = number of cases

Calculation—Grouped Data

Where the arithmetic mean of a considerable number of items is to be computed by "hand," the method outlined above is generally too laborious and too subject to error. With increasing numbers of items, the simple problem of addition can become difficult to the point of physical impossibility. For instance, if the arithmetic mean is to be applied to data containing forty or fifty thousand items, correct addition is next to impossible even with the aid of an adding machine (though not, of course, with electronic data processing equipment).

A more convenient and efficient method of "hand" calculation is to group the data into the form of a frequency distribution, and then compute the arithmetic mean for the distribution.

Long Method

Since no knowledge is available of the actual distribution of the cases within each class interval, it may be assumed that the cases are distributed evenly between the limits of the group.[1] On this assumption the average value for all values in the group would be equal to the midpoint of the group. Thus, the total value for each group may be obtained by multiplying the midpoint of the group by the number of cases in the group.

Thus for the frequency distribution in Table 2.3 the midpoint of the first class interval (28%) is multiplied by the frequency indicated for that group (3) in order to obtain the total value for all cases in the class interval. The products (Table 2.3, column 4) are then added to obtain the total value of all cases in the frequency distribution. The sum is divided by the number of cases (N) to obtain the arithmetic mean.

---

[1] If the class interval is not too large and a sufficient number of cases are available, variation from the assumption will be small and an accurate result will be obtained.

Table 2.3—Computation of Arithmetic Mean by Long Method
Percent of Population Completing High School Education
By County, Middle Atlantic States

Percent Completing High School		Number of Counties	
class interval	midpoint (M.P.)	frequency (f)	frequency X midpoint (f X M.P.)
26–29.9%	28%	3	84
30–33.9	32	29	928
34–37.9	36	39	1404
38–41.9	40	41	1640
42–45.9	44	19	836
46–49.9	48	12	576
50–53.9	52	4	208
54–57.9	56	3	168
		150	5844

Source: U.S. Department of Commerce.

This method is expressed by the following formula:

$$\bar{X} = \frac{\Sigma(f \times M.P.)}{N} = \frac{5844}{150} = 38.96\%$$

This method is known as "the long method" because complex calculations may result when the frequencies and the midpoint values are large.

Short Methods
a. Unit Deviation Method

The second technique used with ungrouped data may readily be applied to grouped data. For the distribution in Table 2.4 an arbitrary starting point (guessed mean) may be selected. Though any value may be taken, the midpoint of one of the class intervals is usually the most convenient choice. For Table 2.4, the midpoint of the third class interval (5.00) may be used as the guessed mean.

Since it has been assumed that the midpoint of each class interval is the average value of all items in that interval, the difference (d) between the midpoint of each group and the guessed mean will represent the *average deviation* of the items in the group from the assumed mean. To obtain the total deviation for all items in the class interval it is necessary to multiply this deviation (d) by the frequency of the group (f). To obtain the total deviation from the guessed mean, this value is to-

Table 2.4–Computation of Arithmetic Mean–Short Unit-Deviation Method

Ratio of Current Assets to Current Liabilities for 221 Industrial Corporations in the United States

Ratio class interval	midpoint (M.P.)	Number of Companies frequency (f)	deviation (d)	frequency × deviation (fd)
.0- 1.99	1	11	-4	-44
2- 3.99	3	53	-2	-106
4- 5.99	5	47	0	0
6- 7.99	7	37	2	74
8- 9.99	9	21	4	84
10-11.99	11	16	6	96
12-13.99	13	13	8	104
14-15.99	15	8	10	80
16-17.99	17	10	12	120
18-19.99	19	1	14	14
20-21.99	21	2	16	32
22-23.99	23	1	18	18
24-25.99	25	0	20	0
26-27.99	27	1	22	22
		221		494

Source: Moody's Investors Service, Moody's Industrials.

taled for all class intervals. Then this total is divided by N to obtain the average deviation about the guessed mean. Expressed mathematically for Table 2.4, this becomes

$$\frac{\Sigma(fd)}{N} = \frac{494}{221} = 2.24$$

The above value is added to the arbitrary starting point (guessed mean) to obtain the true arithmetic mean.[1]

$$\overline{X} = \overline{Z} + \frac{\Sigma(fd)}{N}$$

$$\overline{X} = 5.00 + \frac{494}{221} = 7.24$$

---

[1] See Technical Appendix II for mathematical proof.

where

$\overline{Z}$ = guessed mean
f = frequency of each class interval
d = deviation of midpoint of each group from guessed mean
N = total number of cases

b. Group-Deviation Method

The computation of the arithmetic mean from a frequency distribution may be further simplified after consideration of the characteristics of such a distribution.

If the distribution has class intervals of a uniform size,[1] it will be noted that the deviation of the midpoint of one group from the next is constant and equal to the size of the class interval. In the distribution shown in Table 2.4 there is a constant difference of 2 between the midpoints, and this is equal to the size of each class interval—for example 0 to 1.99....

The deviations of any one group from any other group may then be measured in terms of class intervals. In the distribution on p. 18 (Table 2.5) the midpoint of the third class interval (5.00) has again been selected as the arbitrary or guessed mean $(\overline{Z})$. The midpoint of the first class interval deviates from the guessed mean by the amount of −4 which is the same as −2 class intervals. The deviation column is expressed in terms of class intervals rather than in the original units of the data. The resulting values in the deviation column are numerically smaller and thus the computation is simpler.

The computation is then carried out in the same manner as in the previous method, and the result is an average deviation about the guessed mean, $\dfrac{\Sigma \ (fd')}{N}$, now in terms of class intervals. To convert the value in terms of class intervals to a value in the original terms multiply it by the size of the class interval. The result may then be added to the guessed mean to obtain the arithmetic mean.[2]

---

[1] Wherever possible this should be the rule in compiling such a distribution.

[2] This formula is also variously given by different texts as

$$M = A + \frac{1}{N} \Sigma(f) \, C, \quad A = E + \frac{\Sigma f (V - E)}{n} C,$$

$$M = M' + c, \quad \text{or} \quad \overline{X} = \overline{X}_d + \left(\frac{\Sigma (fd)}{N} \right) i$$

Table 2.5—Computation of Arithmetic Mean—
Short Group-Deviation Method

Ratio of Current Assets to Current Liabilities for 221 Industrial
Corporations in the United States

Ratio		Number of Companies		
class interval	midpoint (M. P.)	frequency (f)	deviation (d')	(fd')
0- 1.99	1	11	-2	-22
2- 3.99	3	53	-1	-53
4- 5.99	5	47	0	0
6- 7.99	7	37	1	37
8- 9.99	9	21	2	42
10-11.99	11	16	3	48
12-13.99	13	13	4	52
14-15.99	15	8	5	40
16-17.99	17	10	6	60
18-19.99	19	1	7	7
20-21.99	21	2	8	16
22-23.99	23	1	9	9
24-25.99	25	0	10	0
26-27.99	27	1	11	11
		221		247

Source: *Moody's Investors Service, Moody's Industrials.*

$$\overline{X} = \overline{Z} + \frac{\Sigma(fd')}{N} C = 5 + \frac{247}{221}(2) = 7.24$$

where

$\overline{Z}$ = guessed mean
f = frequency
d' = deviation from arbitrary origin in terms of class intervals[1]
N = total number of cases
C = size of class interval

Characteristics

1. The value of the arithmetic mean is determined by every item in the distribution. It is a calculated average.
2. It is greatly affected by extreme values.
3. The sum of the deviations about the arithmetic mean is zero.

---

[1] The symbol d' (rather than d) is used to indicate that the computations are in class intervals.

4. The sum of the squares of the deviations from the arithmetic mean is less than those computed about any other point.[1]

5. Its standard error (see Chapter XIV) is less than that of the median.

6. In every case it has a determinate value.

7. The sum of the means equals the mean of the sums, and the difference between the means equals the mean difference.

$$\overline{X}_{1+2} = \overline{X}_1 + \overline{X}_2$$

$$\overline{X}_{1-2} = \overline{X}_1 - \overline{X}_2$$

Advantages

1. The arithmetic mean is the most commonly used, easily understood, and generally recognized average.

2. Its computation is relatively simple.

3. Only total values and the number of items are necessary for its computation.

4. It may be treated algebraically. For example, where averages for sub-groups are available they in turn may be averaged in order to obtain an average for the whole group. It is necessary however, that there be an equal number of items in each sub-group. If the number of items varies among sub-groups, a *weighted average* should be taken.

Disadvantages

1. The arithmetic mean may be greatly distorted by extreme values, and therefore it may not be a typical value.

2. The arithmetic mean cannot be computed from a distribution containing "**open ended**" **class intervals**, that is, when the items are grouped in "**and over**" or "**and under**" class intervals.

Weighted Averages

Given the arithmetic mean of two or more groups of data, it is possible to combine these averages to obtain the over-all arithmetic mean without reading all the individual values.

This calculation is accomplished by means of a weighted average. In general, a weighted arithmetic mean is an average in which the values to be averaged are given varying degrees of importance. In the case of combining averages of several groups of data, this is accomplished by using the number of values in each group as a weight.

$$\overline{X} = \frac{N_1 \overline{X}_1 + N_2 \overline{X}_2 + \dots N_n \overline{X}_n}{N_1 + N_2 + \dots N_n} = \frac{\Sigma(N_i \overline{X}_i)}{N}$$

---

[1] See Technical Appendix X for proof.

where

N_i = the number of values used to compute the arithmetic mean for group i

$\overline{X_i}$ = the arithmetic mean of the values in group i

$N = (N_1 + N_2 + \ldots N_n)$, the total number of values in all groups combined

Thus, if examination grades for two groups of students are such that the first group of 100 students has an arithmetic mean of 70 and the second group of 50 students has an average of 80, then the combined average is

$$\overline{X} = \frac{100(70) + 50(80)}{100 + 50} = \frac{11,000}{150} = 73.33$$

Weighted averages are used for a variety of purposes and, in some instances, the weight may be a value other than the number of items in the group. (See Chapter XVI.) In general, any weighted average may be computed from the following formula.

$$\overline{X} = \frac{W_1 \overline{X}_1 + W_2 \overline{X}_2 + \ldots W_n \overline{X}_n}{N_1 + N_2 + \ldots N_n} = \frac{\Sigma (W_i \overline{X}_i)}{N}$$

where

W_i = a measure of the importance of the average of group i, i.e., the weight for each group

$\overline{X}i$ = the arithmetic mean of group i

$N = (N_1 + N_2 + \ldots N_n)$, the number of values in all groups

ADDITIONAL BIBLIOGRAPHY

DUNCAN, A. J., *Quality Control and Industrial Statistics*, pp. 38–41, Richard D. Irwin, Inc., Homewood, Ill., 1959.

YULE, G. U., & KENDALL, M. G., *An Introduction to the Theory of Statistics*, pp. 102–111, Hafner Publishing Co., New York, 1950.

ZIZEK, FRANK, *Statistical Averages*, pp. 92–127; 138–193. Henry Holt & Co., New York, 1930.

Chapter III

Analysis of the Frequency Distribution—Central Tendency (*continued*)

THE MEDIAN

Definition

The median is the value of the middle item when the items are arranged according to size. If there is an even number of items, the median is taken as the arithmetic mean of the values of the two central items.

The median is an average of position while the arithmetic mean is a calculated average.

Calculation—Ungrouped Data

The median is computed from ungrouped data as follows.

1. Arrange the items according to magnitude (this arrangement is called an **array**).

2. Record the size of the middle value. If there are an even number of items in the array there will be two central values, and the arithmetic mean of these two values is taken as the median.

Calculation—Grouped Data

The median is computed from grouped data as follows.

1. Determine the number of the desired middle item by using

21

the formula, $\dfrac{N}{2}$ where N is the number of items in the distribution.[1]
For the distribution given in Table 3.1, the median item is the seventy-fifth.

$$\frac{N}{2} = \frac{150}{2} = 75$$

2. Find the class interval in which the seventy-fifth item appears by cumulative addition of the frequencies. In Table 3.1, the sum of the frequencies for the first nine class intervals is 61, and the sum for the first ten class intervals is 79. The seventy-fifth item, then, must be in the tenth class interval.

Table 3.1–Computation of Median
Grades on Bookkeeping Examination in a New York
City High School
Grades in Percent   Number of Students

class interval number	class interval	frequency	cumulated frequencies
1	25–29.9%	2	2
2	30–34.9	4	6
3	35–39.9	5	11
4	40–44.9	9	20
5	45–49.9	8	28
6	50–54.9	7	35
7	55–59.9	8	43
8	60–64.9	4	47
9	65–69.9	14	61
10	70–74.9	18	79
11	75–79.9	24	103
12	80–84.9	21	124
13	85–89.9	14	138
14	90–94.9	7	145
15	95–99.9	5	150
		150	

---

[1]There has been considerable discussion as to whether $\dfrac{N}{2}$ or $\dfrac{N+1}{2}$ should be used to locate the number of the median item. If the distribution is regarded as continuous $\dfrac{N}{2}$ should be used, if considered discrete $\dfrac{N+1}{2}$. Most authors use $\dfrac{N}{2}$ for the location of the median.

3. Assume that the values of the items in a class interval are evenly distributed between the upper and lower limits of that class interval, and use the method of **interpolation** to determine the value of the median. In Table 3.1, the upper limit of the ninth class interval is 70%, and there are sixty-one items in the first nine class intervals. The upper limit of the tenth class interval is 75%.[1] and there are seventy-nine items in the first ten class intervals. The median, then, is the percent value which corresponds to the seventy-fifth item. This information can be tabulated as follows.

Value	Item
70%	sixty-first
?	seventy-fifth
75%	seventy-ninth

In order to find the median value by interpolation, the following steps should be observed.

a. Determine the fraction of the distance from 61 to 75 in the tenth class interval.

$$\frac{number\ of\ cases\ needed}{number\ of\ cases\ in\ group} = \frac{75 - 61}{79 - 61} = \frac{14}{18} = \frac{i}{f}$$

b. Find the difference between the limits in the value column.

$$75\% - 70\% = 5\% = C$$

c. Multiply the fraction found in **a** by the difference found in **b**, and add that product to the lower limit. This sum is the median.

$$median = L_{mc} + \frac{i}{f}\ C = 70\% + \left(\frac{14}{18}\right)(5\%) = 73.89\%$$

Characteristics

1. The median is an average of position.

2. The median is affected by the number of items, not by the size of extreme values.

The change of the fifth student's grade from 58 on the first examination to 100 on the second did not affect the value of the median,

---

[1] For the purpose of specifying a class interval by 74.9 is really meant 74.999 etc. For practical purposes 75 is used here.

Table 3.2–Hypothetical Grades for Six Students on
Three Examinations

Student Number	Examination 1	Examination 2	Examination 3
1	50%	50%	50%
2	51	51	51
3	52	52	52
4	54	54	54
5	58	100	58
6	–	–	100
Median	52%	52%	53%
Arithmetic Mean	53%	61.4%	60.8%

but it did change the value of the arithmetic mean. However, the additional value in the third examination caused a change in the median value as well as in the mean value.

3. The sum of the deviations about the median, signs ignored, will be less than the sum of the deviations about any other point.

4. The median is most typical when used to describe distributions whose central values are closely grouped.

5. A value selection at random is just as likely to be located above the median as below. At times, therefore, the median is called the "probable" value.

Advantages

1. The median is easily calculated.

2. It is not distorted in value by unusual items.

3. It is sometimes more typical of the series than are other averages because of its independence of unusual values.

4. The median may be calculated even when the class intervals of the distribution are "open ended."

Disadvantages

1. The median is not so familiar as the arithmetic mean.

2. The items must be arranged according to size before the median can be computed.

3. It has a larger standard error than the arithmetic mean (see Chapter XIV).

4. The median cannot be manipulated algebraically. The average of the medians of sub-groups, for example, is not the median of the group.

QUARTILES, DECILES, PERCENTILES

Just as the median divides a distribution into two parts, the **quartiles** divide it into four parts, the **deciles** divide it into ten parts, and the **percentiles** divide it into one hundred parts. Thus, quartiles, deciles, and percentiles are all averages of position, and they make possible a more minute analysis of the distribution than does the median.

Since the quartile divides the distribution in four parts, there are three quartiles. The second quartile will divide the distribution in half and therefore will be the same as the median. The first (lower) quartile (Q_1) marks off the first one-fourth of the distribution, and the third (upper) quartile (Q_3) marks off the point separating the third from the last quarter.

The percentiles divide the distribution into 100 parts such that each contains one per cent of the cases. A division as fine as this can be used only when there are a considerable number of cases (at least one thousand); otherwise, there would be too few cases in each division to be significant. In many cases, therefore, only certain percentile points are used.

The quartile, decile, and percentile are calculated by the technique used in the computation of the median. For the quartiles, $\frac{N}{4}$ is used to locate Q_1, and $\frac{3N}{4}$ is used to locate Q_3. Similarly, $\frac{N}{10}$ is used to locate the first decile; $\frac{2N}{10}$ is used to locate the second decile; $\frac{3N}{10}$ is used to locate the third; and so on. For the percentiles, $\frac{N}{100}$ is used to locate the first; $\frac{2N}{100}$ to locate the second; $\frac{3N}{100}$ to locate the third; and so on. In general, substitution in the formula, $L + \frac{i}{f}C$, where L is the lower limit of the class interval in which the desired quartile, decile, or percentile is located, will result in the desired values.

As an example of the usefulness of these various measures, consider Table 2.5 on page 18. The ninetieth percentile point for the distribution of ratios of current assets to current liabilities for 221 industrial corporations is 14.225% indicating that only 10% of the corporations had ratios greater than 14.225% while 90% of the corporations had ratios less than that amount.

THE MODE
Definition
The mode is the most frequent or most common value which occurs in a set of data, provided a large number of observations are available.

The value of the mode will correspond to the value of the maximum point (ordinate) of a frequency distribution if it is an "ideal" or smooth distribution.

Computation—Ungrouped Data
There is no difficulty associated with finding the mode for ungrouped data. One need only locate the value (or values) which occurs most frequently in the distribution. Thus, for example, in the array 3, 5, 7, 9, there is no mode; in the array 3, 5, 5, 7, 9, the mode is 5; and in the array 3, 5, 5, 7, 7, 9, there are two modes, 5 and 7. If an array has two modes, the data are called **bimodal**.

Computation—Grouped Data
It is not possible to make an exact mathematical determination of the mode for grouped data. A number of methods may be used, however, to secure reasonably accurate approximations.

Moments-of-Force Method
The midpoint of the modal class interval may not be used as the value of the mode because its value will change if the size of the class interval is changed.

Reducing the size of the class interval will tend to delimit the value of the mode and tend more and more to have it coincide with the midpoint of the group of greatest frequency. This reduction in size of the class interval is, however, decidedly limited by the number of items included in the distribution. If an infinite number of items were available, and an infinitely small class interval used, the midpoint of the class interval of greatest frequency would be the value of the mode.

In practice this ideal situation does not exist, so an approximation somewhat closer than the midpoint of the modal group is necessary.

In spite of the previous midpoint assumption the values within a group are not evenly distributed. On the contrary, there is a tendency for values to gravitate towards the point of greatest density.

In the distribution below (Table 3.3), the modal group contains 43 items and has class limits of .10% and .19%. Since there are a greater number of cases in the class above this one than in the class be-

Table 3.3–Computation of Mode–Moments-of-Force Method

Percent of Population on Old Age Pensions in the United States, by Countries

class interval percent	Number of Counties frequency
.00– .09%	19
.10– .19	43
.20– .29	32
.30– .39	27
.40– .49	17
.50– .59	21
.60– .69	14
.70– .79	9
.80– .89	2
.90– .99	2
1.00–1.09	0
1.10–1.19	0
1.20–1.29	1
	187

Source: United States Bureau of Labor Statistics, Handbook of Labor Statistics.

low[1] it, .00% to .09%, it follows that the true point of greatest concentration will tend towards the upper class interval and will therefore be above the midpoint of the modal group.

The value of the mode may be approximated by use of the formula

$$\text{mode} = L_{mo} + \frac{f_a}{f_a + f_b} \, C = .10\% + \frac{32}{32 + 19} \, (.10\%) = .163\%$$

where

L_{mo} = lower limit of modal group
f_a = frequency of class interval above modal group
f_b = frequency of class interval below modal group
C = size of class interval

Empirical Method

Where the distribution is only moderately skewed another estimation of the value of the mode may be obtained from the relationship that exists between the position of the mean, median, and mode.

---

[1] By *below* in reference to a class interval is meant in the direction of the lowest class interval value.

In a smoothed curve (such as that shown in Figure 3.1) the mode will be located at the highest point in the distribution. If the distribution is skewed to the right, the position of the median will be somewhat to the right of the mode (in the direction of the extreme values) and will divide the area under the curve in half. The mean, since it is affected to the greatest degree by extreme values, will be furthest in the direction of the extreme values.

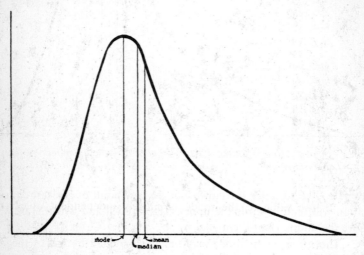

Fig. 3.1—Hypothetical right skewed frequency distribution showing theoretical position of mode, median, and mean.

In a left skewed distribution (such as that shown in Figure 3.2) the same relationship will be obtained but in the opposite direction.

It has been found that *for a moderately skewed distribution, the distance between the mean and the median is one-third of the distance between the mean and the mode.*

Since the values of the mean and the median may be determined exactly, the value of the mode may be approximately determined through this relationship.

$$\text{mode} = \text{mean} - 3(\text{mean} - \text{median})$$

Characteristics

1. By definition the mode is the most usual or typical value. Under certain circumstances it may be considered as the "normal" value.

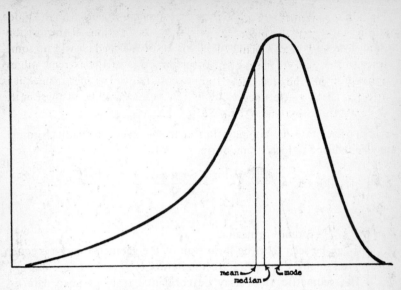

Fig. 3.2—Hypothetical left skewed distribution showing theoretical position of mode, median, and mean.

2. The value of the mode is entirely independent of extreme items.

3. The mode is an average of position.

Advantages

1. It is the most typical value, and therefore the most descriptive average.

2. It is simple to approximate by observation when there are a small number of cases.

3. If there are only a few items, it is not necessary to arrange them in order to determine the mode.

Disadvantages

1. The mode can be approximated *only* when a limited amount of data is available.

2. Its significance is limited when a large number of values is not available.

3. If none of the values are repeated, the mode does not exist.

THE GEOMETRIC MEAN

The geometric mean is the nth root of the product of n items or values, and can be calculated by using the following formula.

$$G_m = \sqrt[n]{X_1 \cdot X_2 \cdot X_3 \cdots X_n}$$

For instance, the geometric mean of $1, $3, and $9 is

$$G_m = \sqrt[3]{1 \times 3 \times 9}$$
$$= \sqrt[3]{27}$$
$$= \$3$$

To facilitate the computation of the geometric mean, its formula may be reduced to logarithmic form.

$$\log G_m = \frac{\log X_1 + \log X_2 + \log X_3 + \cdots + \log X_n}{n}$$

where

G_m = geometric mean

It can be seen that the logarithm of the geometric mean is equal to the average of the logarithms of the items.

The geometric mean may be computed from grouped data by using the technique outlined for the computation of the arithmetic mean by the "long" method (see page 14), except that the logarithms of the midpoints are used in the calculations rather than the actual midpoint values.

Characteristics

1. The geometric mean is a calculated value and depends upon the size of all the values.

2. It is less affected by extreme items than the arithmetic mean.

3. For any series of items it is always smaller than the arithmetic mean.

4. When any single value equals zero the geometric mean will be zero.

Advantages

1. It is a more typical average than the arithmetic mean since it is less affected by extremes.

2. It may be manipulated algebraically.

3. It is particularly useful in the computation of index numbers (see Chapter XVI).

Disadvantages

1. The geometric mean is not widely known.

2. The geometric mean is relatively difficult to compute.

3. It cannot be determined where there are negative values in the series.

THE QUADRATIC MEAN

The quadratic mean is the square root of the mean square of the items (root-mean-square), and can be calculated by using the following formula.

$$Q_m = \sqrt{\frac{\Sigma(X^2)}{N}}$$

The quadratic mean is used in the computation of the standard deviation (see page 37).

THE HARMONIC MEAN

The harmonic mean of a series of values is the reciprocal of the arithmetic mean of the reciprocals of the values. It can be calculated by using the following formula.

$$\frac{1}{H_m} = \frac{\dfrac{1}{X_1} + \dfrac{1}{X_2} + \dfrac{1}{X_3} + \cdots + \dfrac{1}{X_n}}{N}$$

The harmonic mean is used in averaging rates.

ADDITIONAL BIBLIOGRAPHY

CHOU, YA-LUN, *Applied Business and Economic Statistics*, pp. 155–177, Holt, Rinehart & Winston, New York, 1963.

YULE, G. U., & KENDALL, M. G., *An Introduction to the Theory of Statistics*, pp. 111–124, Hafner Publishing Co., New York, 1950.

ZIZEK, FRANK, *Statistical Averages*, pp. 194–247, Henry Holt & Co., New York, 1930.

Chapter IV

Analysis of the Frequency Distribution—Dispersion and Skewness

DISPERSION

The average or typical value is of little use unless the degree of variation which occurs about it is given. For if the scatter about the measure of central tendency is very large, the average is not a *typical* value. It is therefore necessary to develop a quantitative measure of the dispersion (or variation, or scatter) of values about the average.

THE RANGE

The range, the simplest of the measures of dispersion, is the difference between the minimum and maximum values in a series. It is sometimes given in the form of a statement of the minimum and maximum values themselves.

The difference between the two extreme values is indicative of the spread of the series, but quite frequently is misleading because it gives no information about how the items are dispersed. In series A and B in Table 4.1 the range is 30%, but the values are not evenly dispersed over that range.

Characteristics
1. The range is simple and readily understood.
2. It is easily calculated.
3. Its value is dependent on two items only, the highest and lowest values.

Table 4.1
Hypothetical Examination Grades for Ten Students

Student Number	Examination A	Examination B
1	60%	60%
2	60	65
3	61	70
4	63	72
5	65	75
6	65	78
7	66	80
8	67	85
9	68	88
10	90	90

4. It is not necessary to know the distribution of the items between the two extremes in order to obtain the range.

5. Since the range is dependent only upon the two extremes, it is greatly affected by unusual maximum and minimum values.

THE MEAN DEVIATION

The range is dependent for its value entirely upon the two extreme values. Obviously, when these end-values are far-removed from the remainder of the data, a satisfactory measure of dispersion must be dependent upon the position of every value in the series.

A simple method for determining the scatter of a series of values about a given point is to take the average distance of the items from the given point. The smaller the average distance about this point, the smaller the scatter or dispersion of the values.

In a frequency distribution the average distance of the items from a measure of central tendency, such as the arithmetic mean, may be used for this purpose. However, since the sum of the deviations about the arithmetic mean is zero, it is necessary to ignore signs in order to obtain the average distance of items from that measure.

The measure of dispersion found in Table 4.2 (6.37) consists of the average of the deviations of the items from their arithmetic mean. The median (26) could also be used as the measure of central tendency, giving a mean deviation of 6.31.

Characteristics

1. The value of the mean deviation is dependent upon the value of every item in the series.

2. It may be computed about any measure of central tendency.

Table 4.2–Sales Record of Clerk No. 148 in a New York City Department Store for One Month

Date	Number of Sales (X)	Deviation from Monthly Average (x)
June 1	15	11.85
2	27	.15
3	31	4.15
4	27	.15
6	23	3.85
7	23	3.85
8	25	1.85
9	31	4.15
10	29	2.15
11	41	15.15
13	17	9.85
14	30	3.15
15	45	18.15
16	24	2.85
17	26	.85
18	26	.85
20	23	3.85
21	15	11.85
22	37	10.15
23	27	.15
24	18	8.85
25	39	12.15
27	19	7.85
28	18	8.85
29	21	5.85
30	41	13.15
Total	698	165.70
Average	26.85	6.37

3. The mean deviation about the median is less than that calculated about any other point.

Computation–Ungrouped Data

The mean deviation may be computed by using the following formula:

$$MD = \frac{\Sigma |x|}{N} \text{ or } \frac{\Sigma |d|}{N}$$

where

MD = the mean deviation

$\Sigma |x|$ = a sum of the deviations of each value from the arithmetic mean, *signs ignored*

$\Sigma |d|$ = the sum of the deviations from another measure of central tendency such as the median, *signs ignored*

Computation—Grouped Data

When the data are grouped in the form of a frequency distribution the value of the mean deviation may be determined as follows:

1. Obtain the deviation of the midpoint of each class interval from the median (or mean).

2. Multiply the deviations by the number of items (the frequency) in each class interval.

3. Divide the total of the values obtained by the number of cases.

An arithmetically simpler method is to:

1. Select, as an arbitrary origin, the midpoint of the group in which the median (or mean) is located.

2. Obtain the deviations of the midpoints about this more convenient value (see d' in Table 4.3, Column 4), and multiply them by the frequencies (Column 3).

In the illustrated distribution (Table 4.3), the median (73.052 months) is located above the class-interval midpoint used as the arbitrary origin (73). Therefore, all the deviations at the point 73 or be-

Table 4.3—Computation of Mean Deviation

Ages of Pupils in the First Half of First Grade in a New York City Public School

Age in Months class interval	midpoint (M. P.)	Number of Pupils (f)	Deviation from $\overline{Z}$ (in class intervals) (d')	Frequency × deviation (fd')
68–69.9	69	12	2	24
70–71.9	71	33	1	33
72–73.9	73	57	0	0
74–75.9	75	25	1	25
76–77.9	77	9	2	18
78–79.9	79	4	3	12
80–81.9	81	6	4	24
82–83.9	83	2	5	10
84–85.9	85	0	6	0
86–87.9	87	0	7	0
88–89.9	89	2	8	16
		150		162

low are too small by the amount of the difference between the arbitrary origin and the median: .052 months in actual units or .026 in class intervals. There are 102 such deviations (12 + 33 + 57); and, therefore, the total understatement is 102 times .026 class intervals.

In similar fashion, the values above the arbitrary origin are overstated by the amount of .026 class intervals times 48.

The understatement of the 102 values below the arbitrary origin is in part offset by the 48 items overstated, leaving only 54 values understated. The average understatement will then be 54 times .026 divided by the total number of items. If this correction is added to the average deviation about the arbitrary origin the result will be the mean deviation in class intervals. This method is expressed in the following formula:

$$MD' = \frac{\Sigma|fd'|}{N} + \frac{(N_s - N_L)c}{N} = \frac{162}{150} + \frac{(102 - 48).026}{150} = 1.0894$$

where

MD' = mean deviation in class intervals

N_S = number of cases too small or understated

N_L = number of cases too large or overstated

c = difference, in terms of class intervals, between arbitrary origin (midpoint of median or mean group) and median or mean

The mean deviation in actual units will be obtained if the result, which is expressed in terms of class intervals, is multiplied by the size of the class interval.

$$MD = MD' \times C$$

where

C = size of class interval

$$MD = 1.0894 \times 2 = 2.1788 \text{ months}$$

STANDARD DEVIATION AND VARIANCE

The standard deviation is a special form of average deviation from the mean. It is computed by taking the quadratic mean (see page 31) of the deviations from the arithmetic mean of these values. The standard deviation is thus the root-mean-square of the deviations from the arithmetic mean.

$$\sigma = \sqrt{\frac{\Sigma(x^2)}{N}}$$

where

σ = standard deviation[1]
x = deviations from arithmetic mean $(X - \overline{X})$
N = total number of items, $\Sigma(f)$

Table 4.4—Computation of Standard Deviation—Ungrouped Data
Closing Market Price for Twelve Closed-End Investment Companies

Company	Closing Price (X)	Deviation from Mean $(X - \overline{X})$ (x)	(x^2)
Abacus	14	10.1	102.01
Adm Exp	29	4.9	24.01
Am Euro	26	1.9	3.61
Am Int'l	15	−9.1	82.81
Am So. Af.	40	15.9	252.81
Amoskeag	41	16.9	285.61
Bos Pers	17	−7.1	50.41
Carriers	30	5.9	34.81
Consol. Inv	10	−14.1	198.81
Dominick	21	−3.1	9.61
Euro Fund	12	−12.1	146.41
Gen AInv	34	9.9	98.01
Total	289		1288.92
Mean	24.1		107.41

$$\sigma = \sqrt{\frac{\Sigma(x^2)}{N}} = \sqrt{\frac{1288.92}{12}} = \sqrt{107.41} = 10.36$$

Source: Barron's Weekly. (Note: To simplify the illustration fractions were dropped from the stated prices.)

When computed from a sample the formula for the standard deviation is

$$s = \sqrt{\frac{\Sigma(x^2)}{n-1}}$$

When computed in this manner the symbol s is often used rather than σ. This provides an unbiased estimate of the standard deviation of the population from which the sample was drawn. However, when the sample size is large, the difference between σ and s is negligible. For this reason σ is customarily used instead of s for large samples.

The standard deviation is a numerical measure of the degree of

---

[1] The symbol σ is the Greek small letter sigma.

dispersion, variability, or non-homogeneity of the data to which it is applied. At times it is found to be more convenient to eliminate the effect of the square root radical in this formula, and thus the square of the standard deviation (σ^2) is used as a measure of dispersion. This measure is known as the *variance*. Like the standard deviation it is a measure of dispersion of variability of observations but will have a different scale of values than the standard deviation. Either measure (the standard deviation or the variance) may be used as a measure of dispersion.

Computation—Ungrouped Data

1. Get the difference between each actual value and the arithmetic mean.

2. Square the values thus obtained. Obtain the average of the squares.

3. Take the square root of the result.

A more convenient formula for ungrouped data may be derived from an algebraic manipulation of the standard deviation[1]:

$$\sigma = \sqrt{\frac{\Sigma(X^2)}{N} - \left(\frac{\Sigma X}{N}\right)^2}$$

Computation—Grouped Data

Long Method

Where there are a considerable number of items in the series the calculation of the standard deviation can be more readily performed if the data are first grouped into the form of a frequency distribution.

1. The deviation of the midpoint of each group from the arithmetic mean is used as a measure of the average deviation from the mean of all items in that group.

2. The average deviation of each group is squared to obtain the necessary deviation squared.

3. The average deviation squared is multiplied by the frequency indicated for the group in order to obtain the total of the squared deviations for that group.

4. The totals are then added for the entire distribution.

5. The square root of the sum obtained after dividing by N is the standard deviation.

---

[1] See Technical Appendix III for mathematical proof.

$$\sigma = \sqrt{\frac{\Sigma f(x^2)}{N}}$$

or for a sample

$$s = \sqrt{\frac{\Sigma f(x^2)}{n-1}}$$

Table 4.5—Computation of Standard Deviation—Grouped Data—
Long Method
Percent of Tax Delinquency in 151 Cities of Over 50,000 Population
in the United States

(1) Percent of Tax Delinquency (class interval)	(2) (midpoint) (M. P.)	(3) Number of Cities (frequency) (f)	(4) (Deviation from Mean) (Mean = 26.74) (x)	(5) (x^2)	(6) $f(x^2)$
0- 4.99	2.50	1	-24.24	587.5776	587.5776
5- 9.99	7.50	12	-19.24	370.1776	4442.1312
10-14.99	12.50	19	-14.24	202.7776	3852.7744
15-19.99	17.50	24	- 9.24	85.3776	2049.0624
20-24.99	22.50	19	- 4.24	17.9776	341.5744
25-29.99	27.50	19	.76	.5776	10.9774
30-34.99	32.50	16	5.76	33.1776	530.8416
35-39.99	37.50	15	10.76	115.7776	1730.6640
40-44.99	42.50	12	15.76	248.3776	2980.5312
45-49.99	47.50	8	20.76	430.9776	3447.8208
50-54.99	52.50	2	25.76	663.5776	1327.1552
55-59.99	57.50	0	30.76	946.1776	0.
60-64.99	62.50	2	35.76	1278.7776	2557.5552
65-69.99	67.50	2	40.76	1661.3776	3322.7552
		151			27187.4176

$$\sigma = \sqrt{\frac{\Sigma f(x^2)}{N}} = \sqrt{\frac{27{,}187.4176}{151}} = 13.42\%$$

Source: Dun & Bradstreet's Municipal Review.

Short Method

The computation of the standard deviation may be simplified by
the following method.

1. Instead of using the arithmetic mean, compute the standard
deviation about a conveniently selected point. For this purpose the
midpoint of any group may be selected. Since the quadratic mean of
the deviations about the arithmetic mean is smaller than the quadratic
mean of deviations about any other point, the resulting value will be
larger than the true standard deviation.

2. Subtract from it a correction value to obtain the necessary

result. The value of the correction factor may be determined by an algebraic manipulation of the formula[1]:

$$\sigma = \sqrt{\frac{\Sigma f(x^2)}{N}}$$

The resulting formula is

$$\sigma = \sqrt{\frac{\Sigma f(d^2)}{N} - \left(\frac{\Sigma fd}{N}\right)^2}$$

where

σ = standard deviation

d = deviation of midpoint of each class interval from that of arbitrary group

Where the class intervals are uniform in size, the calculation may be further simplified by carrying out all computations in terms of class intervals and then multiplying the final results by the size of the class interval.

The formula will then read[2]

$$\sigma = C\sqrt{\frac{\Sigma f(d'^2)}{N} - \left(\frac{\Sigma fd'}{N}\right)^2}$$

where

σ = standard deviation

d' = deviation of midpoint of class interval from arbitrary origin in terms of class intervals

f = frequency of values in class interval

C = size of class interval

The application of this formula to the computation of the standard deviation of the minimum line rates for national advertising for 249 daily newspapers is shown in Table 4.6.

Correction for Grouping

The value computed by the grouping method will be subject to the assumption that all the values are located at the midpoint of each

---

[1] See Technical Appendix III for mathematical proof.

[2] When computed from a sample this formula becomes

$$s = \sqrt{\frac{\Sigma f(d')^2 - (\Sigma fd')^2/n}{n-1}}$$

Table 4.6–Computation of Standard Deviation–Short Method
Minimum Line Rate for National Advertising for 249 Daily
Newspapers in Cities of 25,000 to 50,000 in the United States

(1) Rate per line in dollars	(2) Number of newspapers (f)	(3) Deviation from $\bar{Z}$ In Class Intervals (d')	fd'	fd'^{2}
\$.01–.019	2	−5	−10	50
.02–.029	4	−4	−16	64
.03–.039	23	−3	−69	207
.04–.049	30	−2	−60	120
.05–.059	40	−1	−40	40
.06–.069	45	0	0	0
.07–.079	35	1	35	35
.08–.089	25	2	50	100
.09–.099	12	3	36	108
.10–.109	9	4	36	144
.11–.119	6	5	30	150
.12–.129	10	6	60	360
.13–.139	3	7	21	147
.14–.149	1	8	8	64
.15–.159	1	9	9	81
.16–.169	3	10	30	300
	249		120	1970

Source: Editor and Publisher, International Year Book

$$\sigma = C \sqrt{\frac{\Sigma f(d'^2)}{N} - \left(\frac{\Sigma fd'}{N}\right)^2}$$

$$= \$.01 \sqrt{\frac{1970}{249} - \left(\frac{120}{249}\right)^2} \qquad = \$.01 \sqrt{7.9116 - .2322}$$

$$= \$.01 \sqrt{(7.6794)} \qquad = (2.7711) \,\$.01 = \$.0277$$

class interval. It is dependent in part on the size of the class interval used. The error in the assumption is constant when

1. The distribution is "continuous" (see page 6).
2. The distribution tapers off gradually in both directions.

Under these conditions the corrected standard deviation is

$$\sigma^2 = (\sigma'^2 - 1/12)\,C^2$$

where

σ' = standard deviation in class interval units
C = size of class interval

Check on Computation—The Charlier Check

A simple check may be used in order to determine the accuracy of the computations before the actual substitution is made in the formula for the standard deviation.

If an additional value $\Sigma f(d' + 1)^2$ is computed it will be seen that

$$\Sigma f(d' + 1)^2 = \Sigma f(d'^2 + 2d' + 1)$$
$$= \Sigma fd'^2 + 2\Sigma fd' + \Sigma f$$

But since

$$\Sigma f = N,$$
$$\Sigma f(d' + 1)^2 = \Sigma fd'^2 + 2\Sigma fd' + N$$

These are the values used in the computation of the standard deviation. If the equation is fulfilled the preliminary values are correct. If the computation check is applied to the problem shown above the result is

$$\Sigma fd'^2 + 2\Sigma fd' + N = 1970 + 2\,(120) + 249$$
$$\Sigma f(d' + 1)^2 = 2459$$

Table 4.7—Charlier Check—Applied to Data of Table 4.6

(1) Rate per line in dollars	(2) Number of newspapers (f)	(3) (From Table) d'	(4) d' + 1	(5) (d' + 1)²	(6) f(d' + 1)²
$.01–.019	2	−5	−4	16	32
.02–.029	4	−4	−3	9	36
.03–.039	23	−3	−2	4	92
.04–.049	30	−2	−1	1	30
.05–.059	40	−1	0	0	0
.06–.069	45	0	1	1	45
.07–.079	35	1	2	4	140
.08–.089	25	2	3	9	225
.09–.099	12	3	4	16	192
.10–.109	9	4	5	25	225
.11–.119	6	5	6	36	216
.12–.129	10	6	7	49	490
.13–.139	3	7	8	64	192
.14–.149	1	8	9	81	81
.15–.159	1	9	10	100	100
.16–.169	3	10	11	121	363
	249				2459

Characteristics of Standard Deviation

1. The standard deviation is affected by the value of every item.

2. Greater emphasis is placed on extremes than in the mean deviation; this because all values are squared in the computation.

3. In a normal or bell-shaped distribution the mean deviation is .7979 σ. In a moderately skewed distribution this relationship is still approximately true.

4. a. If a distance equal to one standard deviation is measured off on the X axis on both sides of the arithmetic mean in a normal distribution, 68.26% of the values will be included within the limits indicated.

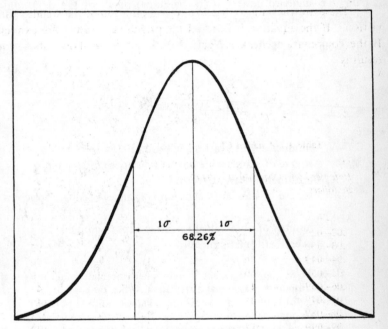

Fig. 4.1–"Normal" distribution showing the percentage of the area included within one standard deviation measured both plus and minus about the arithmetic mean.

b. If two standard deviations are measured off 95.46% of the values will be included.

c. Three standard deviations measured off will include 99.73% of the values.

The above percentages are exact only where the distribution is perfectly normal.[1] In the case of a moderately skewed distribution the percentages are approximations, and are often considered as about 68%, 95%, or all the values.

The exact percentage of cases included for any number of standard deviations measured from the arithmetic mean in one direction only may be found in Table 12.4.

5. Since 3 standard deviations on either side of the mean include almost all (99.7%) of the cases in moderately skewed or normal distributions, the standard deviation usually is about 1/6 of the range.

Standard Deviation (Variance) of Group Data

1. Given the standard deviation (or variance) of several groups of data, the standard deviation (or variance) of all of the data combined may be computed from[2]

$$N\sigma^2 = N_1 \sigma_n^2 + N_2 \sigma_2^2 + \ldots + N_n \sigma_n^2 + N_1 (\overline{X}_1 - \overline{X})^2 +$$
$$N_2 (\overline{X}_2 - \overline{X})^2 + \ldots + N_n (\overline{X}_n - \overline{X})^2$$

or

$$N\sigma^2 = \Sigma (N_i \sigma_i^2) + \Sigma [N_i (\overline{X}_i - \overline{X})^2]$$

where

N_i = number of observations in group i

σ_i^2 = square of standard deviation (variance) of group i

$\overline{X}_i$ = arithmetic mean of group i

σ^2 = square of standard deviation (variance) of all groups combined.

N = number of observations in all groups combined

$\overline{X}$ = overall arithmetic mean

---

[1] The above percentages apply only when the underlying frequency distribution is a normal distribution. When the form of the underlying distribution is not known or is known not to be normal, information about the percentage of the observations contained within a range of $\pm T$ standard deviations of the mean can be obtained from the **Tchebycheff Inequality**. This inequality states that within a range of $\pm T$ standard deviations about the mean there will be contained *not less than*

$$1 - \frac{1}{T^2}$$

proportion of the observations.

[2] See Technical Appendix XI for proof.

2. The variance of a sum (say $X_1 + X_2$) may be computed from

$$\sigma_{x_1 + x_2}^2 = \sigma_{x_1}^2 + \sigma_{x_2}^2 + 2r\,\sigma_{x_1}\,\sigma_{x_2}$$

where $\sigma_{x_1}^2$ is variance of x_1, etc., or if the values x_1 and x_2 *are uncorrelated* $(r = 0)$.[1]

$$\sigma_{x_1 + x_2}^2 = \sigma_{x_1}^2 + \sigma_{x_2}^2$$

3. The variance of a difference (say $x_1 - x_2$) may be computed from

$$\sigma_{x_1 - x_2}^2 = \sigma_{x_1}^2 + \sigma_{x_2}^2 - 2r\sigma_{x_1}\,\sigma_{x_2}$$

or if the values X_1 and X_2 are uncorrelated

$$\sigma_{x_1 - x_2}^2 = \sigma_{x_1}^2 + \sigma_{x_2}^2$$

RELATIVE MEASURES OF DISPERSION

The measures outlined above are *absolute* measures of dispersion and therefore the resulting values cannot always be compared with other similar measures with significance. The standard deviation in months of the ages of students in a given grade of a New York City school, for example, cannot be compared to the dispersion of the Intelligence Quotients in the same class where the standard deviation is expressed as a percentage because of the difference in units.

A measure of dispersion must also be compared to the size of the average about which it is measured. For instance, a variation of five dollars in the price of a share of stock, the average price of which is ten dollars, does not have the same value in relation to its average price as a share which has the same variation but an average price of one hundred dollars.

To relate the measure of dispersion to its average and to convert it to a percentage, the standard deviation is divided by the arithmetic mean. Stating this measure in percentage form solves the problem presented by the differing units. The resulting measure developed by Pearson is known as the **coefficient of variation** or V (V^2 is sometimes called the **relative variance**).

$$V = \frac{\sigma}{X} 100$$

---

[1] r is the coefficient of correlation, described on page 88.

Other comparative coefficients of dispersion may be computed by using other measures of dispersion. With the mean deviation, for example,

$$V_{MD} = \frac{MD}{\text{median (or mean)}}$$

and with the quartile deviation,

$$V_Q = \frac{\dfrac{Q_3 - Q_1}{2}}{\dfrac{Q_3 + Q_1}{2}} = \frac{Q_3 - Q_1}{Q_3 + Q_1}$$

SKEWNESS

Skewness is a term for the degree of distortion from symmetry exhibited by a frequency distribution.

When a distribution is perfectly symmetrical with one mode, the values of the mean, median, and mode coincide. In an asymmetrical (skewed) distribution their values will be different. Since the arithmetic mean is most affected by extremes it will be farthest from the mode. The mode is not affected at all by unusual values; therefore the greater the degree of skewness the greater the distance between the mean and the mode.

It follows that this distance between mean and mode may be used as a measure of skewness, since the greater the lack of symmetry the larger the discrepancy between them. However, because the measure of skewness is used mostly for comparative purposes, the problem of differing units comes up again. A second difficulty arises in that the distance between mean and mode will be larger in a widely dispersed distribution than in one with a narrow dispersion. Both difficulties may be removed by dividing the distance by the standard deviation.[1]

$$S_K = \frac{\text{mean} - \text{mode}}{\sigma}$$

Since the distance between the mean and mode in moderately skewed distributions is three times the distance between the mean and the median (see page 28), for this type of distribution the formula may

---

[1] This measure is called the **Pearsonian Coefficient** of skewness. Another measure of skewness may be employed. This is described in Chapter XV.

be rewritten as follows

$$S_K = \frac{3 \, (\text{mean} - \text{median})}{\sigma}$$

When the distribution is symmetrical the values of the mode and the mean will coincide. Under these circumstances the coefficient of skewness will be zero.

Where the curve is right skewed the extremely large values will *increase* the value of the mean, while the value of the mode will remain unaffected. The coefficient will then be positive. If the distribution is skewed to the left, the extreme cases will *reduce* the value of the mean, and the coefficient of skewness will be negative.

Another measure is based on the position of the quartiles. In a symmetrical distribution the quartiles are equidistant from the median. In a skewed distribution the quartiles will differ in their distances from the median.

The greater the lack of symmetry the larger the discrepancy between the two distances of the quartiles from the median. If this is divided by the quartile deviation (the measure of dispersion based on the quartiles) the result is a coefficient of skewness.

$$S_K = \frac{(Q_3 - \text{median}) - (\text{median} - Q_1)}{QD}$$

In a symmetrical distribution the distances will be equal and S_K will equal zero. Where the distribution is right skewed the right quartile (Q_3) will be a greater distance from the median than Q_1. The opposite will be true where the curve is left skewed. The resulting coefficient will then be negative.

KURTOSIS

The "peakedness" of the frequency distribution is another characteristic which can be measured. The measurement of this characteristic is outlined in Chapter XVII.

ADDITIONAL BIBLIOGRAPHY

CHOU, YA-LUN, *Applied Business and Economic Statistics,* pp. 184–207, Holt, Rinehart & Winston, 1964.

TUTTLE, A. M., *Elementary Business and Economic Statistics,* pp. 230–259, McGraw-Hill Book Co., New York, 1957.

YULE, G. U., & KENDALL, M. G., *An Introduction to the Theory of Statistics,* pp. 125–150, Hafner Publishing Co., New York, 1950.

Chapter V

Time Series Analysis— Trend

The time series is an arrangement of statistical data in accordance with its time of occurrence.

CLASSIFICATION OF MOVEMENTS

The analysis of the time series consists of the description and measurement of the various changes or movements as they appear in the series during a period of time. These changes or movements may be classified as

1. **Secular trend**, or the growth or decline occurring within the data over a long period of time. The period should be not less than ten years, when economic data are involved.

2. **Seasonal variation**, or the more or less regular movement within each twelve month period. This movement occurs year after year and is caused by the changing seasons.

3. **Cyclical movement**, or the swing from prosperity through recession, depression, recovery, and back again to prosperity. This movement varies in time, length, and intensity.

4. **Residual, accidental, or random** variations, including such unusual disturbances as wars, disasters, strikes, fads, or other non-recurring factors.

Periodic variations similar to those encountered in economics are found in other fields. The weather, for instance, has seasonal, cyclical, accidental, and even secular movement. Sunspots have cyclical movement, and there are other examples.

MEASUREMENT OF TREND

Four methods are commonly used for measuring trends. These are

1. freehand
2. semi-average
3. moving average
4. least squares (see Chapter VI)

Freehand Method

To fit (i.e., to describe) a trend by the freehand method draw a line through a graph of the data in such a way as to describe what appears to the eye to be the movement over the whole period. A line of trend fitted by this method is shown in Figure 5.1. The drawing of this

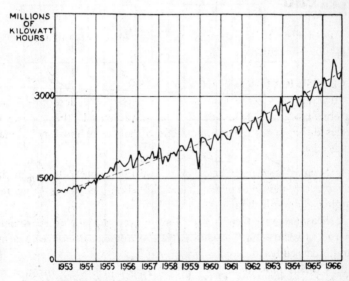

Source: Standard and Poor's.

Fig. 5.1–Average daily output of electric power in the United States, 1953–1966. Trend indicated by a freehand line.

line need not be strictly freehand but may be accomplished with the aid of a transparent straight edge or a "French" curve.

Advantages

1. The method is simple.
2. The method may better describe the trend than would a mathematical equation.

3. If drawn with care, the trend line fitted by this method will be a close approximation to a mathematically fitted trend.

Disadvantages

1. The results vary according to personal estimate.

2. Considerable practice is required to make a good fit.

Semi-Average Method

In this method the data are split into two equal parts, and the figures in each half are averaged. The averages thus obtained are plotted at the center of their respective periods (see Figure 5.2) and a straight line is drawn through the two points.

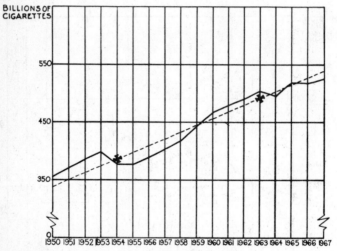

Source: United States Department of Commerce.

Fig. 5.2—Consumption of cigarettes in the United States, 1950–1967. Trend indicated by semi-average method.

In the example shown below (Table 5.1) the data are split in half (1950–1958 and 1959–1967). The two values obtained by averaging the figures in each half (385.38 for the first half and 492.80 for the second half) are plotted at the midpoints of their respective periods (the middle of 1954 for the first group and the middle of 1963 for the second). With the aid of a ruler a straight line is then drawn through the two points.

Table 5.1–Computation of Trend–Semi-Average Method
Consumption of Cigarettes in the United States, 1950–1967

Year	Consumption (Billions of cigarettes)	Totals	Arithmetic means
1950	355.10		
1951	369.71		
1952	385.43		
1953	396.70		
1954	378.43	3468.46 ÷ 9 =	385.38
1955	376.05		
1956	387.26		
1957	402.73		
1958	417.05		
1959	442.78		
1960	468.55		
1961	479.39		
1962	488.44		
1963	502.96	4435.24 ÷ 9 =	492.80
1964	495.07		
1965	516.96		
1966	516.14		
1967	524.95		

Advantages

1. The method is simple.

2. The result is entirely objective, i.e., it is not dependent upon individual estimate.

Disadvantages

1. The semi-average method makes use of the arithmetic mean, which as we have seen before is greatly affected by extreme values. For this reason the semi-average trend line may be pulled out of its true position by such unusual occurrences as strikes, etc.

2. The method is only useful in the fitting of straight line trends.

Moving Average Method

In the moving average method the trend is described by smoothing out fluctuations in the data by means of a moving average.

The moving average is a series of successive averages secured from a series of items by grouping a number of items and securing the group average. The first item in each group is dropped and the next item in the series included to obtain the next average. To obtain a three item moving average, in the illustration below, the first three numbers

(3, 5, and 7) are added (the total is entered in Column 2 next to the middle item of the group). The first number (3) is then replaced by the next number in the column of figures (in this case 10) and the process is continued until the entire series has been included. Each total is then divided by three and the result is placed in Colunn 3.

Table 5.2–Moving Average

(1) Values	(2) 3 Item Moving Total	(3) 3 Item Moving Average
3		
5	15	5.00
7	22	7.33
10	29	9.67
12	36	12.00
14	41	13.67
15	46	15.33
17		

The fluctuations caused by the business cycle in an economic time series may be removed or partially eliminated by including in the moving average a number of items (years, for example) equal to the length of the cycle which is evident in the data. The cyclical fluctuations will thus be smoothed out and a better measure of trend obtained. A twelve-month moving average is especially useful in handling monthly statistics subject to a seasonal variation.

Table 5.3 illustrates the application of a moving average to the consumption of cigarettes in the United States. This table demonstrates the procedure to be followed in fitting a moving average consisting of an odd number of items. The method consists in obtaining successive totals (in this case 7 items) by consecutively dropping the first item and adding the next in the series. The total for the first seven items in Table 5.3 is 2613.63. This sum is placed next to the middle item of the group (1952). The second figure in Column 3 (2648.68) is obtained by dropping the figure for 1949 and adding the figure for 1956. This process is continued until all the figures in Column 2 have been included. Each of the figures in Column 3 is then divided by 7 to obtain the seven-year moving average entered in Column 4.

When an even number of items is included in the moving average (as 6 years in Table 5.4, p. 54) the center point of the group will be between two years. It is, therefore, necessary to adjust or to shift (known technically as to center) these averages so that they coincide

Table 5.3—Computation of Trend—Moving Average Method

Consumption of Cigarettes in the United States, 1949–1967

(1) Year	(2) Consumption (Billions of cigarettes)	(3) Seven-Year Moving Total	(4) Seven-Year Moving Average (Col. 3 divided by 7)
1949	352.21		
1950	355.10		
1951	369.71		
1952	385.43	2613.63	373.38
1953	396.70	2648.68	378.38
1954	378.43	2696.31	385.19
1955	376.05	2743.65	391.95
1956	387.26	2801.00	400.14
1957	402.73	2872.85	410.41
1958	417.05	2973.81	424.83
1959	442.78	3086.20	440.89
1960	468.55	3201.90	457.41
1961	479.39	3294.24	470.61
1962	488.44	3394.15	484.88
1963	502.96	3467.51	495.36
1964	495.07	3523.91	503.42
1965	516.96		
1966	516.14		
1967	524.95		

Source: United States Department of Commerce.

with the years. Columns 3 and 4 (six-year moving total and six-year moving average) may be obtained by the methods outlined above. To center the values a two-item moving average is taken of the even-item moving average.

A two-year moving average is taken of the six-year average. The resulting average is located between the two six-year moving average values and coincides with the years. The end result (a two-year moving average of a six-year moving average) is known as the six-year moving average *centered*.

Advantages

1. Only simple computations are involved.
2. It may replace the fitting of complex mathematical curves.

Disadvantages

1. It cannot be brought up to date. Depending upon the number of items included, the last point in the trend must occur several years

Table 5.4–Computation of Trend–Moving Average Method
(Even Year Period)

Consumption of Cigarettes in the United States, 1949–1967

(1) Year	(2) Consumption (Billions of Cigarettes)	(3) Six-Year Moving Total	(4) Six-Year Moving Average	(5) Two-Year Moving Total of Col. 4	(6) Six-Year Moving Average Centered
1949	352.21				
1950	355.10				
1951	369.71				
		2237.58	372.93		
1952	385.43			749.83	374.92
		2261.42	376.90		
1953	396.70			759.16	379.58
		2293.53	382.26		
1954	378.43			770.03	385.02
		2326.60	387.77		
1955	376.05			780.81	390.41
		2358.22	393.04		
1956	387.26			793.76	396.88
		2404.30	400.72		
1957	402.73			816.46	408.23
		2494.42	415.74		
1958	417.05			848.70	424.35
		2597.76	432.96		
1959	442.78			882.78	441.39
		2698.94	449.82		
1960	468.55			916.35	458.18
		2799.17	466.53		
1961	479.39			946.06	473.03
		2877.19	479.53		
1962	488.44			971.43	485.72
		2951.37	491.90		
1963	502.96			991.73	495.87
		2998.96	499.83		
1964	495.07			1007.25	503.63
		3044.52	507.42		
1965	516.96				
1966	516.14				
1967	524.95				

Source: United States Department of Commerce.

before the end of the data. A five-year moving average ends three years before the end of the data, a six- and seven-year average four years, etc.

2. The concept of trend involves the idea of smooth growth or decline. The moving average is usually irregular in appearance.

3. A moving average fitted where the trend is that of a concave (upturning) curve will be higher than the true trend at all points, and lower in the case of a convex trend.

4. The moving average is computed by the use of the arithmetic mean. This form of average is greatly affected by extreme values. Because of this fact the moving average will be distorted by such unusual events as strikes, disasters, etc.

5. The number of items giving the smoothest moving average is equal to the number of years included in the average length of the business cycle in the data. Since this average length must be estimated by

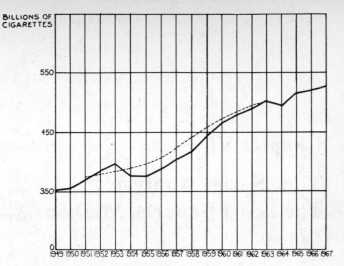

Source: *United States Department of Commerce.*

Fig. 5.3–Consumption of cigarettes in the United States, 1949–1967. Trend indicated by a six-year moving average.

the statistician the estimation will vary from person to person and, therefore, the method is not completely objective.

ADDITIONAL BIBLIOGRAPHY

DAVIS, H. T., *Analysis of Economic Time Series,* Principia, Bloomington, Ind., 1941.

SASULY, M., *Trend Analysis of Statistics*, Brookings Institute, Washington, D.C., 1934.

STEINER, P. O., *An Introduction to the Analysis of Time Series*, Holt, Rinehart & Winston, New York, 1956.

Chapter VI

Time Series Analysis
The Least Squares Method—
Linear

FORMULAE FOR STRAIGHT LINES

If a straight line is drawn on a graph its formula may be read by inspection.

The formula for line d in Figure 6.1, for example, is clearly

$$Y = X$$

Similarly, for lines e and f, $Y = 2X$ and $Y = 3X$

The value of the coefficient of X indicates the number of units the value of Y will increase for each unit increase in X. This constant is denoted by the letter b. The equation may be written more generally as

$$Y = bX$$

The greater the value of the constant b the more rapid the rise in the line. The value of b therefore is the measure of the slope of the line. If the line of trend is downward, indicating a decrease in Y for each unit increase in X, the b value will be negative.

The Y Intercept

If the values of X and Y for line g in Figure 6.2 are obtained the result will be

$$Y = 1 + 1X$$

The formula for line h is

$$Y = 1 + 2X$$

56

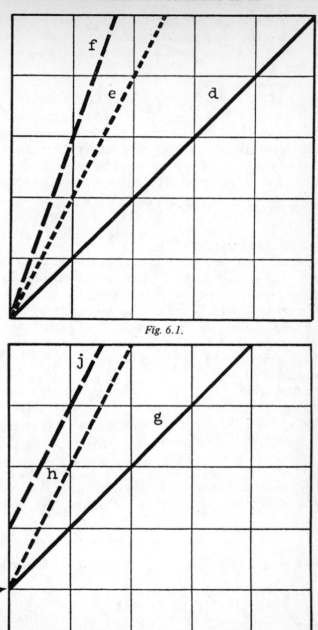

Fig. 6.1.

Fig. 6.2.

The value of the new constant is equal to the value indicated on the Y or vertical axis at the point at which the line crosses it, when X is equal to zero. This point is indicated by the arrow in Figure 6.2 for lines g and h.

The formula for line j is

$$Y = 2 + 2X$$

The constant (usually denoted by the letter a) is sometimes called the **Y intercept** because the value is determined by the point at which the line crosses the Y axis, while b indicates the **slope** of the line.

The formula for any straight line may be written generally as

$$Y = a + bX$$

THE LEAST SQUARES METHOD

If a straight line trend is assumed, the values of a and b must be determined. It is necessary to decide which line best describes the data. The principle of **least squares** aids in determining the line that best describes the trend of the data. The principle states that the line of best fit to a series of values is the line about which the sum of the squares of the deviations (the differences between the line and the actual values) is a minimum. There can be only one line having this property.

Least Squares Line

The least squares line for a given series is obtained by using a set of normal equations. These equations are derived mathematically (see Technical Appendix IV) but for working purposes they may be obtained by multiplying the type equation, in this case[1]

$$Y = a + bX$$

through by the coefficients of each unknown (a and b). The coefficient of the first unknown (a) is 1. Multiplying the type equation through by 1 we have

$$Y = a + bX$$

The formula must be summed up for all points. The summation results in

$$\Sigma(Y) = \Sigma a + b\Sigma(X)$$

---

[1] The method outlined here is not a derivation but rather a short method for obtaining the necessary normal equations.

But, the sum of a equals the number of items times the constant

$$\Sigma a = Na$$

Therefore

$$\text{(I)} \ \Sigma(Y) = Na + b\,\Sigma(X)$$

The coefficient of the second unknown (b) is X. Multiplying the type equation through by X we obtain

$$XY = aX + bX^2$$

This sums up to

$$\text{(II)} \ \Sigma(XY) = a\,\Sigma(X) + b\,\Sigma(X^2)$$

By the use of these two equations the values of the two unknowns can be determined and the trend line fitted.

APPLICATION OF THE LEAST SQUARES METHOD

The application of the least squares method for the determination of the trend for the production of aluminum for fifteen years in the United States is shown in Table 6.1. The equation, since it is as-

Table 6.1—Computation of Trend—Least-Squares Method

Annual Production of Aluminum in the United States, 1951–1965

(1)	(2)	(3) Production of Aluminum (Short tons)	(4)	(5)
Year	X	Y	XY	X^2
1951	0	1.13	0	0
1952	1	1.24	1.24	1
1953	2	1.62	3.24	4
1954	3	1.75	5.25	9
1955	4	1.90	7.60	16
1956	5	2.02	10.10	25
1957	6	2.01	12.06	36
1958	7	1.86	13.02	49
1959	8	2.31	18.48	64
1960	9	2.34	21.06	81
1961	10	2.24	22.40	100
1962	11	2.58	28.38	121
1963	12	2.82	33.84	144
1964	13	3.11	40.43	169
1965	14	3.40	47.60	196

$\Sigma(X) = 105$ $\Sigma(Y) = 32.33$ $\Sigma(XY) = 264.70$ $\Sigma(X^2) = 1015$

Source: Statistical Abstract of the United States.

sumed to be linear (straight line), must be of the type

$$Y = a + bX$$

from which the two normal equations are obtained.

(I) $\Sigma(Y) = Na + b\Sigma(X)$
(II) $\Sigma(XY) = a\Sigma(X) + b\Sigma(X^2)$

In order to solve for a and b the following values are necessary

$$\Sigma(X); \Sigma(Y); \Sigma(XY); \Sigma(X^2); N$$

Time is invariably placed on the X axis and for this reason constitutes the X variable, with production as the Y variable.

The sum for X or $\Sigma(X)$ is obtained by adding the *numbers* of
each year; for $\Sigma(Y)$ the production figures for all years are totaled.

The numbers of the years (1951, 1952, etc.) are inconvenient for
calculating purposes and, since the numbering system for the years is
arbitrary, the usual numbers are replaced with a simpler series of con-
secutive numbers. The correspondence between the two series must be
noted by indicating the original number of the year to which the number zero is now assigned. This is known as the **origin year**. The new
numbering system is shown in Column 2 of Table 6.1.

By substituting the values obtained in the two normal equations
(N = the number of years)

(I) $\Sigma(Y) = Na + b\Sigma(X)$
(II) $\Sigma(XY) = a\Sigma(X) + b\Sigma(X^2)$

we obtain

(I) $32.33 = 15a + 105b$
(II) $264.70 = 105a + 1015b$

When the equations are solved simultaneously, a is found to equal 1.196
and b, .137, The normal equation can now be written with numerical
coefficients and the formula for the line of trend written as

$$Y = 1.196 + .137\,X$$

In interpreting the equation it is necessary to know the origin year and
the units used in the enumeration of the original values.

The equation as finally stated will then read

Trend of Annual Production of Aluminum
in the United States 1951-1965

$$Y = 1.196 + .137\,X$$
Year of origin: 1951
Units: short tons

Graphic Presentation of Trends

To obtain the various trend values of Y (in order that the trend line may be drawn on the graph) the various values of X indicated for each year on the work sheet are substituted in the equation. For 1953 the X value indicated in the work sheet (Table 6.1) is 2. Therefore

$$Y = 1.196 + .137\,(2)$$
$$Y = 1.196 + .274$$
$$Y = 1.470 \text{ for the year } 1953$$

This process can be repeated until the trend value for each year is obtained. Since two points determine a straight line, in practice all that is needed to plot the line are two values for two different years, preferably years at the beginning and end of the period to ensure accuracy.

SHORT METHOD FOR COMPUTING TREND—ODD NUMBER OF YEARS

This procedure may be simplified when an odd number of years is used for the trend computation. The middle year may be taken as the origin year, thereby assigning to it an X value of zero. A minus sign is generally given to the X values for the years previous to the origin year and a plus sign to those following the origin year.

Applying this technique to the problem worked out above it will be found (see Column 2 of Table 6.2) that the sum of the X values will be zero, since they will consist of two like arithmetic progressions equal in amount but opposite in sign. This will modify the normal equations.

$$\text{(I)} \quad \Sigma(Y) = Na + b\Sigma(X)$$
$$\text{(II)} \quad \Sigma(XY) = a\Sigma(X) + b\,\Sigma(X^2)$$

Since $\Sigma(X) = 0$, the equations are simplified to

$$\text{(I)} \quad \Sigma(Y) = Na$$
$$\text{(II)} \quad \Sigma(XY) = b\Sigma(X^2)$$

and the need for a simultaneous solution is now eliminated.

By substituting the values of Table 6.2 in the two simplified normal equations we obtain

$$\text{(I)} \quad 32.33 = 15a$$
$$a = 2.155$$

$$\text{(II)} \quad 38.39 = 280b$$
$$b = .137$$

Table 6.2–Computation of Trend–Least-Squares Straight Line–
Odd Number of Years–Short Method

Annual Production of Aluminum in the United States, 1951–1965

(1)	(2)	(3) Production of Aluminum (Short tons)	(4)	(5)
Year				
	X	Y	XY	X^2
1951	-7	1.13	-7.91	49
1952	-6	1.24	-7.44	36
1953	-5	1.62	-8.10	25
1954	-4	1.75	-7.00	16
1955	-3	1.90	-5.70	9
1956	-2	2.02	-4.04	4
1957	-1	2.01	-2.01	1
1958	0	1.86	.00	0
1959	1	2.31	2.31	1
1960	2	2.34	4.68	4
1961	3	2.24	6.72	9
1962	4	2.58	10.32	16
1963	5	2.82	14.10	25
1964	6	3.11	18.66	36
1965	7	3.40	23.80	49

$\Sigma(X) = 0$ $\Sigma(Y) = 32.33$ $\Sigma(XY) = 38.39$ $\Sigma(X^2) = 280$

Source: Statistical Abstract of the United States.

The resulting equation is written as follows:

Trend of Annual Production of Aluminum
in the United States 1951-1965

$$Y = 2.155 + .137X$$
Origin: 1958
Units: short tons

Since a represents the value of Y when X equals zero, when a different zero point is used for X the a value is changed. The difference in the a value in the two trend equations (i.e., those obtained respectively by the long and short method) is due to the different points of origin. In the short method, 1958 was taken as the origin year; in the long method 1951. The value of X was zero in 1958 for the short method, and zero in 1951 in the long method.

The equations can be shown to give the same results by using the corresponding X values for any year in both formulae.

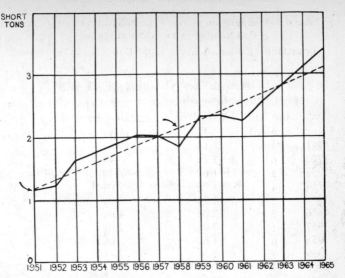

SHORT
TONS

Fig. 6.3–Annual production of aluminum in the United States, 1951-1965. Trend indicated by a least-squares straight line.

For 1956, for example, the long method gives

$$Y = 1.196 + .137 (5) = 1.881$$

while the short method gives

$$Y = 2.155 + .137 (-2) = 1.881$$

Shifting the Origin

A change in origin is merely a change in the starting point for the computation of the trend figure.

The b value (the measure of slope) is not affected by a shift in origin because no matter where the starting point be taken the line of trend will always have the same slope. In both equations in the previous section the b value was .137 in spite of the differing origins (1951 and 1958).

The a value indicates the value of Y when X equals zero. The following method may be used to change the point of origin of the first equation (1951 origin) to a 1958 origin.

Summary of steps for shifting origin:

1. Original equation $Y = 1.196 + .137 X$ (origin 1951)
2. Substitute value for X for new $X = 7$ since the 1958 trend value
    origin year   is desired

3. Obtain Y (trend) value $Y = 2.155$

4. Replace old a value with this $Y = 2.155 + .137\,X$ (origin 1958)
   new amount

SHORT METHOD–EVEN NUMBER OF YEARS

In applying the short method to a time series with an even number of years the fact that there is no middle year presents a difficulty. However, the middle point of the series may be used by assigning an origin between the two center years (as January 1, 1961 in Table 6.3).[1]

Table 6.3–Computation of Trend–Least Squares Straight Line
Short Method–Even Year Period
Trend Expansion of F. W. Woolworth Co. 1956–1965

(1)	(2)	(3)	(4)		
	Y	X	X'		
Years	Number of Stores	(In Years)	(In Half Years)	$X'Y$	X'^2
1956	2101	-4.5	-9	-18909	81
1957	2121	-3.5	-7	-14847	49
1958	2152	-2.5	-5	-10760	25
1959	2319	-1.5	-3	- 6957	9
1960	2430	- .5	-1	- 2430	1
1961	2502	.5	1	2502	1
1962	2529	1.5	3	7587	9
1963	3108	2.5	5	15540	25
1964	3129	3.5	7	21903	49
1965	3160	4.5	9	28440	81

$\Sigma(Y) = 25551$ $\Sigma(X) = 0$ $\Sigma(X') = 0$ $\Sigma(X'Y) = 22069$ $\Sigma(X'^2) = 330$

Source: Fairchilds Financial Manual.

The first year on either side of the origin will now be one half year away from the point of origin and will be given the number +.5 or -.5, as the case may be. The second year on either side will be numbered +1.5 and -1.5. This numbering system is shown in Table 6.3, Column 3.

Since decimals or fractions are cumbersome, the trend equation may be obtained more readily by working in terms of half years, eliminating decimal values (Column 4). These values may now be substituted in the two simplified normal equations

$$\text{(I)} \quad \Sigma(Y) = Na$$
$$\text{(II)} \ \Sigma(X'Y) = b\,\Sigma(X'^2)$$

---

[1] When a figure for a year is indicated in a time series it represents the middle of the year, or July 1 of that year.

resulting in

$$(I) \quad 25551 = 10a$$
$$a = 2555.1$$
$$(II) \quad 22069 = 330b$$
$$b = 66.88$$

The equation then reads

Trend of Number of Stores in Woolworth Chain 1956-1965

$$Y = 2555.1 + 66.88 \, X'$$

Origin: January 1, 1961
Unit: number of stores
X' in $\frac{1}{2}$ years

The equation in its present form is difficult to handle. For this reason it should be converted to the standard form with July 1 of the year as its origin.

To shift the origin to the center of 1961 (July 1) it is necessary to obtain the trend value (Y) at the new point of origin and use that value for a.

The trend value for $\frac{1}{2}$ year later may be obtained by substituting +1 for X' in the equation (since X' is in terms of half years).

$$Y = 2555.1 + 66.88 \times 1$$
$$Y = 2621.98$$

Y may now be used as the new a value and the equation written

$$Y = 2621.98 + 66.88 \, X'$$
Origin: July 1, 1961
Unit: number of stores
X' in $\frac{1}{2}$ years.

Finally, an adjustment must be made to convert X' (in half years) to X (in years).

The symbol b used as the coefficient of X represents the increase per half year (in this case 66.88 stores). To obtain the increase per year, b is doubled (133.76). The equation is now written:

Trend of Number of Stores in Woolworth Chain 1956-1965

$$Y = 2621.98 + 133.76 \, X$$
Origin: 1961
Unit: number of stores.

LEAST SQUARES METHOD—LINEAR

Advantages

1. The method expresses trend in the form of a mathematical formula which can be easily interpreted.

2. Results obtained by the method are definite and independent of any subjective estimate on the part of the statistician.

3. The resulting equation is in convenient form for extrapolation (extension into future or past).

Disadvantages

1. The technique demands mathematical calculations.

2. The method is based on the assumption that the data follow a trend that can be expressed by a linear equation.

ADDITIONAL BIBLIOGRAPHY

DAVIS, H. T., *Analysis of Economic Time Series*, Principia, Bloomington, Ind., 1941.

SASULY, M., *Trend Analysis of Statistics,* Brookings Institute, Washington, D.C., 1934.

STEINER, P. O., *An Introduction to the Analysis of Time Series*, Holt, Rinehart & Winston, New York, 1956.

Chapter VII

Time Series Analysis—
Non-Linear Trends

The straight line trend does not satisfactorily describe the trend of data which have a varying rate of growth. For example, in Figure 7.1 only a curved line can accurately describe the trend of the data.

METHODS OF FITTING NON-LINEAR TRENDS
The Potential Series
The parabola is the simplest type of curve used to describe the trend of data. The formula for a curve of this type is

$$Y = a + bX + cX^2$$

The fitting of a parabola follows closely the method of fitting a linear equation explained in the previous chapter. Here, however, there are three unknowns (a, b, and c) in the equations and three normal equations are needed for its solution.[1]

Method
1. Write down the equation for a parabola

$$Y = a + bX + cX^2$$

---

[1] In order to solve for a given number of unknowns in an equation it is necessary to have the same number of equations involving these unknowns.

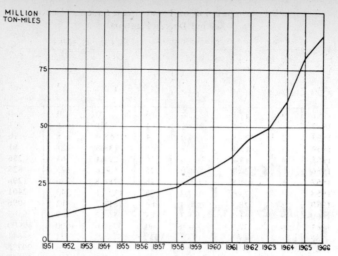

Source: Civil Aeronautics Administration.

Fig. 7.1—Air express and freight ton-miles flown, 1951–1966.

2. Multiply each term of the equation by the coefficient of each unknown and sum up.

　　a. For normal equation I multiply the type equation by the coefficient of a (1) and sum up

$$\text{(I)} \quad \Sigma(Y) = Na + b\Sigma(X) + c\Sigma(X^2)$$

　　b. For equation II multiply the equation by the coefficient of b (X) and sum up

$$\text{(II)} \quad \Sigma(XY) = a\Sigma(X) + b\Sigma(X^2) + c\Sigma(X^3)$$

　　c. For normal equation III multiply the equation by the coefficient of c (X^2), and sum up

$$\text{(III)} \quad \Sigma(X^2 Y) = a\Sigma(X^2) + b\Sigma(X^3) + c\Sigma(X^4)$$

3. From the values of X and Y, $\Sigma(X)$, $\Sigma(Y)$, $\Sigma(XY)$, $\Sigma(X^2)$, $\Sigma(Y^2)$, $\Sigma(X^2 Y)$, and N can be calculated in the way outlined in the previous chapter and their values substituted in the normal equations shown above. A simultaneous solution is then used to obtain the desired values for the unknown constants.[1]

---

[1] For review of the method of simultaneous solutions where there are more than two unknowns see any standard textbook on Elementary Algebra, as Harding, A. M. & Mullins, G. W., *College Algebra.*

Table 7.1—Computation of Trend—Least Squares Method
—Second Degree Parabola

Air Express and Freight Ton-Miles Flown in the U.S. 1951–1966

(1)	(2)	(3)	(4)	(5)	(6)	(7)	(8)
Years	X	Exports (Millions of ton-miles) Y	XY	X^2	X^2Y	X^3	X^4
1951	0	11.7	0	0	0	0	0
1952	1	13.1	13.1	1	13.1	1	1
1953	2	14.5	29.0	4	58.0	8	16
1954	3	15.4	46.2	9	138.6	27	81
1955	4	18.4	73.6	16	294.4	64	256
1956	5	20.0	100.0	25	500.0	125	625
1957	6	21.8	130.8	36	784.8	216	1296
1958	7	23.9	167.3	49	1171.1	343	2401
1959	8	28.3	226.4	64	1811.2	512	4096
1960	9	31.7	285.3	81	2567.7	729	6561
1961	10	37.1	371.0	100	3710.0	1000	10000
1962	11	45.3	498.3	121	5481.3	1331	14641
1963	12	39.2	590.4	144	7084.8	1728	20736
1964	13	60.6	787.8	169	10241.4	2197	28561
1965	14	76.3	1068.2	196	14954.8	2744	38416
1966	15	89.6	1344.0	225	20160.0	3375	50625
	120	556.9	5731.4	1240	68971.2	14400	178312

Source: Standard and Poors Statistics.

The fitting procedure is outlined in Table 7.1.

$$\text{(I)} \ \Sigma(Y) = Na + b\Sigma(X) + c\Sigma(X^2)$$
$$556.9 = 16a + 120b + 1240c$$
$$\text{(II)} \ \Sigma(XY) = a\Sigma(X) + b\Sigma(X^2) + c\Sigma(X^3)$$
$$5731.4 = 120a + 1240b + 14400c$$
$$\text{(III)} \ \Sigma(X^2Y) = a\Sigma(X^2) + b\Sigma(X^3) + c\Sigma(X^4)$$
$$68971.2 = 1240a + 14400b + 178312c$$

Solving Equations I and II to eliminate a

(II) $5731.40 = 120a + 1240b + 14400c$
(IV) $4176.75 = 120a + 900b + 9300c$ (Equation I × 7.5)
(V) $1554.65 = 340b + 5100c$ (II − IV)
(III) $68971.20 = 1240a + 14400b + 178312c$
(VI) $43159.75 = 1240a + 9300b + 96100c$ (Equation I × 77.5)
(V) $25811.45 = 5100b + 82212c$ (III − VI)
(VII) $23319.75 = 5100b + 76500c$ (Equation V × 15)
  $2491.70 = 5712c$ (V − VII)
  $c = .43622$

Substituting the value of c in Equation V

$$1554.65 = 340b + 5100\,(.43622)$$
$$= 340b + 2224.72$$
$$b = -1.97079$$

Substituting the values of b and c in Equation II

$$5731.4 = 120a + 1240\,(-1.97079) + 14400\,(.43622)$$
$$120a = 1893.6116$$
$$a = 15.78009$$

The final trend equation then reads

Trend of Air Express and Freight
United States 1961-1966
$$Y = 15.78009 - 1.97079X + .43622X^2$$
Origin: 1951
Units: Million ton-miles

The trend values for the various years may now be obtained by substituting the appropriate values of X (as indicated in Table 7.1, Colum 2). Thus for 1961, substituting 10 for X

$$Y = 15.78009 - 1.97079(10) + .43622(10)^2$$
$$Y = 15.78009 - 19.7079 + 43.622$$
$$Y = 39.69419$$

The method can be further simplified when the data consist of an odd number of items. The middle year is selected as the year of origin, so that $\Sigma(X)$ and $\Sigma(X^3)$ will be equal to zero. The normal equations are therefore reduced to

$$\text{(I)} \quad \Sigma(Y) = Na + c\Sigma(X^2)$$
$$\text{(II)} \quad \Sigma(XY) = b\Sigma(X^2)$$
$$\text{(III)} \quad \Sigma(X^2 Y) = a\Sigma(X^2) + c\Sigma(X^4)$$

The values of a and c are then obtained simultaneously in the usual way, while b is obtained directly.

A more flexible curve than the second degree parabola may be obtained by using a curve of a higher degree.[1]

The third degree equation has the formula

$$Y = a + bX + cX^2 + dX^3$$

---

[1] The "degree" of an equation corresponds to the largest exponent in the equation.

and the general formula for this type of curve is

$$Y = a + bX + cX^2 + dX^3 + eX^4 + \ldots$$

These more elaborate forms of equation tend to follow the data more closely but they must be used with care if they are to describe the trend of the figures rather than the cyclical or seasonal movement.

The solutions for curves of a higher degree may be arrived at by simultaneous equations, as with second degree curves. The normal equations for the more complex formulae are obtained as before from the type equation and the values of the various unknown coefficients a, b, c, d, etc. calculated by the method previously described.

Exponential Series.

Occasionally neither the straight line nor the potential curve will be appropriate for describing the trend of a particular series. This occurs, for instance, where the trend is geometric in nature. One curve descriptive of the geometric type of trend has the formula

$$Y = ab^x$$

This type of trend appears when the Y values tend to form a geometric progression.[1] If plotted on semi-log paper (with logarithmic ruling on Y axis) the trend will become linear, and the resulting curve is therefore referred to as the **semi-logarithmic curve.** (See Chapter XX, p. 202).

If a geometric progression is formed by the Y values when the X values are arranged geometrically the formula is

$$Y = aX^b$$

This type of curve will be linear on logarithmic paper (logarithmic rulings on both the X and Y axes).

The formulae of the exponential type such as those above may be fitted readily by reducing them to logarithmic form.

$$Y = ab^x$$

reduced to logarithms reads

$$\log Y = \log a + X \log b$$

The normal equations may then be obtained as above.

There are a number of special exponential curves of some im-

---

[1] A geometric progression is a series in which the values increase at a constant ratio, as in the series 1, 2, 4, 8, 16, etc.

portance for trend purposes. One of the more important curves of this type is known as the Gompertz curve. The formula is

$$Y = ab^{c^x}$$

ADDITIONAL BIBLIOGRAPHY

DAVIS, H. T., *Analysis of Economic Time Series*, Principia, Bloomington, Ind., 1941.

SASULY, M., *Trend Analysis of Statistics*, Brookings Institute, Washington, D. C., 1934.

STEINER, P. O., *An Introduction to the Analysis of Time Series*, Holt, Rinehart & Winston, New York, 1956.

Chapter VIII
Seasonal and Cyclical Analysis

Seasonal variation is the technical term given the more or less regular movements within the year which occur year after year.

In every time series with seasonal variation each month has a typical value position in relation to the year as a whole. The problem of seasonal variation is to determine this typical (or average) position of each month.

METHODS OF MEASURING SEASONAL VARIATION

The most generally used methods for measuring the seasonal variation occurring within a time series are

1. simple average method
2. link relative method
3. ratio to moving average method
4. ratio-to-trend method

THE SIMPLE AVERAGE METHOD

1. Average (using the arithmetic mean) the values for each month for all the years (see Table 8.1, p. 74). The result is the typical value for each of the twelve months.

2. Adjust for trend. Each of the averages just computed will be distorted by the secular trend of the data. If the trend is upward, December will be higher than it should be in relation to the rest of the months since it occurs later along the trend line.

The increase per month due to trend may be determined by fitting a least squares line to the yearly figures and dividing the b value (slope) by 12. The resulting value will then represent the amount each

Table 8.1—Average Weekly Freight Car Loadings in U.S. 1956–1967

Year	Jan.	Feb.	March	Apr.	May	June	July	Aug.	Sept.	Oct.	Nov.	Dec.	Aver.
						(Unit 1000 Cars)							
1956	678	692	701	742	765	786	599	740	788	817	735	660	725
1957	643	668	678	674	712	740	690	749	713	727	627	555	682
1958	543	528	537	528	549	622	552	644	642	682	615	547	582
1959	548	573	509	633	686	703	559	542	553	584	600	572	596
1960	597	573	580	622	640	613	574	592	582	630	545	470	586
1961	480	489	501	527	555	582	543	593	588	645	577	509	549
1962	518	530	548	562	574	589	511	576	568	608	551	477	551
1963	482	516	532	560	587	596	539	578	577	628	554	493	554
1964	517	526	526	554	586	601	537	589	603	644	594	510	566
1965	535	519	547	575	604	585	555	595	579	626	587	530	570
1966	526	524	558	557	609	593	544	594	593	631	581	513	569
1967	512	514	532	555	558	555	492	558	551	586	552	496	538
Total	6579	6652	6749	7089	7425	7565	6695	7350	7337	7808	7118	6332	
Averages	548.3	554.3	562.4	590.8	618.8	630.4	557.9	612.5	611.4	650.7	593.2	527.7	

Source: Standard and Poors Statistics.

Table 8.2–Computation of Index of Seasonal Variation
Simple Average Method
Freight Car Loadings in the United States, 1956–1967

(1) Month	(2) Average for month (from Table 8.1)	(3) Trend Correction	(4) Corrected Average	(5) Index of Seasonal Variation
January	548.3	–	548.25	.93
February	554.3	−.22	554.08	.94
March	562.4	−.44	561.96	.96
April	590.8	−.66	590.09	1.00
May	618.8	−.88	617.87	1.05
June	630.4	−1.10	629.30	1.07
July	557.9	−1.32	556.58	.95
August	612.5	−1.54	610.96	1.04
September	611.4	−1.76	609.64	1.04
October	650.7	−1.98	648.72	1.10
November	593.2	−2.20	591.00	1.01
December	527.7	−2.42	525.28	.89
Total			7046.73	
Average			587.23	

monthly average is distorted by the trend as compared to the previous month.

Thus to reduce the February average to the level of the first month, January, the amount of the trend increment is subtracted from that average. To reduce March to the January level, it is necessary to subtract from it twice the trend increment, for April 3 times, etc. (see Table 8.2, Column 3).

3. The resulting corrected averages may then be expressed as a percentage of the average of the entire period (587.23). These values are known as the indices of seasonal variation. The figure of 93% for January means that the figure for January is typically 7% below the average for the year.

LINK RELATIVE METHOD

The first step in the **link relative** method is to express the value for each month as a percentage of the previous month. From the data for bituminous coal production in the United States 1951–1966 the figure for January, 1951 (51.53) is divided into the figure for February, 1951 (39.99), the figure for February, 1951 (39.99) is divided into that

Table 8.3–Link Relatives–Bituminous Coal Production in the United States, 1951–1966

Year	Jan.	Feb.	March	April	May	June	July	Aug.	Sept.	Oct.	Nov.	Dec.
1951	. . .	77.60	111.81	93.68	103.33	100.39	78.27	138.42	91.04	120.59	95.22	89.42
1952	113.48	87.60	93.67	95.44	93.25	86.29	82.01	132.54	137.21	69.84	125.40	104.05
1953	93.52	86.88	106.30	101.59	100.62	103.45	90.49	115.14	101.79	98.96	87.42	104.54
1954	91.78	87.27	106.05	89.75	102.38	105.02	90.33	120.69	102.88	106.25	101.39	102.64
1955	95.31	97.22	104.56	92.85	110.75	93.87	101.41	117.76	94.92	102.50	104.36	104.85
1956	99.86	93.57	102.31	92.76	109.46	89.40	77.77	143.87	91.53	119.12	92.35	89.14
1957	113.34	89.29	107.89	98.18	102.17	91.64	87.19	125.57	94.64	111.59	84.21	96.51
1958	104.02	83.39	102.01	92.54	102.20	111.39	70.19	141.64	107.37	108.79	86.56	114.36
1959	91.67	93.94	103.28	99.15	101.14	103.61	66.29	123.43	108.25	107.22	103.08	112.66
1960	89.84	96.07	111.71	89.43	103.68	92.69	75.21	144.21	94.53	102.33	94.60	98.48
1961	101.23	88.91	103.16	97.46	118.10	91.46	84.33	139.79	93.56	110.95	94.38	94.51
1962	108.25	87.54	109.62	94.26	107.97	101.69	58.85	176.28	87.61	117.96	92.48	88.90
1963	112.34	92.47	98.82	113.18	107.72	94.95	71.14	150.69	95.32	111.30	86.51	100.64
1964	107.89	85.77	105.71	102.06	101.86	105.89	76.90	131.10	103.17	104.96	91.19	103.55
1965	92.28	94.62	113.08	97.77	100.46	102.81	79.13	135.65	93.79	107.48	99.49	100.50
1966	91.54	95.20	119.57	62.64	150.92	99.19	76.74	415.32	92.70	103.70	95.51	103.28

for March, 1951 (44.71) etc. The resulting percentage figures as 77.60% for February, 111.81% for March, etc. are called link relatives (see Table 8.3).

The link relatives for each month (all the Januarys, etc.) are then averaged. The typical position of each month in relation to the previous month is thus obtained. For instance, June production is typically 99.8% of May. Since the arithmetic mean of the monthly averages would be distorted by unusual monthly values, the median is used as the averaging method since it is less disturbed by extreme values.

The median (or typical) link relatives show the relation of each month to the month before but not to the rest of the months. To do this, it is necessary to establish a relationship between the various links or convert them into a series of chain relatives. This is accomplished by arbitrarily setting the value of January as 100%. The median link for February is 89.1%, which indicates that February is typically that percent of January and therefore is 89.1% of 100%. The median link for March establishes its chain relative as 105.9% of the chain for February (89.1%) or 94.3%. This computation is continued by multiplying each of the median link relatives by the chain for the preceding month. This process is repeated for each month, and the process is continued by establishing a second median link for January. The result may be seen in Table 8.4, Column 2.

Table 8.4—Computation of Seasonal Variation (Link Relative Method)

Month	(1) Median Link Relative	(2) Chain Relatives	(3) Adjusted Chain Relatives
January	99.9	100.0	100.0
February	89.1	89.1	89.3
March	105.9	94.3	94.8
April	94.9	89.5	90.2
May	102.9	92.0	93.0
June	99.8	91.9	93.0
July	78.0	71.7	73.0
August	139.1	99.7	101.3
September	94.8	94.5	96.3
October	107.4	101.4	103.5
November	94.5	95.8	98.1
December	101.6	97.4	99.9
January	99.9	97.3	100.0

A discrepancy exists between the chain relative for the first January and that for the second. The difference is due to the trend, which makes each succeeding January higher or lower than that of the preceding year.

The difference between the two values (2.9) represents the trend increment. To adjust the chain relatives for the effect of trend, increasing multiples of one-twelfth of the discrepancy from each chain value— starting with 1/12 for February, 2/12 for March—must be subtracted out. The chain relatives will then conform to the same base as January[1] (see Table 8.4, Column 2).

The end result is an index of seasonal variation with a base of January.

RATIO TO MOVING AVERAGE METHOD

1. The seasonal variations in the data are smoothed by means of a twelve-month moving average. The differences between the actual values and this moving average are due to seasonal movements.

2. The ratio of each value to the corresponding moving average value for each month is obtained.

3. The ratios are then averaged for each month of all the years, using either the mean or median for this purpose.

4. The resulting averages are the indices of seasonal variation.

RATIO–TO–TREND METHOD

The ratio-to-trend method measures the seasonal variation and in addition the combined cyclical and residual variations.

Express each actual monthly value as a percentage of its corresponding trend value, as computed by the least squares method[2] (see Table 8.5, Column 4). These values represent the original data expressed in percentage form with trend removed.

The ratios $\left(\dfrac{\text{actual}}{\text{trend}}\right)$ are then averaged for each month over the

---

[1] The trend discrepancy may be distributed on various other bases (see Mills, F. C., *Statistical Methods*).

[2] The monthly trend values may be readily obtained by fitting the least squares line to the annual averages and dividing the b value by 12. Since the annual equation has its origin at the middle of the year (July 1) it will be necessary to shift the origin ½ month forward to center it on the month. This may be accomplished by adding ½ of the b value to the a value.

Table 8.5—Computation of Seasonal and Cyclical Variation—Ratio to Trend Method

Production Workers
Confectionery and Related Products
United States Total 1947–1965

(1)	(2)	(3)	(4)	(5)	(6)
Year and Month	Production Workers (Thousands of Men)	Trend		Index of Seasonal	Cyclical and Residual
	A	T	A/T	1 + S	C + R
1964					
January	62.00	56.79	1.09	1.02	.07
February	62.50	56.68	1.10	.98	.12
March	60.60	56.56	1.07	.98	.09
April	58.00	56.45	1.03	.93	.10
May	57.80	56.33	1.03	.90	.13
June	58.50	56.22	1.04	.92	.12
July	56.40	56.10	1.01	.89	.12
August	61.60	55.99	1.10	.95	.15
September	65.40	55.87	1.17	1.08	.09
October	67.40	55.76	1.21	1.14	.07
November	67.70	55.64	1.22	1.14	.08
December	67.20	55.53	1.21	1.11	.10
1965					
January	62.60	55.41	1.13	1.02	.11
February	61.30	55.30	1.11	.98	.13
March	61.60	55.18	1.12	.98	.14
April	57.70	55.07	1.05	.93	.12
May	57.20	54.95	1.04	.90	.14
June	57.30	54.84	1.05	.92	.13
July	54.60	54.72	1.00	.89	.11
August	61.70	54.61	1.13	.95	.18
September	65.90	54.49	1.21	1.08	.13
October	67.90	54.38	1.25	1.14	.11
November	68.20	54.26	1.26	1.14	.12
December	67.00	54.15	1.24	1.11	.13

Source: *Employment and Earnings Statistics for the United States 1909–1966, U.S. Department of Labor.*

entire period of years. If used in averaging the arithmetic mean may be distorted by extreme values; therefore these values are excluded before averaging.[1]

---

[1] If the median is used for the averaging process it may not be typical when there are few items in the group averaged.

The extreme or unusual values may be located by means of a **multiple frequency table**. The multiple frequency table is a frequency distribution of the ratios (A/T) with one column for each month (illustrated in Figure 8.1).

	Jan.	Feb.	Mar.	Apr.	May	June	July	Aug.	Sept.	Oct.	Nov.	Dec.
1.22–1.25										//	///	//
1.18–1.21									/	////	/	/
1.14–1.17	/								//	//	///	//
1.10–1.13	/	(///)	/					//	///	////	///	////
1.06–1.09	//		/		(//)			/	ℕℕ /	ℕℕ //	ℕℕ /	////
1.02–1.05	///	//	//	//		//		///	ℕℕ		//	ℕℕ
.98–1.01	ℕℕ //	ℕℕ //	///			/	//	ℕℕ	//			/
.94– .97	ℕℕ	ℕℕ /	ℕℕ //	////	//	/	/	ℕℕ				
.90– .93		/	ℕℕ	ℕℕ ///	////	ℕℕ /	///	///				
.86– .89				////	ℕℕ ////	ℕℕ ///	ℕℕ ////					
.82– .85				//	/	////						

Fig. 8.1–Multiple Frequency Table of ratios of production working employment trend values for each month, 1947-1965.

When the average is computed the unusual figures (indicated by circles in the multiple frequency table) are excluded from the calculation.

The resulting average ratios of actual values to trend will show the typical relation of each month to trend and can be used as seasonal indices (see Table 8.6). For example, the value of 98% for March means that that month is typically 2% below the trend for the month.

If the seasonal index for each month is subtracted from the month's ratio of actual to trend, the seasonal variation will be eliminated from the series, leaving only the combined cyclical and residual (random) fluctuations.

Ratio-to-Trend—Summary

1. Fit a trend line to the data.

2. Compute the ratio of each actual value to its respective trend (A/T).

3. Average the ratios for each month, using the arithmetic mean for averaging. First, however, eliminate extreme values located by means of the multiple frequency table. The resulting figures are the indices of seasonal variation.

4. Subtract the respective indices of seasonal variation from the

Table 8.6—Computation of Seasonal Index—Ratio-to-Trend Method

Month	Monthly* Total (1947–1965)	Number of* Months Used	Monthly Average
January	19.45	19	1.02
February	15.72	16	.98
March	18.59	19	.98
April	17.75	19	.93
May	15.28	17	.90
June	17.54	19	.92
July	16.96	19	.89
August	18.04	19	.95
September	20.61	19	1.08
October	21.70	19	1.14
November	21.64	19	1.14
December	21.02	19	1.11

*Does not include "extreme" months, see multiple frequency table.

ratios for each month. The resulting series represents the cyclical and residual fluctuations occurring in the series.

ADDITIONAL BIBLIOGRAPHY

DAVIS, H. T., *Analysis of Economic Time Series,* Principia, Bloomington, Ind., 1941.
SASULY, M., *Trend Analysis of Statistics,* Brookings Institute, Washington, D. C., 1934.
STEINER, P. O., *An Introduction to the Analysis of Time Series*, Holt, Rinehart & Winston, New York, 1956.

Chapter IX

Correlation and Regression Analysis—Linear

It is often desirable to observe and measure the association which occurs between two or more statistical series. It may, for instance, be desirable to know whether there is a relationship between changes in the cost of living and changes in wages; the grades and intelligence quotients of a group of students; the amount of electric current passed through a solution and the amount of substance deposited by electrochemical reaction; the length of time elapsed and the amount of academic material retained in the memory; and many other similar associated (correlated) series.

The relationship, or more accurately the association, between series may be established and measured by means of the correlation technique.

THE SCATTER DIAGRAM

If two related (associated) series are plotted graphically with one variable placed on the X axis and the other on the Y axis, the result is known as the **scatter diagram**.[1]   If there is a definite relationship between the associated variables the result will be seen on a chart as a definite line of movement or "path" as in Figure 9.1.

If the relationship were perfect it is obvious that for every given value on the X axis, there would always be indicated a certain value on

---

[1] The independent variable is placed on the X axis while the dependent variable is placed on the Y axis. The dependent variable is that which it is desired to estimate from the other variable.

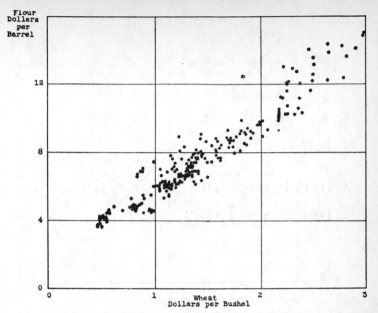

Source: *Wheat prices—Daily Trade-Bulletin. Flour prices—Northwestern Miller.*

Fig. 9.1—Scatter Diagram of relation between wheat prices and flour prices, by months.

the *Y* axis. In a situation like this all the points would coincide with a curve or line instead of forming a path across the face of the scatter diagram. In Figure 9.2 the figures of a bond yield index are plotted against those for the bond price index. Inasmuch as one series was computed from the other the relationship, of course, is perfect.

When the series are imperfectly associated the values of *Y* when a given value of *X* is selected will not lie on a definite line or curve. In accordance with the more or less imperfect relationship the points will depart from the indicated line or curve, creating a scatter. If there is a high degree of association the scatter will be confined to a narrow path. The less perfect the relationship between the two sets of data, the greater will be the departures from the indicated line or course. These departures are known as a **scatter**.

LINE OF REGRESSION

The trend of this movement may be defined by means of a least squares line or curve. The resulting curve is known as the **line of re-**

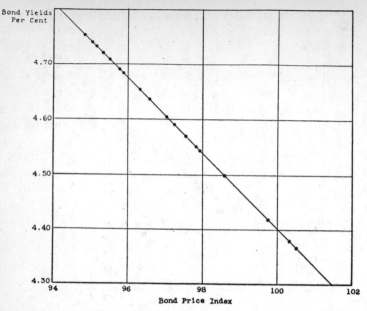

Fig. 9.2—Scatter Diagram of relationship between standard statistics index of bond yields and index of bond prices.

gression.[1]   If the trend of the data is linear (non-linear regressions are treated in Chapter X), the resulting equation will be of the type

$$Y = a + bX$$

The values of a and b are obtained from the normal equations by the least squares procedure described in Chapter VI.[2]

$$\text{(I)} \ \Sigma(Y) = Na + b\Sigma(X)$$
$$\text{(II)} \ \Sigma(XY) = a\Sigma(X) + b\Sigma(X^2)$$

STANDARD ERROR OF ESTIMATE

The line of regression is used to estimate a theoretical value of Y for a given value of X. If the relationship is not perfect the actual

---

[1] The techniques described below for deriving an equation of relationship are often referred to as *regression analysis*.

[2] Other methods of fitting such curves may be used, generally with less useful results.

values will not usually coincide with the theoretical values, because of the scatter. If the scatter is definitely measured the variations may then be allowed for, and a range established within which a given proportion of values will fall.

The measure used for this purpose, the **standard error of estimate**, is similar to the standard deviation. The standard deviation measures the variation or scatter about the arithmetic mean, while the standard error of estimate is a measure of the variation or scatter about the line of regression.

The standard deviation is the average (quadratic mean) of the deviations about the arithmetic mean, while the standard error of estimate is the average (quadratic mean) of the deviations about the line of regression.

$$S_y = \sqrt{\frac{\Sigma(d^2)}{N}}$$

where

S_y = standard error of estimate
d = deviation of actual values (Y)
from theoretical (Y_c), or $(Y - Y_c)$

The standard error of estimate may be used in the same manner as the standard deviation. One standard deviation measured off plus and minus about the arithmetic mean includes approximately 68% of the cases; and one standard error of estimate will also include about 68% of the cases when measured off plus and minus about the line of regression.[1]

Number of Standard Errors	Percent of Cases Included
± .6745 S_y	50%
± 1.0000	68.26%
± 1.9600	95%
± 2.5800	99%
± 3.0000	99.73%

---

[1] It is assumed that there is a normal or approximately normal distribution of the values about the line of regression. It is also assumed that there is uniformity of variation about the entire range of the line of regression (homoscedasticity). For a more complete discussion of these percentage figures see Chapter XII.

Table 9.1—Computation of Coefficient of Correlation—Ungrouped Data

Circulation and Minimum Line Rates for National Advertising in 39 Daily Newspapers in New England

(1) News-papers	(2) Circula-tion (thou-sands) (X)	(3) Rate per Line (in cents) (Y)	(4) XY	(5) X^2	(6) Y^2	(7) Theoret-ical Re-gression Values (Y_c)	(8) $(Y - Y_c)$ d	(9) d^2
#1	166	33	5478	27556	1089	34	−1	1
2	192	42	8064	36864	1764	38	+4	16
3	301	57	17157	90601	3249	55	+2	4
4	149	30	4470	22201	900	31	−1	1
5	111	25	2875	13225	625	25	+0	0
6	108	23	2484	11664	529	25	−2	4
7	446	75	33450	198916	5625	78	−3	9
8	381	65	24765	145161	4225	68	−3	9
9	399	70	27930	159201	4900	71	−1	1
10	158	32	5056	204964	1024	33	−1	1
11	451	79	35629	203401	6241	79	+0	0
12	133	27	3591	17689	729	29	−2	4
13	108	22	2376	11664	484	25	−3	9
14	154	30	4620	23716	900	32	−2	4
15	231	47	10857	53361	2209	44	+3	9
16	150	32	4800	22500	1024	31	+1	1
17	403	70	28210	162409	4900	71	−1	1
18	149	32	4768	22201	1024	31	+1	1
19	343	65	22295	117649	4225	62	+3	9
20	247	50	12350	61009	2500	47	+3	9
21	117	25	2925	13689	625	26	−1	1
22	231	47	10857	53361	2209	44	+3	9
23	217	43	9331	47089	1849	42	+1	1
24	196	42	8232	38416	1764	39	+3	9
25	166	33	5478	27556	1089	34	−1	1
26	124	25	3100	15376	625	27	−2	4
27	182	35	6370	33124	1225	36	−1	1
28	166	33	5478	27556	1089	34	−1	1
29	112	28	3136	12544	784	26	+2	4
30	177	35	6195	31329	1225	36	−1	1
	6468	1252	322227	1725088	60650			125

Source: Editor and Publisher, International Yearbook.

For the data in Table 9.1 the standard deviation is

$$\sigma_y = \sqrt{\frac{\Sigma(Y^2)}{N} - \left(\frac{\Sigma(Y)}{N}\right)^2} = \sqrt{\frac{60650}{30} - \left(\frac{1252}{30}\right)^2} = 16.74 \text{¢}$$

The procedure for obtaining the least squares line of regression and the standard error of estimate is outlined below.

(I) $\Sigma(Y) = Na + b\Sigma(X)$

(II) $\Sigma(XY) = a\Sigma(X) + b\Sigma(X^2)$

$$\begin{array}{ll}
\text{(I)} & 1252 = 30a + 6468b \\
\text{(II)} & 322227 = 6468a + 1{,}725{,}088.0b \\
\text{(III)} & 269931 = 6468a + 1{,}394{,}500.8b \quad \text{(I} \times 215.6) \\
\hline
& 52296 = 330{,}587b \qquad\qquad \text{(II–III)} \\
& b = .1582
\end{array}$$

Substituting the value of b in Equation I

$$1252 = 30a + 6468(.1582)$$
$$30a = 228.7624$$
$$a = 7.6254$$

The line of regression

$$Y_c = 7.6254 + .1582X$$

where

$$Y_c = \text{minimum rate per line in cents}$$
$$X = \text{circulation in thousands}$$

The standard error may now be computed by obtaining the theoretical regression values from this equation (Table 9.1, Column 7), and the difference between these and the actual values (Column 8).

For instance the regression value for paper #3 with a circulation (X) of 301 is obtained as follows[1]

$$Y_c = 7.6254 + .1582(301)$$
$$Y_c = 55¢$$

The standard error of estimate may then be computed.

$$S_y = \sqrt{\frac{\Sigma(d^2)}{N}} = \sqrt{\frac{125}{30}}$$

$$= \sqrt{4.17} = 2.04¢$$

On the basis of the equation obtained:

$$Y = 7.6254 + .1582X$$

and with a standard error of estimate of 2.04¢, a newspaper that has a circulation of 400,000 can be expected to have a minimum linage rate for national advertising of between 65¢ to 77¢ (3 standard errors—

---

[1] The symbol Y_c is used to differentiate the theoretical regression value from the actual value.

99.7% chances out of 100). 95 papers out of 100 with this circulation would have a minimum rate between 67¢ and 75¢.

The range of values is obtained by substituting the circulation value for X

$$Y_c = 7.6254 + .1582(400)$$

resulting in a theoretical value of 71¢ for the estimated linage rate. If 3 standard errors of estimate are then measured about this value, 99.7% of all papers can be expected to fall within this range while if 1.96 times the value of the standard error of estimate is added to and subtracted from this value 95% of the newspapers would be included within this range.

COEFFICIENT OF CORRELATION

The standard error of estimate is a measure of the degree of association between series. The larger the value of the standard error the greater the scatter about the line of regression and the poorer the relationship.

Standard errors cannot always be directly compared because the standard error is expressed in terms of the original unit of the Y variables, and the units are different. For instance, the standard error in the association of the intelligence quotient and examination grades cannot be compared with the association of the intelligence quotient and the number of words misspelled in a spelling test.

Very often one of two variables has a more limited range. For instance, if there are only 20 words in the spelling test and the number spelled correctly is taken as the score, the variation about the regression line is limited by this range, while a wider variation is allowed within the limits of 0 to 100% on an arithmetic test. If the standard error of estimate is divided by the standard deviation (of the Y values) the resulting ratio can be put into percentage form. Both measures are in the same units, the factor of dispersion is made a constant, and thus both difficulties arising from differing dispersions and units are overcome.

$$\frac{S_y}{\sigma_y}$$

When the relationship is perfect there will be no deviations from the line of regression. S_y will then be equal to zero and so will the ratio. If the relationship is poor the value of the standard error will be larger, the limit of its value being that of the standard deviation. The ratio will thus attain as its other limit 1, or 100%.

A perfect relationship is indicated by a ratio equal to zero and an imperfect relationship by a value of 1. Since this inverts the usual manner of thinking in regard to such subjects a more readily understandable value can be obtained by subtracting the ratio from 1.

$$1 - \frac{S_y}{\sigma_y}$$

The new measure has a value of 1 for a perfect relationship and a value of zero for a wholly imperfect relationship. A similar measure is termed the **coefficient of correlation** and is used as the comparative measure of association. The formula is

$$r = \sqrt{1 - \frac{S_y^2}{\sigma_y^2}}$$

The coefficient of correlation has the same limits as the value outlined above, zero and 1. The value of r in Table 9.1 is

$$r = \sqrt{1 - \frac{(2.04)^2}{(16.74)^2}} = .9925$$

Although the relationship may be good or even perfect it may be inverse; that is to say, an increase in the value of X results in a decrease in the value of Y. In that case the line of regression slopes downward. The value of b, the coefficient of slope (also called the coefficient of regression), is then negative.

The sign of the coefficient of slope (or regression) is attached to r to indicate whether it is positive or negative.

The problem of the measurement of association may be divided into three parts.

1. The determination of the form of relationship—the line of regression.
2. The measurement of variation about the established form of relationship—the standard error of estimate.
3. The reduction of measurement of association to a relative basis—the coefficient of correlation.

PRODUCT MOMENT METHOD–UNGROUPED DATA

By the use of algebraic manipulation a much less arduous method for computing r, S_y, and the line of regression[1] may be evolved from the fundamental formula for the coefficient of correlation.

$$r = \sqrt{1 - \frac{S_y^{\,2}}{\sigma_y^{\,2}}}$$

The formula for the simpler method is[2]

$$r = \frac{p}{\sigma_x \sigma_y}$$

where, for ungrouped data,

$$p = \frac{\Sigma(XY)}{N} - \left(\frac{\Sigma X}{N}\right)\left(\frac{\Sigma Y}{N}\right)$$

$$\sigma_x = \sqrt{\frac{\Sigma(X^2)}{N} - \left(\frac{\Sigma X}{N}\right)^2}$$

$$\sigma_y = \sqrt{\frac{\Sigma(Y^2)}{N} - \left(\frac{\Sigma Y}{N}\right)^2}$$

S_y can be obtained from the original formula for r.

$$r = \sqrt{1 - \frac{S_y^{\,2}}{\sigma_y^{\,2}}}$$

From this equation

$$r^2 = 1 - \frac{S_y^{\,2}}{\sigma_y^{\,2}}$$

---

[1] It is assumed that the regression line is linear.
[2] For the derivation of the formula see Technical Appendix V. This formula is written in a variety of equivalent forms in various textbooks. For instance,

$$r = \frac{\Sigma(X - \overline{X})(Y - \overline{Y})}{\sqrt{\Sigma(X - \overline{X})^2 \, \Sigma(Y - \overline{Y})^2}}$$

$$r = \frac{\Sigma(XY) - n\overline{XY}}{n S_x S_y}$$

$$r = \frac{n\Sigma(XY) - (\Sigma X)(\Sigma Y)}{\sqrt{n\Sigma(X^2) - (\Sigma X)^2} \, \sqrt{n\Sigma(Y^2) - (\Sigma Y)^2}}$$

$$S_y^2 = \sigma_y^2 (1 - r^2)$$
$$S_y = \sigma_y \sqrt{1 - r^2}$$

Where the equation has its origin at the point of averages[1] the line of regression may be determined from the formula[2]

$$y = r \frac{\sigma_y}{\sigma_x} x$$

since y and x are in terms of deviation from their respective means. Because the line of regression must pass through the point of averages, the a (Y intercept) value will be zero when that point is used as the origin. With the origin now at the point of averages the customary equation (here expressed in terms of deviations from the means) is resolved into

$$y = a + bx$$

where

$$a = 0$$

and

$$b = r \frac{\sigma_y}{\sigma_x}$$

therefore

$$y = r \frac{\sigma_y}{\sigma_x} x$$

The more usual form of equation with an origin at zero (or in terms of actual values) may be determined by a transformation. Since

$$y = Y - \overline{Y}$$
$$x = X - \overline{X}$$
$$Y - \overline{Y} = r \frac{\sigma_y}{\sigma_x} (X - \overline{X})$$

For Table 9.1

$$p = \frac{\Sigma(XY)}{N} - \frac{\Sigma(X)}{N} \frac{\Sigma(Y)}{N} = \frac{322,227}{30} - \left(\frac{6468}{30}\right)\left(\frac{1252}{30}\right) = 1743.92$$

$$\sigma_x = \sqrt{\frac{\Sigma(X^2)}{N} - \left(\frac{\Sigma(X)}{N}\right)^2} = \sqrt{\frac{1,725,088}{30} - \left(\frac{6468}{30}\right)^2} = 104.97$$

---

[1] The point of averages is the point defined by the arithmetic means of X and Y (denoted by the symbols $\overline{X}$ and $\overline{Y}$).
[2] For mathematical proof see Technical Appendix VI.

$$\sigma_y = \sqrt{\frac{\Sigma(Y^2)}{N} - \left(\frac{\Sigma(Y)}{N}\right)^2} = \sqrt{\frac{60,650}{30} - \left(\frac{1252}{30}\right)^2} = 16.74$$

$$r = \frac{p}{\sigma_x \sigma_y} = \frac{1743.92}{(104.97)(16.74)} = .9925$$

$$S_y = \sigma_y \sqrt{1 - r^2} = 16.74\sqrt{1 - (.9925)^2} = 2.04$$

$$y = r\frac{\sigma_y}{\sigma_x}x$$

$$y = .9925\,\frac{16.74}{104.97}x$$

$$y = .1582\,x$$

$$Y - \overline{Y} = b\,(X - \overline{X})$$

$$Y - 41.73 = .1582\,(X - 215.6)$$

$$Y = 7.63 + .1582\,X$$

PRODUCT MOMENT METHOD (GROUPED DATA)– THE CORRELATION TABLE

Where electronic computers[1] are not available and there are very many items in the series to be analyzed for association, the procedure outlined above is unsatisfactory. Its use invites error in proportion to the number of computations while the work of calculation is very great.

Large groups of data may be handled more efficiently by first grouping them into the form of a double frequency distribution or **correlation table**. Table 9.2, pp. 94-95, is an example for the association between the prices of wheat and flour.

The items are located in the various boxes by reference to the class intervals in which their X and Y values fall. Thus an item with an X value of $.80 and a Y value of $6.50 will be located in the box indicated by the X class interval $.80–.89 and the Y class interval $6.00–6.99.

The calculations may be more readily carried out if, instead of trying to use the midpoints of the class intervals as the actual values, an

---

[1]There are "library" programs available for this and similar computations.

arbitrary origin is selected for both X and Y values, and the deviations from this arbitrary origin are measured in terms of class intervals, as in the short methods for calculating the mean and the standard deviation.

The necessary values for the determination of r by the product moment method may now be readily secured.

$$r = \frac{p}{\sigma_x \sigma_y}$$

where

$$p = \frac{\Sigma [f(d_x d_y)]}{N} - \frac{\Sigma(fd_x)}{N} \cdot \frac{\Sigma(fd_y)}{N}$$

$$\sigma_x = \sqrt{\frac{\Sigma(fd_x^2)}{N} - \left(\frac{\Sigma(fd_x)}{N}\right)^2}$$

$$\sigma_y = \sqrt{\frac{\Sigma(fd_y^2)}{N} - \left(\frac{\Sigma(fd_y)}{N}\right)^2}$$

The sum of $fd_x d_y$ is determined as follows.

1. Obtain the product of the indicated value of d_x and d_y for each box (inserted in the lower left-hand corner of the box).

2. Multiply the values obtained by the frequency entered in the box (result entered in upper right-hand corner of box).

3. Add up for all boxes in the table.

It is to be noted that all calculations for r are carried out in terms of class intervals, not actual values. The application of this technique for the series in Table 9.2 is shown below.

$$p = \frac{\Sigma [f(d_x d_y)]}{N} - \frac{\Sigma(fd_x)}{N} \cdot \frac{\Sigma(fd_y)}{N} = \frac{6319}{240} - \left(\frac{1119}{240}\right)\left(\frac{999}{240}\right)$$

$$= 26.3292 - (4.6625)(4.1625) = 6.9215$$

$$\sigma_x = \sqrt{\frac{\Sigma(fd_x^2)}{N} - \left(\frac{\Sigma(fd_x)}{N}\right)^2} = \sqrt{\frac{7217}{240} - \left(\frac{1119}{240}\right)^2}$$

$$= \sqrt{30.0708 - 21.7389} = 2.8865$$

$$\sigma_y = \sqrt{\frac{\Sigma(fd_y^2)}{N} - \left(\frac{\Sigma(fd_y)}{N}\right)^2} = \sqrt{\frac{5697}{240} - \left(\frac{999}{240}\right)^2}$$

$$= \sqrt{23.7375 - 17.3264} = 2.5320$$

Table 9.2—Correlation Table—Wheat and Flour Prices by Months.

X = Wheat Price per Bushel in Dollars

Y = Flour Price per Barrel in Dollars

Class Interval	Mid-Point	Deviation d	Frequency f	fd	fd²	.40 / -.59	.60 / -.79	.80 / -.99	1.00 / -1.19	1.20 / -1.39	1.40 / -1.59	1.60 / -1.79	1.80 / -1.99	2.00 / -2.19	2.20 / -2.39	2.40 / -2.59	2.60 / -2.79	2.80 / -2.99	Total	f(dxdy)
15.00 -15.99	15.5	12	1	12	144													144 / 1 / 144		144
14.00 -14.99	14.5	11	5	55	605											110 / 1 / 110	242 / 2 / 121	264 / 2 / 132		616
13.00 -13.99	13.5	10	5	50	500										90 / 1 / 90	200 / 2 / 100	110 / 1 / 110	120 / 1 / 120		520
12.00 -12.99	12.5	9	10	90	810										486 / 6 / 81	180 / 2 / 90	198 / 2 / 99			864
Mid-Point						5	7	9	11	13	15	17	19	21	23	25	27	29		
Deviation d						0	1	2	3	4	5	6	7	8	9	10	11	12		
Frequency f						20	6	25	37	52	24	15	15	13	18	6	5	4	240	
fd						0	6	50	111	208	120	90	105	104	162	60	55	48	1119	
fd²						0	6	100	333	832	600	540	735	832	1458	600	605	576	7217	

Y = Flour Price per Barrel in Dollars	midpoint	x	f	fx	fx²	Σ
11.00 –11.99	11.5	8	5	40	320	360
10.00 –10.99	10.5	7	14	98	686	840
9.00 – 9.99	9.5	6	17	102	612	726
8.00 – 8.99	8.5	5	28	140	700	790
7.00 – 7.99	7.5	4	46	184	736	764
6.00 – 6.99	6.5	3	54	162	486	576
5.00 – 5.99	5.5	2	16	32	64	86
4.00 – 4.99	4.5	1	34	34	34	33
3.00 – 3.99	3.5	0	5	0	0	0
Total			240	999	5697	6319

Scatter cell entries (each cell: product / f / value), by Y-row:

- 11.5: 80 / 1 / 80
- 10.5: 216 / 3 / 72 ; 504 / 8 / 63
- 9.5: 64 / 1 / 64 ; 336 / 6 / 56 ; 240 / 5 / 48
- 8.5: 420 / 10 / 42 ; 140 / 4 / 35 ; 40 / 1 / 40
- 7.5: 36 / 1 / 36 ; 330 / 11 / 30 ; 21 / 1 / 21
- 6.5: 30 / 1 / 30 ; 200 / 8 / 25 ; 72 / 3 / 24
- 5.5: 80 / 4 / 20 ; 240 / 12 / 20 ; 45 / 3 / 15
- 4.5: 352 / 22 / 16 ; 300 / 25 / 12 ; 8 / 1 / 8
- 3.5: 84 / 7 / 12 ; 180 / 20 / 9 ; 60 / 10 / 6 ; 30 / 5 / 6 ; 16 / 4 / 4 ; 28 / 14 / 2 ; 2 / 1 / 2 ; 5 / 5 / 1 ; 0 / 15 / 0 ; 0 / 5 / 0

$$r = \frac{p}{\sigma_x \sigma_y} = \frac{6.9215}{(2.8865)(2.5320)} = .9470$$

The standard error of estimate may now be computed from

$$S_y = \sigma_y \sqrt{1 - r^2}$$

by first converting σ_y to the original units of its series through multiplying by the size of the Y class interval (in this case, $1.00).

$$S_y = \$2.5320 \sqrt{1 - (.9470)^2} = \$.8133$$

the line of regression now may be obtained, using x and y in *original*, not class interval, values

$$y = r\frac{\sigma_y}{\sigma_x}x$$

$$y = .9470 \left(\frac{2.5320}{.5773}\right) = 4.1535x$$

but since

$$y = Y - \overline{Y}$$

$$x = X - \overline{X}$$

$$Y - 7.6625 = 4.1535 (X - 1.4325)$$

$$Y = 1.7126 + 4.1535X$$

COEFFICIENTS OF DETERMINATION AND ALIENATION

When the dependent variable (Y) is causally related to the independent variable (X) and both series consist of simple elements of equal variability, r^2 measures the proportion of the variance[1] in Y that is explained by X. The measure (r^2) is then termed the **coefficient of determination** (the phrase **index of determination** is used where the correlation is curvilinear).

Just as the coefficient of correlation is a measure of the *degree* of association between two series, the **coefficient of alienation** is a measure of the *lack* of association.

$$\text{Coefficient of alienation } (k) = \sqrt{\frac{S_y^2}{\sigma_y^2}} = \sqrt{1 - r^2}$$

---

[1]Variance is the technical term for the square of the standard deviation.

The square of the coefficient of alienation may be interpreted in a similar fashion to the square of the coefficient of correlation and is known as the **coefficient of non-determination.**

$$1 = r^2 + k^2$$
and
$$k^2 = 1 - r^2$$

CORRECTION FOR SMALL NUMBER OF CASES

When the number of cases is small, the coefficient of correlation must be adjusted for exaggeration of its value and the standard error must be adjusted for an underestimate.

The correction formulae are

$$\overline{S}_y^{\,2} = S_y^{\,2}\,\frac{N}{N-2}$$

$$\overline{r}^2 = 1 - (1 - r^2)\frac{N-1}{N-2}$$

where $\overline{S}_y$ and $\overline{r}$ are the corrected values.[1]

CORRELATION FROM RANKS

The method of measuring correlation from the rank or position of the various items has proven particularly useful in education and psychology.

In this method the data are numbered according to their position, as in Table 9.3, p. 98, and the following formula (Spearman's)[2] is used

$$\rho = 1 - \frac{6\Sigma(D^2)}{N(N^2 - 1)}$$

where

ρ is the measure of correlation[3]

---

[1] For more complete analysis of the formulae for S_y and r see Ezekiel, M., *Methods of Correlation Analysis*, p. 121.

[2] See Kelley, T. L., *Statistical Method*, pp. 191–194, for derivation of this formula.

[3] The symbol ρ is the Greek small letter rho. It is usually used as the index of correlation (see p. 103) and its use in Spearman's formula must be distinguished. Spearman's coefficient is sometimes designated r_r.

D is the difference between the two ranks given for each individual

N equals number of individuals

For Table 9.3

$$\rho = 1 - \frac{6(32)}{15(225 - 1)} = .953$$

Table 9.3—Illustrating Rank Method of Measuring Association
Hypothetical Grades of Fifteen Students on Two Examinations

Student Number	Examination 1 Grade	Examination 1 Rank	Examination 2 Grade	Examination 2 Rank	Difference in Rank (D)	(D²)
# 1	100%	1	90%	3	2	4
2	98	2	95	1	−1	1
3	95	3	89	4	1	1
4	91	4	87	5	1	1
5	90	5	93	2	−3	9
6	85	6	86	6	0	0
7	83	7	80	7	0	0
8	82	8	79	8	0	0
9	81	9	76	10	1	1
10	80	10	77	9	−1	1
11	70	11	72	11	0	0
12	65	12	60	14	2	4
13	63	13	62	13	0	0
14	60	14	50	15	1	1
15	50	15	63	12	−3	9
						32

In case of ties in rank either one of two methods of assigning ranks may be used. In the **bracket method** all are assigned the same rank; but the next individual is given the rank that would have been assigned if the ties had received successive ranks. In the **mid-rank method,** a rank equal to that of the middle of the tie is assigned to all items with identical values. (See Table 9.4)

Assuming that the original data constitute a normal distribution, the following relationship may be established between r (the coefficient of correlation) and ρ.

$$r = 2 \sin\left(\frac{\pi}{6}\rho\right)$$

Table 9.4 – Ranking Methods

Student	Grade	Ranks Bracket Method	Mid Rank Method
A	100%	1	1
B	95	2	3
C	95	2	3
D	95	2	3
E	94	5	5
F	92	6	6.5
G	92	6	6.5
H	90	8	8

Spearman's Footrule

When only a rough approximation of the correlation based on ranks is desired, "Spearman's footrule" formula may be used.

$$R = 1 - \frac{6\Sigma G}{N^2 - 1}$$

where G represents the *positive* differences in rank.

CORRELATION AND THE TIME SERIES

When two time series are correlated, if there is a similar upward trend the higher items toward the end of the first series will be associated with the higher items in the second. Because of the trend there will be an exaggeration of the correlation. In either case the long-time relationship would tend to overshadow the short-time movements upon which attention particularly centers.

The following methods may be used to overcome the difficulty

1. The deviations from trend may be correlated.

2. The first differences (deviation of each item from previous items in series) may be correlated.

3. The two series may be adjusted for trend.

ADDITIONAL BIBLIOGRAPHY

DUNCAN, A. J., *Quality Control and Industrial Statistics*, pp. 674–703, Richard D. Irwin, Inc., Homewood, Ill., 1965.

EZEKIEL, M., & FOX, F. A., *Methods of Correlation and Regression Analysis*, pp. 55–150, John Wiley & Sons, New York, 3rd ed., 1959.

YULE, G. U., & KENDALL, M. G., *An Introduction to the Theory of Statistics*, pp. 199–280, Hafner Publishing Co., New York, 1950.

WALKER, H. M., & LEV, J., *Statistical Inference*, pp. 230–249, Henry Holt & Co., New York, 1953.

Chapter X

Correlation and Regression Analysis—Non-Linear, Multiple, Partial

A straight line of regression does not always satisfactorily describe the association between variables. Frequently the relationship is too complex to be described by means of a simple straight line and a curve must be used.

For instance, if the association between rainfall and crop yield is examined it is found that beyond a certain point an increase in the amount of rainfall will not result in an increase in crop yield. On the contrary, a point could be established beyond which the yield would decrease as the rainfall increases up to complete extinction of the crop.

TYPES OF REGRESSION CURVES
The types of curves which may be used are similar to those described in the chapter on trend (see Chapter VII).

The two most important types of curves are

1. potential curves of the type

$$Y = a + bX + cX^2 + \ldots$$

2. exponential curves of the type

$$Y = ab^x$$

In terms of logarithms exponential curves may be divided into

a. logarithmic

$$\log Y = a + b \log X$$

$$\log Y = a + b \log X + c(\log X)^2$$

b. semi-logarithmic

$$\log Y = a + bX$$
$$\log Y = a + bX + cX^2$$

$$Y = a + b \log X$$
$$Y = a + b \log X + c(\log X)^2$$

Figure 10.1 illustrates the typical appearance of some of these curves. Comparison of the scatter diagram with these typical curves may suggest the likeliest curve for the regression line.

The normal equations for any of the curves may be derived by the same method as outlined for trend curves (Chapter VI).

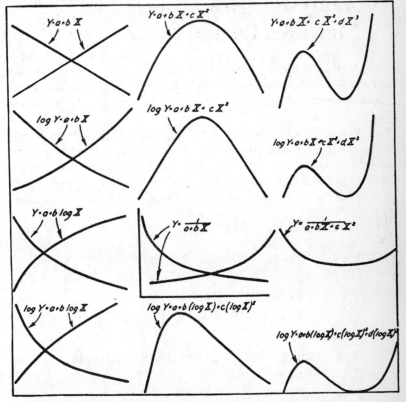

Source: Ezekiel, Mordecai, Methods of Correlation Analysis.

Fig. 10.1 – Curves illustrating a number of different types of mathematical functions.

For a second degree curve of the potential group

$$Y = a + bX + cX^2$$

the normal equations are

(I) $\Sigma(Y) = Na + b\Sigma(X) + c\Sigma(X^2)$

(II) $\Sigma(XY) = a\Sigma(X) + b\Sigma(X^2) + c\Sigma(X^3)$

(III) $\Sigma(X^2 Y) = a\Sigma(X^2) + b\Sigma(X^3) + c\Sigma(X^4)$

For a semi-logarithmic equation of the type

$$\log Y = a + bX + cX^2$$

the normal equations are

(I) $\Sigma(\log Y) = Na + b\Sigma(X) + c\Sigma(X^2)$

(II) $\Sigma(X \log Y) = a\Sigma(X) + b\Sigma(X^2) + c\Sigma(X^3)$

(III) $\Sigma(X^2 \log Y) = a\Sigma(X^2) + b\Sigma(X^3) + c\Sigma(X^4)$

.. Before fitting an equation of the type

$$Y = \frac{1}{a + bX}$$

the equation is first converted to

$$\frac{1}{Y} = a + bX$$

or $Y' = a + bX$

where Y' is the reciprocal of Y.

The fitting of the curve may now be carried out in the usual fashion by obtaining the normal equations.

(I) $\Sigma(Y') = Na + b\Sigma(X)$

(II) $\Sigma(XY') = a\Sigma(X) + b\Sigma(X^2)$

NON–LINEAR STANDARD ERROR OF ESTIMATE

The standard error of estimate about the curve may be determined as before (see page 85) from the formula

$$S_y = \sqrt{\frac{\Sigma(d^2)}{N}}$$

or from[1]

$$S_y^2 = \frac{\Sigma(Y^2) - a\Sigma(Y) - b\Sigma(XY) - c\Sigma(X^2 Y) - d\Sigma(X^3 Y) - \ldots}{N}$$

For curves other than the potential series this formula may be adapted by using the log of X or Y or its reciprocal as called for by the type of equation. Thus for a curve of the type

$$\log Y = a + bX + cX^2$$

the formula would read

$$S_y^2 = \frac{\Sigma(\log Y)^2 - a\Sigma(\log Y) - b\Sigma(X \log Y) - c\Sigma(X^2 \log Y)}{N}$$

THE INDEX OF CORRELATION—RHO (ρ)

When computed about a curve the measure of correlation corresponding to the coefficient of correlation is known as the **index of correlation**. It is assigned the Greek letter rho (ρ) as a symbol.

$$\rho = \sqrt{1 - \frac{S_y^2}{\sigma_y^2}}$$

The measure ρ is equal in value to r, the coefficient of correlation, when the regression is linear. If the regression is non-linear, a curve will more closely approximate the data. As a result the deviations about the curve will tend to be smaller than those about the straight line. The standard error, therefore, will be smaller and will result in a larger value for rho. *Rho is always either equal to or larger than r.* Rho may be computed from the following formula

$$\rho^2 = \frac{a\Sigma(Y) + b\Sigma(XY) + c\Sigma(X^2 Y) + \ldots - Nc_y^2}{\Sigma(Y^2) - Nc_y^2}$$

where

$$c_y^2 = \left(\frac{\Sigma Y}{N}\right)^2$$

The **index of alienation** and the **indices of determination** and **nondetermination** are computed in the same manner and have the same meaning as the linear indices (see page 96).

---

[1] The derivation of this formula follows the same lines as that for the linear standard error outlined in Technical Appendix V.

It is important to remember that the value for a given index of correlation (rho) may be compared to that for another association only when the same type of curve is used to describe both regressions.

Non-linear regression curves must be used with extreme care. Since the more complex the curve the higher the index of correlation, an ultimate value of 100% may be reached if one so complex as to pass through all of the points is used. The regression line and the index would then be meaningless.

CORRELATION RATIO

The **correlation ratio** is only rarely used. In this measure the regression passes through the means of all the columns when the scatter diagram is divided into columns.

The regression is not defined by an equation.

Since the data must be divided into groups, the correlation ratio can be computed from the following formula[1]

$$\eta = \sqrt{1 - \frac{\sigma_{ay}^2}{\sigma_y^2}}$$

where σ_{ay} is the standard deviation of the various values about the means of their respective columns.

The value of η is dependent upon the number of columns in the correlation table as compared to the number of cases used. With only one item in each column the correlation as computed by the correlation ratio would be perfect but meaningless. A correction for fineness of grouping can be made by use of the formula

$$\eta'^2 = \frac{\eta^2 - \frac{(\kappa - 3)}{N}}{1 - \frac{(\kappa - 3)}{N}}$$

where κ is the number of arrays or columns in the correlation table.

Since a curve[2] passing through the means of the columns will be most descriptive of the data, the deviations about this curve will be smallest. The correlation ratio, therefore, will be either larger than or equal to the index of correlation or the coefficient of correlation. If

---

[1] The symbol η is the Greek small letter eta.

[2] "Curve" here does not mean a curve with a mathematical meaning; it describes a series of straight lines joining the means.

the relationship is essentially linear the means will all coincide with a straight line and η and r will be equal.

Since ρ will also equal r if the regression is definitely linear:

$$\eta = \rho = r$$

but if it is non-linear:

$$\eta > \rho > r$$

Thus where the regression is basically non-linear, η would be larger than ρ. A test for linearity of regression has been devised on this basis.

$$\zeta = \eta^2 - r^2$$

where ζ[1] is the test for linearity.

When $\zeta = 0$, the regression is linear; when $\zeta \neq 0$ a non-linear regression is called for.

METHOD OF SUCCESSIVE ELIMINATION

Although in scientific experiments it is usually possible to control all the different variables and to allow only the factors being studied to vary, this is not often possible in other fields. In the social sciences and business, especially, numerous uncontrollable factors vary simultaneously. The relationship of one of these numerous factors to a particular dependent variable excluding the others may be studied by the method of successive elimination.

The effectiveness of advertising as gauged by the number of returns linked to a keyed advertisement is dependent upon the circulation of the newspaper in which it was inserted and the size of the advertisement[2] among other factors. If it is desired to study the effectiveness of various sizes of advertisements, allowing for or, more exactly, excluding the effect of the differing circulation of the various papers in which the advertisements appeared, the relationship between circulation and returns is studied by either linear or non-linear correlation. The line of regression is determined by the usual methods. Since this variable is not the only factor determining the number of returns the points (actual values) will be scattered about the line of regression.

---

[1] The symbol ζ is the Greek small letter zeta.
[2] Assuming that the advertising copy is the same for all advertisements.

The theoretical values based on circulation may now be determined from the line of regression and the difference between the actual number of returns and these theoretical values calculated. These values are termed the residuals (z').[1]

$$z' = Y - Y_c$$

The residuals may now be correlated with the size of the advertisement to determine their relationship. The line of regression will express the variation in the returns if the circulation of the paper were held constant.

Another effective method is to use the line of regression to adjust the given data so that the returns are expressed as though all were obtained from a paper of a given circulation (say, 100,000). The returns for a paper of smaller or larger circulation are increased or diminished in relation to the difference on the regression line between the theoretical value for its circulation and that for a paper of 100,000 circulation. The b value (in a linear regression) expresses the increase in returns per unit of circulation and can be used for this purpose.

MULTIPLE CORRELATION

The fluctuations in a given series are seldom dependent upon a single factor or cause. The measurement of the association between such a series and several of the variables associated with the dependent variable which may be affecting the fluctuations is known as **multiple correlation**.

Multiple correlation is the measurement of the relationship between a dependent variable and two or more independent variables. The procedure is similar to that for simple correlation, with the exception that other variables are added to the regression equation. For two independent variables the regression equation, if linear, is of the type

$$X_1 = a + b_{12.3} X_2 + b_{13.2} X_3$$

where X_1 is the dependent or estimated variable (replacing the symbol Y previously used) and X_2 and X_3 are the independent variables.

The coefficient of regression or slope, $b_{12.3}$, indicates the number of units change in the dependent variable for a given unit change in

---

[1] Some of these values will be negative, showing below average returns for a paper of the specified circulation.

X_2, while $b_{13.2}$ indicates the change in X_1 for a unit change in X_3. However, in the computation of these coefficients of regression the associations of each of the other independent variables with the dependent variable is taken into consideration. The coefficients, therefore, indicate the *net* relationship between the dependent and an independent variable, allowing for the other factors or variables which are also considered in the equation. The subscripts after the period indicate the other variables included. Thus $b_{12.345}$ would give the net regression of variable X_2 in relationship to X_1, allowing for X_3, X_4, and X_5. The last named coefficients are therefore known as the **coefficients of net regression.**

The values for the coefficients may be obtained in the usual manner by making use of the previously outlined method (see page 58) of obtaining the normal equations. For a linear relationship with three independent variables

$$X_1 = a + b_{12.34} X_2 + b_{13.24} X_3 + b_{14.23} X_4$$

the normal equations would be

(I) $\quad \Sigma(X_1) = Na + b_{12.34} \Sigma(X_2) + b_{13.24} \Sigma(X_3) + b_{14.23} \Sigma(X_4)$

(II) $\Sigma(X_1 X_2) = a\Sigma(X_2) + b_{12.34} \Sigma(X_2^2) + b_{13.24} \Sigma(X_2 X_3) +$
$\qquad b_{14.23} \Sigma(X_2 X_4)$

(III) $\Sigma(X_1 X_3) = a\Sigma(X_3) + b_{12.34} \Sigma(X_2 X_3) + b_{13.24} \Sigma(X_3^2) +$
$\qquad b_{14.23} \Sigma(X_3 X_4)$

(IV) $\Sigma(X_1 X_4) = a\Sigma(X_4) + b_{12.34} \Sigma(X_2 X_4) + b_{13.24} \Sigma(X_3 X_4) +$
$\qquad b_{14.23} \Sigma(X_4^2)$

By assuming the origin of the equation to be at the point of averages these equations reduce to[1]

(I) $p_{12} = b_{12.34} \sigma_2^2 + b_{13.24} p_{23} + b_{14.23} p_{24}$

(II) $p_{13} = b_{12.34} p_{23} + b_{13.24} \sigma_3^2 + b_{14.23} p_{34}$

(III) $p_{14} = b_{12.34} p_{24} + b_{13.24} p_{34} + b_{14.23} \sigma_4^2$

The equations may now be solved simultaneously to arrive at the desired values for $b_{12.34}$, $b_{13.24}$, and $b_{14.23}$.

The standard error of estimate may now be computed from

---

[1] See Technical Appendix VII–*a* may be obtained from the first normal equation

$$\Sigma(X) = Na + b_{12.34} \Sigma(X_2) + b_{13.24} \Sigma(X_3) + b_{14.23} \Sigma(X_4)$$

$$S_{1.234} = \sqrt{\frac{\Sigma d^2}{N}} \text{ or more readily from}[1]$$

$$S_{1.234}^2 = \sigma_1^2 - b_{12.34}p_{12} - b_{13.24}p_{13} - b_{14.23}p_{14}$$

and the coefficient of multiple correlation from

$$R_{1.234} = \sqrt{1 - \frac{S_{1.234}^2}{\sigma_1^2}}$$

or

$$R_{1.234}^2 = \frac{b_{12.34}p_{12} + b_{13.24}p_{13} + b_{14.23}p_{14}}{\sigma_1^2}$$

The coefficients of multiple alienation, determination, and non-determination may now be computed and applied as in simple correlation.

The same technique may be used for non-linear multiple correlation, using the general equation

$$X_1 = a + f(X_2) + f(X_3) + f(X_4) + \ldots$$

where $f(X_2)$ indicates any function of X_2 such as $bX_2 + c(X_2)$ (a parabola), etc. As the complexity of the function increases the amount of computation required soon becomes so great as to render the use of the technique difficult without the use of electronic computers.[2]

PARTIAL CORRELATION
Coefficient of Partial Correlation
When it is desired to compute the separate or net effect or importance of each independent variable the technique of partial correlation may be used.

The **coefficient of partial correlation** is a relative measure of the association between the dependent variable and a given independent variable, eliminating the effect of the other independent variables.

---

[1] See Technical Appendix VIII.

[2] A less arduous technique for curvilinear multiple correlation is outlined in Ezekiel, M., *Methods of Correlation Analysis*. The advent of the electronic computer now makes possible the solution of far more complex non-linear multiple regressions than formerly.

There are a number of formulae which may be used to compute the association.

1. $r_{13.24} = \sqrt{b_{13.24} \cdot b_{31.24}}$

2. $r_{12.3} = \dfrac{r_{12} - r_{13} r_{23}}{\sqrt{1 - r_{13}^2} \sqrt{1 - r_{23}^2}}$

3. $r_{14.23} = 1 - \dfrac{(1 - R_{1.234}^2)}{(1 - R_{1.23}^2)}$

4. $r_{14.23} = \sqrt{1 - \dfrac{S_{1.234}^2}{S_{1.23}^2}}$

Coefficient of Part Correlation

A similar coefficient somewhat easier to compute is called the **coefficient of part correlation** $(_{12}r_{34})$.

The differences between these two coefficients may be shown by comparing the coefficient of partial correlation with that resulting from the correlation of[1]

$$(X_2 - b_{23.4}X_3 - b_{24.3}X_4) \text{ with } (X_1 - b_{13.4}X_3 - b_{14.3}X_4)$$

while the coefficient of part correlation may be compared to the correlation of

$$X_2 \text{ with } (X_1 - b_{13.24}X_3 - b_{14.23}X_4)$$

The coefficient of part correlation may be computed from

$$_{12}r_{34}^2 = \frac{b_{12.34}^2 \, \sigma_2^2}{b_{12.34}^2 \, \sigma_2^2 + \sigma_1^2 (1 - R_{1.234}^2)}$$

The subscripts to the right of the letter r indicate the variables excluded.

Beta Coefficients

The relative importance of the individual independent variables in a multiple correlation in determining the dependent variable may be determined through resort to the **beta (β) coefficients**.

---

[1] Suggested by Ezekiel in *Methods of Correlation Analysis*.

The coefficients of regression of the multiple correlation regression equation indicate the increase in the dependent variable resulting from a unit increase in the indicated independent variable. The various independent variables are, however, often expressed in different units making a direct comparison of the coefficients impossible. The coefficients of multiple regression may be made comparable by dividing each variable by its own standard deviation. Thus the original regression equation

$$X_1 = a + b_{12.3}X_2 + b_{13.2}X_3$$

becomes

$$\frac{X_1}{\sigma_1} = a + b_{12.3}\frac{X_2}{\sigma_2} + b_{13.2}\frac{X_3}{\sigma_3}$$

and therefore the beta coefficients replacing the coefficients of regression are

$$\beta_{12.3} = b_{12.3}\frac{\sigma_2}{\sigma_1}$$

$$\beta_{13.2} = b_{13.2}\frac{\sigma_3}{\sigma_1}$$

The beta coefficients are comparable measures and indicate the increase in standard deviations in the dependent variable resulting from an increase of *one standard deviation* in each independent variable.

TYPES OF CORRELATION—SUMMARY

I. **Simple correlation**, with two variables, one considered dependent, the other independent.
   a. **Linear Correlation**: The rate of change of one variable is constant with respect to the other.
      1. Direct: An increase in one variable is accompanied by an increase in the other.
      2. Inverse: An increase in one variable is accompanied by a decrease in the other.
   b. **Non-Linear (Curvilinear) Correlation**: The change in one variable with respect to the other is at constantly changing rate, decreasing or increasing.

II. **Multiple Correlation**, with more than two variables. One is considered dependent, the others independent. Multiple correlation may be
   a. Linear, where some variables may be inversely and some directly associated.

b. Non-Linear

c. Joint, where the relationship between various independent and dependent variables changes with a change in
another independent variable.

III. Partial Correlation, which measures the association between
an independent and a dependent variable. It allows for the
variation associated with specified other independent variables.

ADDITIONAL BIBLIOGRAPHY

DUNCAN, A. J., *Quality Control and Industrial Statistics,* pp. 704–738, Richard
D. Irwin, Inc., Homewood, Ill., 1965.

EZEKIEL, M., & FOX, K. A., *Methods of Correlation and Regression Analysis,*
pp. 151–278, John Wiley & Sons, New York, 3rd ed., 1959.

WALKER, H. M., & LEV, J. *Statistical Inference,* pp. 315–386, Henry Holt &
Co., New York, 1953.

YULE, G. U., & KENDALL, M. G., *An Introduction to the Theory of Statistics,*
pp. 281–339, Hafner Publishing Co., New York, 1950.

Chapter XI

Correlation of Attributes

The previous section dealt with the measurement of the association between two measured characteristics of a given set of items. Two quantitative values were determined for each item and a coefficient of correlation was computed for the various paired values. Thus if the height and weight of a group of individuals are measured the coefficient of correlation can be computed. It will show the degree to which the greater heights are associated with the greater weights.

It is not always possible to use measurements for various characteristics. For instance, many classifications are qualitative such as light and dark, good and poor, etc. The association between heights and weights may be determined by classifying the individuals as light or heavy and tall or short and by cross-classifying each individual in a fourfold table of the type shown below, where a, b, c, and d represent the number of cases with each of the given pairs of characteristics.

	Short	Tall
Light	a	b
Heavy	c	d

If the association is perfect, that is, if all tall people are heavy and all short people are light, all of the cases will be located in two boxes (cells)—in this case a and d. If there is absolutely no association, that is, if it is a matter of indifference insofar as weight is concerned if a person is tall or short, the cases will be distributed at random throughout the four boxes. Since there is an equal likelihood of a case appearing in any box there will tend to be an equal number in each box.

This assumes an equal number of short and tall and light and heavy persons in the sample. Otherwise, the distribution is proportionally random.

Similarly, qualitative characteristics of a group of individuals or items may be arranged in a table when more than two alternative attributes are dealt with. A table of this type follows:

	A	B	C	D	Total
A' B' C' D'		n_{rc}			n_{r1} n_{r2} n_{r3} n_{r4}
Total	n_{c1}	n_{c2}	n_{c3}	n_{c4}	N

where $ABCD$ and A', B', C', D' are two sets of qualitative specifications.[1]

In this type of table if the association is perfect there will be only one entry in each row and column of cells for quality A will always accompany quality A', B will accompany quality B', etc. If there is no correlation there will be a tendency towards an equal distribution of cases among various boxes.

It is possible on the basis of these facts to compute a coefficient of association which will serve as a comparative measure of correlation.

COEFFICIENT OF CONTINGENCY

This coefficient of association (the **coefficient of mean-square contingency**) is based upon a comparison of the number of cases actually occurring in a given cell or box and the number of cases which would be likely to occur in the cell if chance alone were involved. It is a comparison of the actual distribution and the distribution occurring when there is no association.

If n_r is the number of cases in a given row, n_c is the number in a given column, and n_{rc} is the number in a given cell

$$n_{rc} - \frac{n_r n_c}{N}$$

---

[1] The qualitative specifications should be arranged in order where there are more than two such as poor, fair, good, etc.

will be the difference between the actual number of cases and the theoretically expected number of cases due to chance alone. But the ratio of the square of this value to the theoretical number of cases is χ^2 (test for goodness of fit)[1] or

$$\chi^2 = \Sigma \left[\frac{\left(n_{rc} - \dfrac{n_r n_c}{N}\right)^2}{\dfrac{n_r n_c}{N}} \right]$$

Pearson's **mean-square contingency,** ϕ^2, is obtained by dividing this value by N^2

$$\phi^2 = \frac{\chi^2}{N}$$

and the coefficient of mean-square contingency is

$$cc = \sqrt{\frac{\phi^2}{1 + \phi^2}} = \sqrt{\frac{\chi^2}{N + \chi^2}}$$

A simpler form of this formula shown by Yule is

$$C = \sqrt{\frac{S - N}{S}}$$

where

$$S = \Sigma \left(\frac{n_{rc}^2}{\dfrac{n_r n_c}{N}} \right)$$

or

$$S = N\Sigma \left(\frac{n_{rc}^2}{n_r n_c} \right)$$

Steps in Computation

1. Square the value in each cell (n_{rc}^2)

2. For each box multiply the number in that row by the number in that column $(n_r n_c)$.

---

[1] See page 131. χ is the Greek letter chi.
[2] ϕ is the Greek letter phi.

3. Divide the squared value in each box (n_{rc}^2) by the corresponding $n_r n_c$.

4. Sum up for all rows and multiply by N. The resulting value is S.

5. Substitute in

$$C = \sqrt{\frac{S - N}{S}}$$

Limits of Coefficient of Contingency

The maximum value obtainable for the coefficient of contingency is dependent upon the number of columns and rows in the table analyzed. The maximum possible values of the coefficient of contingency for various size tables are given below.

Number of Columns and Rows	Maximum Value of CC
2 × 2	.71
3 × 3	.82
4 × 4	.87
5 × 5	.89
6 × 6	.91
7 × 7	.93
8 × 8	.94

FOURFOLD TABLES (2 × 2 CLASSIFICATIONS)

When two variants are classed in alternate categories, A and not A, B and not B, the following coefficients of correlation are occasionally used.

1. Yule's coefficient of association and coefficient of colligation.[1]

The coefficient of association is

$$Q = \frac{ad - bc}{ad + bc}$$

The coefficient of colligation is[2]

$$\omega = \frac{\sqrt{ad} - \sqrt{bc}}{\sqrt{ad} + \sqrt{bc}}$$

---

[1] Objections have been offered to the use of both of these measures.

[2] ω is the Greek letter omega.

where a, b, c, d represent the frequencies contained within the various cells as follows

a	b
c	d

When the relationship is perfect all of the cases will be concentrated in two of the boxes, either a and d or b and c, and therefore both Q and ω will be equal to 1.00 or -1.00. When there is no association and the distribution of cases is equal to the expected random distribution, both Q and ω will equal zero.

    2. **Pearson's cosine method.**

    The coefficient of correlation by the cosine method for a four-fold table is

$$r = \cos \frac{\sqrt{bc}}{\sqrt{ad} + \sqrt{bc}} \pi$$

    This coefficient will vary from $r = 0$ to $r = 1.00$. When the association is perfect there will be frequencies in only two of the squares (either a and d or b and c) and therefore $\dfrac{\sqrt{bc}}{\sqrt{ad} + \sqrt{bc}}$ will equal zero and r will equal 1.00. When there is a proportionate distribution of cases, $ad = bc$, the fraction will equal .50 and $r = .00$.

BISERIAL COEFFICIENT OF CORRELATION

    When the table is of $2 \times N$ classifications, i.e., when there are only two possible categories of one attribute and a continuous variable, the **biserial coefficient of correlation** may be computed. This type of classification is common since classification by sex and other similar twofold categories appear frequently.[1] It assumes that the distribution of the attribute is approximately normal.

    The formula is

$$\text{biserial } r = \frac{(\overline{X}_p - \overline{X}_q)pq}{.3989h\sigma}$$

where

        $\overline{X}_p$ = the mean value of the p category

---

[1] Division of observations into two categories or a scale with only two values is called a *dichotomy*.

$\overline{X}_q$ = the mean value of the q category

p = percent of cases in p category

q = percent of cases in q category

σ = standard deviation of combined categories (p and q)

h = height of ordinate of normal curve at a distance from the mean including $\dfrac{p-q}{2}$ of the area of the curve.

The value h is computed by determining the number of standard deviations from the mean that include $\dfrac{p-q}{2}$ of the area of the normal curve (from the table of the area of the normal curve, pp. 128-129) and using that number of standard deviations in the table of the ordinates of the normal curve to determine the height of the curve (in percent of the maximum ordinate) at that point.

Another similar measure of correlation is the **point biserial coefficient of correlation.** This coefficient is free of the requirement of an assumption of normality for the dichotomous variable. The formula is

$$\text{point biserial } r = \frac{\overline{X}_1 - \overline{X}_2}{\sigma} \sqrt{pq}$$

where these terms have the same meaning as for the biserial coefficient of correlation.

ADDITIONAL BIBLIOGRAPHY

PEATMAN, J. G., *Introduction to Applied Statistics*, pp. 125–147, Harper & Row, New York, 1963.

WALKER, H. M., & LEV, J., *Statistical Inference*, pp. 261–287, Henry Holt & Co., New York, 1953.

YULE, G. U., & KENDALL, M. G., *An Introduction to the Theory of Statistics*, pp. 49–65, Hafner Publishing Co., New York, 1950.

Chapter XII
The Normal Curve

PROBABILITY

When events must occur in one of two ways, and there are many such events, they *tend* to be divided into two groups, the first consisting of favorable (or desired) occurrences, the second of unfavorable occurrences.

The chances are even that whenever a coin is flipped a "head" will appear. Analysis of this statement should clarify the nature of probability and the theory of sampling. Since only a head or a tail can appear, and each is equally likely, the probability of success in the appearance of the desired face can be stated as one half, or

$$p = \frac{1}{2}$$

In general

$$p = \frac{a}{n}$$

where

    p = probability of success
    a = number of equally likely ways in which a favorable outcome can appear
    n = total number of possible events

In general if an event can happen in a ways and not happen in b equally likely ways, the probability of its occurrence is

$$p = \frac{a}{a + b} = \frac{a}{n}$$

where $a + b = n$

The probability of failure is therefore

$$q = \frac{b}{n}$$

where

      b = the possible number of unfavorable results
      q = probability of failure

If a coin is tossed one hundred times, the probability of either heads or tails for each toss is ½. This ratio of probability may be expressed for the group as (½) (100).

The probability of occurrence of one or another of a series of *mutually exclusive* events may be determined by adding the probabilities of the individual occurrences. Thus if a card is drawn from a deck of cards, the probability of drawing either a space or a club is

$$\frac{1}{4} + \frac{1}{4}$$

To ascertain the probability of all of a series of mutually independent events, the respective probabilities are multiplied. Thus the probability of drawing a spade and then a club from a deck[1] is

$$\frac{1}{4} \times \frac{1}{4}$$

The probability that an event will either happen or fail to happen is certainty, to which is assigned the value 1.

The probability of the appearance of either a head or a tail in the toss of a coin is certainty

$$\left(\frac{1}{2}h + \frac{1}{2}t \right) = 1$$

For a toss of two coins the probability of the occurrence of heads and tails may be determined from

$$\left(\frac{1}{2}h + \frac{1}{2}t \right)^2 \text{ or } \frac{1}{4}h^2 + \frac{1}{2}ht + \frac{1}{4}t^2$$

---

[1]Providing that the first card is returned to the deck before the second is drawn so that the drawing of one card will have no effect on that of the next.

so that the

probability of all heads $= \dfrac{1}{4}$

probability of all tails $= \dfrac{1}{4}$

probability of a head and a tail $= \dfrac{1}{2}$

More generally the probability of occurrence of an event a given number of times in n trials can be found by the expansion of

$$n \, (p + q)^n$$

Figure 12.1 graphically presents the distribution of the number of heads appearing theoretically in tosses of ten coins. This is known as a **binomial distribution.**[1]

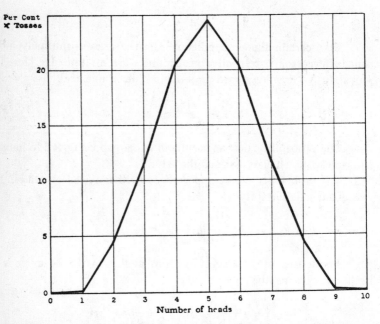

Fig. 12.1 – *Theoretical distribution of number of heads appearing in tosses of 10 coins.*

---

[1] Another name for this type of distribution is the Bernoulli Distribution. It is to be noted that this is a discrete distribution.

THE BINOMIAL DISTRIBUTION

The individual terms of the binomial distribution can be computed from

$$\Pr(r) = \frac{n!}{r!\,(n-r)!}\,p^r q^{n-r}$$

where

n = number of events

r = number of type specified

p = probability of event of type specified

q = probability of event other than type specified $(1 - p)$

For instance, if 6 coins are tossed, the probability of exactly two heads can be computed from

$$\Pr(2) = \frac{6!}{2!\,4!}\,(1/2)^2\,(1/2)^4 = \frac{720}{(2)\,(24)}\,(1/4)\,(1/16) = .2344$$

As n is increased in the binomial distribution the number of points to be plotted will increase and the curve will become smoother and smoother ultimately approaching in form the "normal" or **Gaussian**[1] **distribution** as seen in Figure 12.2[2] (p. 125). This type of curve frequently occurs when data exhibiting chance variation are plotted.

The mean of a binomial distribution is

$$\overline{X} = np$$

where

n = number of trials

p = probability of success

The standard deviation is[3]

$$\sigma = \sqrt{npq}$$

---

[1] The formula for the Gaussian distribution is

$$y = \frac{N}{\sigma\sqrt{2\pi}}\,e^{-x^2/2\sigma^2}$$

[2] When p is not equal to .5 (or 50%) and n is small, the binomial distribution is skewed but nevertheless approaches the shape of the normal distribution as n is increased.

[3] On a relative or percentage basis this becomes

$$\overline{X}_\% = p \text{ and } \sigma_\% = \sqrt{\frac{pq}{n}}$$

where $\qquad$ q = the probability of failure

The binomial distribution approaches the normal or Gaussian distribution previously referred to as n approaches infinity. The standard deviation will bear the same relation to the distribution as outlined previously (see page 43) and these values may be used to approximate the binomial distribution by using the normal distribution as the basis of the estimation.

Number of Standard Deviations (Measured plus and minus from the mean)	Percent of Cases Included
.6745 σ	50.00%
1.0000 σ	68.26%
1.9600 σ	95.00%
2.5800 σ	99.00%
3.0000 σ	99.73%

Thus, if 100 coins are tossed, the normal distribution assumption may be used to approximate the results

$$\overline{X} = np = 100\,(^1/_2) = 50$$

$$\sigma = \sqrt{npq} = \sqrt{100\,(^1/_2)\,(^1/_2)} = \sqrt{25} = 5$$

It can now be stated that there is a 95% probability that if 100 coins are tossed, between 50 plus and minus 1.96σ heads (or 50 $\pm 9.80 = 40.2$ and 59.8) or 41 to 59 heads (since the values are discrete) will occur.

THE POISSON DISTRIBUTION

It has been indicated that the binomial distribution, even when p is not .5, tends to approach the form of the normal distribution as n increases.

However, when p is very small and n is very large another form of discrete probability distribution called the **Poisson Distribution** may be used. Since p is very small this distribution deals with the occurrences of rare events.

The formula for the terms of the Poisson distribution is

$$\Pr(r) = \frac{e^{-\overline{X}}\overline{X}^r}{r!}$$

where

$\qquad$ e is a constant equal to 2.71828 ...
$\qquad$ $\overline{X}$ is the arithmetic mean

r is the number of events

$\Pr(r)$ is the probability of r events

The mean of a Poisson distribution is

$$\overline{X} = np$$

and its standard deviation is

$$\sigma = \sqrt{np}$$

Thus the variance of a Poisson distribution is equal to its arithmetic mean.

Tables[1] of the values of the terms of the Poisson distribution have been prepared to facilitate its use. These tables are entered through the value of $\overline{X}$ (or np) and probabilities for 0, 1, 2, etc., occurrences are given.

As an illustration, assume that a manufacturer of plastic film produces a product which frequent sampling discloses has contained a defect in the form of an average of one pin hole per square yard. What is the probability that the next square yard examined will contain no pin holes, 1 pin hole, 2 pin holes, etc?

Entry into the table for $\overline{X} = 1$ provides the following probabilities

Number of Defects per Sq. Yard	Probability
0	.37
1	.37
2	.180
3	.060
4	.015
5	.003

GENERALIZATION OF CURVES

When only a limited number of items are available the frequency distribution compiled from them is generally irregular in form. If the

---

[1] A condensed set of such tables may be found in Arkin, H. and Colton, R. R., *Tables for Statisticians,* 2nd ed., Barnes & Noble Inc., 1963, pp. 129–139 and a very extended set in Molina, E. C., *Tables of Poisson's Exponential Limit,* Van Nostrand Co., New York, 1942.

number of items is increased sharply the distribution will show a tendency to eliminate irregularities and be smoothed.

The sample is frequently used to generalize about the underlying data (technically the **population** or **universe**) and therefore it is often desirable to smooth the irregular curve obtained from the sample. By smoothing, the curve of the sample is put into the more general form of the underlying data, or in other words into an "ideal" distribution. The ideal distribution represents the distribution which would appear if an infinite number of cases were used rather than a sample.

Where it is believed that the underlying data can be best described by means of a normal curve,[1] data can be smoothed by this curve by use of special tables of the ordinates and area of the normal curve (see Tables 12.2 and 12.5).[2]

AREA METHOD OF FITTING NORMAL CURVE

In a normal distribution the percentage of the cases (area under the curve) included within any number of standard deviations measured from the mean can be determined from tables of areas of the normal curve.

If a distance equal to one standard deviation is measured in both directions from the mean it will include 68.26% of the cases, two standard deviations will include 95.46% of the cases, etc. (see p. 122). Similarly, the percentage of cases included within a distance equal to any number of standard deviations measured by only one direction from the mean will give one-half of the percentages shown above.

The mean of the normal curve is located at the center of the distribution. If a distance from the mean to any given point on the X axis is calculated *in terms of standard deviations*, the area included within this distance may now be determined by reference to Table 12.2 (pp. 128–129). For example, in Figure 12.2 the distance from the mean

---

[1] See Chapter XVII for methods of determining the type of underlying curve.

[2] Various other types of curves may be used to smooth a frequency distribution. Two groups of curves frequently used for this purpose are the Pearson system of curves and the Gram Charlier series.

Since the technique of fitting the curves is beyond the scope of this book, the reader is referred to Elderton, W. P., *Frequency Curves and Correlation* for the Pearson curve, and Camp, B. W., *Mathematical Part of Elementary Statistics* for the Gram Charlier Group.

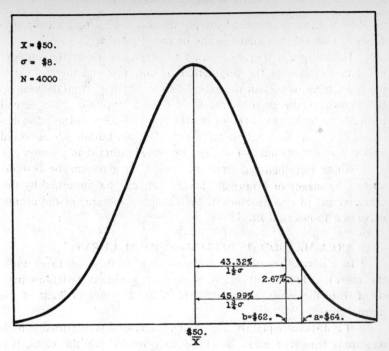

Fig. 12.2.

$50 to point *b* $62 is $12. The standard deviation of the distribution is given as $8. Thus the distance from the mean to this point in terms of standard deviations is 1.5 standard deviations (12 ÷ $8). By reference to the figure it will be seen that 43.32% of the cases are included between the mean and $62. If another point on the *X* axis is now selected, say $64, it can be found that since this point is 1.75 standard deviations away from the mean $\left(\dfrac{\$64 - \$50}{\$8} \right)$, 45.99% of the cases will be included within these limits. Since there are 45.99% of the cases between $64 and the mean and 43.32% of the cases between $62 and the mean, 2.67% (45.99% - 43.32%) of the frequencies must occur between $64 and $62. As there are 4000 cases in all, 106.8 of them should theoretically be between $62 and $64 in value.

In a similar manner the percentage of cases included within the limits of any one class interval may now be determined, and the number of cases or the theoretical frequency can be found by applying that

Table 12.1 – Fitting of Normal Curve – Area Method
Variation of Thickness in 600 Brass Washers Manufactured by the ABC Co.*

(1) Thickness (In Inches)	(2) Mid-Points	(3) Number of Washers (f)	(4) Deviation of Class Limit from Mean (x)	(5) Column (4) In Terms of Class Intervals $\left(\frac{x}{\sigma}\right)$	(6) Percent of Area between Class Limit and Mean	(7) Percent of Area in Class Interval	(8) Theoretical Frequency (f)
.0180–.01839	.0182	6	−.0022	−2.59	49.52%	1.22%	7.3
.0184–.01879	.0186	30	−.0018	−2.12	48.30	3.25	19.5
.0188–.01919	.0190	42	−.0014	−1.65	45.05	6.95	41.7
.0192–.01959	.0194	66	−.0010	−1.18	38.10	11.98	71.9
.0196–.01999	.0198	94	−.0006	−.71	26.12	16.64	99.8
.0200–.02039	.0202	120	−.0002 } .0002 }	−.24 } .24 }	9.48 } 9.48 }	18.96	113.8
.0204–.02079	.0206	102	.0006	.71	26.12	16.64	99.8
.0208–.02119	.0210	60	.0010	1.18	38.10	11.98	71.9
.0212–.02159	.0214	54	.0014	1.65	45.05	6.95	41.7
.0216–.02199	.0218	14	.0018	2.12	48.30	3.25	19.5
.0220–.02239	.0222	12	.0022	2.59	49.52	1.22	7.3
		600				99.10	594.2

*Hypothetical data based on smaller distribution given by W. A. Shewhart, *Economic Control of Quality* of Manufactured Product.

percentage to the total number of cases. If the theoretical frequencies are then plotted, the result will be a normal curve.

The determination of the theoretical frequency (for one of the class intervals) in the distribution of the variations in the thickness of 600 brass washers manufactured by the ABC Co. as shown in Table 12.1, p. 126, may be used as an example of this procedure. The fitting of this curve will make it possible to determine the variation to be expected in large numbers of these washers as manufactured by this company.

The following measures were computed from the sample:

$$\overline{X} = .0202 \text{ inches}$$
$$\sigma = .00085 \text{ inches}$$
$$n = 600$$

The third class interval has a lower limit of .0188 and an upper limit of .01919. The distance between the lower limit and the mean (.0202 − .0188) is .0014. Since the standard deviation is .00085 inches, in terms of standard deviations, the distance is 1.65 standard deviations. Reference to the area table indicates that 45.05% of the cases are included within this distance. Between the lower limit of the next class interval and the mean (.0202 − .0192 = .0010) there is a distance equal to 1.18 class deviations. The area table indicates that 38.10% of the cases are included within this distance. It can then be seen that there must be 6.95% of the cases between .0188 and .01919 or between the upper and lower limits of group three. Since the total frequency is given as 600 the theoretical frequency for this particular class interval will be 6.95% of 600. The frequency is then plotted at the midpoint of the group. The same process is repeated for all other class intervals as shown in Table 12.1.

FITTING THE NORMAL CURVE—ORDINATE METHOD

The normal curve may also be fitted by reference to a table of ordinates of the probability curve (Table 12.5, p. 132). The table gives the ordinates of the normal curve at any distance (in terms of standard deviations) from the mean as a percentage of the maximum ordinate. The maximum ordinate occurs at the center of the distribution.

The formula for the normal curve is

$$Y = Y_o e^{-x^2/2\sigma^2}$$

Table 12.2–Normal Curve Area Table

$\dfrac{x}{\sigma}$	.00	.01	.02	.03	.04	.05	.06	.07	.08	.09
0.0	.0000	.0040	.0080	.0120	.0159	.0199	.0239	.0279	.0319	.0359
0.1	.0398	.0438	.0478	.0517	.0557	.0596	.0636	.0675	.0714	.0753
0.2	.0793	.0832	.0871	.0910	.0948	.0987	.1026	.1064	.1103	.1141
0.3	.1179	.1217	.1255	.1293	.1331	.1368	.1406	.1443	.1480	.1517
0.4	.1554	.1591	.1628	.1664	.1700	.1736	.1772	.1808	.1844	.1879
0.5	.1915	.1950	.1985	.2019	.2054	.2088	.2123	.2157	.2190	.2224
0.6	.2257	.2291	.2324	.2357	.2389	.2422	.2454	.2486	.2518	.2549
0.7	.2580	.2612	.2642	.2673	.2704	.2734	.2764	.2794	.2823	.2852
0.8	.2881	.2910	.2939	.2967	.2995	.3023	.3051	.3078	.3106	.3133
0.9	.3159	.3186	.3212	.3238	.3264	.3289	.3315	.3340	.3365	.3389
1.0	.3413	.3438	.3461	.3485	.3508	.3531	.3554	.3577	.3599	.3621
1.1	.3643	.3665	.3686	.3708	.3729	.3749	.3770	.3790	.3810	.3830
1.2	.3849	.3869	.3888	.3907	.3925	.3944	.3962	.3980	.3997	.4015
1.3	.4032	.4049	.4066	.4083	.4099	.4115	.4131	.4147	.4162	.4177
1.4	.4192	.4207	.4222	.4236	.4251	.4265	.4279	.4292	.4306	.4319

1.5	.4332	.4345	.4357	.4370	.4382	.4394	.4406	.4418	.4430	.4441
1.6	.4452	.4463	.4474	.4485	.4495	.4505	.4515	.4525	.4535	.4545
1.7	.4554	.4564	.4573	.4582	.4591	.4599	.4608	.4616	.4625	.4633
1.8	.4641	.4649	.4656	.4664	.4671	.4678	.4686	.4693	.4699	.4706
1.9	.4713	.4719	.4726	.4732	.4738	.4744	.4750	.4758	.4762	.4767
2.0	.4773	.4778	.4783	.4788	.4793	.4798	.4803	.4808	.4812	.4817
2.1	.4821	.4826	.4830	.4834	.4838	.4842	.4846	.4850	.4854	.4857
2.2	.4861	.4865	.4868	.4871	.4875	.4878	.4881	.4884	.4887	.4890
2.3	.4893	.4896	.4898	.4901	.4904	.4906	.4909	.4911	.4913	.4916
2.4	.4918	.4920	.4922	.4925	.4927	.4929	.4931	.4932	.4934	.4936
2.5	.4938	.4940	.4941	.4943	.4945	.4946	.4948	.4949	.4951	.4952
2.6	.4953	.4955	.4956	.4957	.4959	.4960	.4961	.4962	.4963	.4964
2.7	.4965	.4966	.4967	.4968	.4969	.4970	.4971	.4972	.4973	.4974
2.8	.4974	.4975	.4976	.4977	.4977	.4978	.4979	.4980	.4980	.4981
2.9	.4981	.4982	.4983	.4984	.4984	.4984	.4985	.4985	.4986	.4986
3.0	.49865	.4987	.4987	.4988	.4988	.4988	.4989	.4989	.4989	.4990
3.1	.49903	.4991	.4991	.4991	.4992	.4992	.4992	.4992	.4993	.4993

Table 12.3.–Fitting of Normal Curve–Ordinate Method

Variation of Thickness in 600 Brass Washers Manufactured by the ABC Co.

(1) Thickness (In Inches)	(2) Mid-Points	(3) Number of Washers (f_o)	(4) Deviation of Midpoint from Mean (x)	Column 4 In Terms of Standard Deviations ($\frac{x}{\sigma}$)	Percent of Maximum Ordinate (From Ordinate Table)	Theoretical Frequency (f)
.0180-.01839	.0182	6	-.0020	-2.37	6.03%	6.84
.0184-.01879	.0186	30	-.0016	-1.90	16.45	18.67
.0188-.01919	.0190	42	-.0012	-1.42	36.49	41.42
.0192-.01959	.0194	66	-.0008	-.95	63.68	72.28
.0196-.01999	.0198	94	-.0004	-.47	89.54	101.62
.0200-.02039	.0202	120	.0000	.00	100.00	113.50
.0204-.02079	.0206	102	.0004	.47	89.54	101.62
.0208-.02119	.0210	60	.0008	.95	63.68	72.28
.0212-.02159	.0214	54	.0012	1.42	36.49	41.42
.0216-.02199	.0218	14	.0016	1.90	16.45	18.67
.0220-.02239	.0222	12	.0020	2.37	6.03	6.84
		600				

where[1]

$$Y_o, \text{ the maximum ordinate, } = \frac{N}{\sigma\sqrt{2\pi}} = \frac{N}{2.506628\,\sigma}$$

The value of the maximum ordinate (Y_o) for the distribution can now be computed.

$$\sigma \text{ (class interval units)} = 2.109$$

$$Y_o = \frac{N}{2.506628\sigma} = \frac{600}{2.109(2.506628)} = 113.5$$

For the midpoint of the first group which is 2.37 standard deviations (.0020 inches) from the mean we find that an ordinate erected at this point would be 6.03% of the maximum ordinate or 6.84. The same procedure may be used on the other class intervals as shown in Table 12.3.

TESTING THE GOODNESS OF FIT—THE CHI SQUARE TEST

A test to determine the goodness of fit of the actual data to the theoretical distribution has been devised by Karl Pearson.

The test involves the calculation of χ^2 (chi square)

$$\chi^2 = \Sigma\left(\frac{(f_o - f)^2}{f}\right)$$

Table 12.4—The Chi Square Test for Goodness of Fit—
Data of Table 12.1

(1)	(2)	(3)	(4)	(5)	(6)
Thickness (In Inches)	Number of Washers (f_o)	Theoretical Frequency (f)	($f_o - f$)	($f_o - f$)^2	$\frac{(f_o - f)^2}{f}$
.0180–.01839	6 ⎫	7.3 ⎫	9.2	84.64	3.158
.0184–.01879	30 ⎭	19.5 ⎭			
.0188–.01919	42	41.7	.3	.09	.002
.0192–.01959	66	71.9	− 5.9	34.81	.484
.0196–.01999	94	99.8	− 5.4	29.16	.292
.0200–.02039	120	113.8	6.2	38.44	.338
.0204–.02079	102	99.8	2.2	4.84	.048
.0208–.02119	60	71.9	−11.9	141.61	1.970
.0212–.02159	54	41.7	12.3	151.29	3.628
.0216–.02199	14 ⎫	19.5 ⎫	− .8	.64	.024
.0220–.02239	12 ⎭	7.3 ⎭			

$$\chi^2 = 9.944$$

---

[1]In the application of this formula the standard deviation in class interval units, not original units, must be used.

Table 12.5–Normal Curve Ordinates
(As decimal of maximum ordinate)

$\frac{x}{\sigma}$	.00	.01	.02	.03	.04	.05	.06	.07	.08	.09
0.0	1.00000	.99995	.99980	.99955	.99920	.99875	.99820	.99755	.99681	.99596
0.1	.99501	.99397	.99283	.99159	.99025	.98881	.98728	.98565	.98393	.98211
0.2	.98020	.97819	.97609	.97390	.97161	.96923	.96676	.96421	.96156	.95882
0.3	.95600	.95309	.95000	.94701	.94384	.94059	.93725	.93384	.93034	.92677
0.4	.92312	.91939	.91558	.91169	.90774	.90371	.89960	.89543	.89119	.88688
0.5	.88250	.87805	.87354	.86897	.86433	.85963	.85488	.85006	.84518	.84025
0.6	.83527	.83023	.82514	.82000	.81481	.80957	.80429	.79896	.79358	.78816
0.7	.78270	.77721	.77167	.76609	.76048	.75484	.74916	.74345	.73771	.73194
0.8	.72615	.72033	.71448	.70861	.70272	.69680	.69087	.68492	.67896	.67297
0.9	.66698	.66097	.65495	.64892	.64288	.63683	.63078	.62472	.61866	.61260
1.0	.60653	.60047	.59440	.58834	.58228	.57623	.57018	.56414	.55811	.55209
1.1	.54607	.54007	.53409	.52811	.52215	.51621	.51028	.50437	.49848	.49260
1.2	.48675	.48092	.47511	.46933	.46357	.45783	.45212	.44644	.44078	.43516
1.3	.42956	.42399	.41845	.41294	.40747	.40202	.39661	.39123	.38589	.38058
1.4	.37531	.37007	.36488	.35971	.35459	.34950	.34445	.33944	.33447	.32954

1.5	.28251	.28702	.29158	.29618	.30082	.30550	.31023	.31499	.31980	.32465
1.6	.23978	.24385	.24797	.25213	.25634	.26059	.26489	.26923	.27361	.27804
1.7	.20148	.20511	.20879	.21250	.21627	.22007	.22392	.22782	.23176	.23575
1.8	.16762	.17081	.17404	.17732	.18064	.18400	.18741	.19086	.19436	.19790
1.9	.13806	.14083	.14364	.14649	.14938	.15232	.15529	.15831	.16137	.16447
2.0	.11258	.11496	.11737	.11982	.12230	.12483	.12740	.13000	.13265	.13534
2.1	.09090	.09290	.09495	.09702	.09914	.10129	.10347	.10569	.10795	.11025
2.2	.07265	.07433	.07604	.07779	.07956	.08137	.08320	.08508	.08698	.08892
2.3	.05750	.05888	.06030	.06174	.06321	.06471	.06624	.06780	.06939	.07101
2.4	.04505	.04618	.04734	.04852	.04972	.05096	.05221	.05349	.05480	.05613
2.5	.03494	.03586	.03679	.03775	.03873	.03972	.04074	.04179	.04285	.04394
2.6	.02683	.02757	.02831	.02908	.02986	.03066	.03148	.03232	.03317	.03405
2.7	.02040	.02098	.02157	.02217	.02279	.02343	.02408	.02474	.02542	.02612
2.8	.01536	.01581	.01627	.01674	.01723	.01772	.01823	.01876	.01929	.01984
2.9	.01145	.01179	.01215	.01252	.01289	.01328	.01367	.01408	.01449	.01492
3.0										.01111
4.0										.00034

134 THE NORMAL CURVE

where

f_o = the observed or actual frequencies
f = the theoretical frequencies

For the problem outlined above chi square may be calculated as in Table 12.4. If the value of f for any class interval is very small, several groups must be combined in order not to obtain disproportionate values of χ^2. See Table 12.4. Thus the 1st and 2nd class intervals are combined as well as the last two.

Chi square may be evaluated by reference to a set of χ^2 tables.[1] In tables of χ^2, N represents the number of "degrees of freedom." See Table 12.6. This is based on the number of groups (N_k), and equals

Table 12.6–Section of χ^2 Table

Degree of Freedom (N)	p = .99	p = .05	p = .01
1	.00016	3.84	6.64
2	.020	5.99	9.21
3	.115	7.82	11.34
4	.297	9.49	13.28
5	.554	11.07	15.09
6	.872	12.59	16.81
7	1.239	14.07	18.48
8	1.646	15.51	20.09
9	2.088	16.92	21.67
10	2.558	18.31	23.21
11	3.053	19.68	24.73
12	3.571	21.03	26.22
13	4.107	22.36	27.69
14	4.660	23.69	29.14
15	5.229	25.00	30.58

N_k - 3. The value p in the table is the probability of obtaining a fit, due to chance, *as poor as or worse than* the one actually obtained. If this probability is small the likelihood that the disparities between observed and actual data are due to chance is small.[2]

---

[1] An extended set of these tables is to be found in Arkin, H., and Colton, R. R., *Tables for Statisticians*, Barnes & Noble, Inc., New York, 1963.

[2] Generally, if the indicated value of p is less than some specified value, usually .05 or .01, the discrepancies are accepted as too large to be accidental. (See Chapter XIII.)

Thus the value of χ^2 in the problem above for 6 degrees of freedom indicates a value for p of about .10.   In other words there are about 10 chances out of 100 that the fit obtained would be as bad as or worse than the one shown; 10 chances out of 100 that the variations due to chance fluctuations with no real departure from normality occurring in the sample might be either as bad as or worse than those actually occurring.   There is no evidence that the normal curve is a bad fit.

ADDITIONAL BIBLIOGRAPHY

BURR, IRVING W., *Engineering Statistics and Quality Control*, pp. 180–211, McGraw-Hill Book Co., New York, 1953.

DUNCAN, A. J., *Quality Control and Industrial Statistics*, pp. 13–79, Richard D. Irwin, Inc., Homewood, Ill., 1959.

GOLDBERG, SAMUEL, *Probability, An Introduction*, pp. 45–131, Prentice-Hall, Inc., Englewood Cliffs, N. J., 1960.

WALKER, H. M., & LEV, J., *Statistical Inference*, pp. 8–42, Henry Holt & Co., 1953.

YULE, G. U., & KENDALL, M. G., *An Introduction to the Theory of Statistics*, pp. 169–198, Hafner Publishing Co., New York, 1950.

Chapter XIII

Chi Square and Tests of Hypotheses

TESTING HYPOTHESES

The chi square test was described in the previous chapter as a test of goodness of fit for a frequency distribution. However, a broader and more important application of the chi square test is its use in testing hypotheses by comparing observed or experimental data to theoretical frequencies based on a hypothesis.

The use of chi square in a hypothesis test is applicable where the data can be represented in the form of a *contingency table*. A contingency table consists of a cross-tabulation of categories or classes of observations with the frequency for each cross-classification shown. An example of a contingency table is shown in Table 13.1.

Table 13.1 – Responses of a Sample of Respondents to a Question According to Educational Level

	EDUCATIONAL LEVEL			
Reply	*Elementary School*	*Secondary School*	*College*	*Total*
Yes	66	22	12	100
No	32	36	13	81
No Answer	20	23	16	59
Total	118	81	41	240

The columns classify the respondents according to their educational level while the rows classify the respondents according to the manner in which they answered the question. The contingency table used as an illustration contains 3 columns and 3 rows (a 3 x 3 table), but as many columns and rows may be used as needed.

The purpose of the technique is to test the independence of the cross-classification in the contingency table, that is, to establish whether the classification principle used for the columns is related to that of the rows. For the data of Table 13.1, the aim is to determine whether the educational level of the respondents affected their responses to the question.

The Hypotheses

Since the objective is to test a hypothesis, it is necessary to state the hypothesis explicitly. In tests of this type, the hypothesis is always that the classification of the columns and rows are related, or for the data of Table 13.1, that the differences in educational level *did* affect the responses.

To establish the necessary theoretical frequencies to provide the basis for a chi square computation, the *null hypothesis* is used. This null hypothesis is the opposite of the original hypothesis and thus assumes independence of classification or that the classification of the rows is unrelated to that of the columns.[1] If the null hypothesis can be disproved, the original hypothesis must be true, since there is no other alternative. For Table 13.1, the null hypothesis is that the characteristics described by the column headings (educational level) are unrelated to those of the row headings (response).

The Method

If the classification for the columns is independent of that for the rows and the null hypothesis is true, the frequencies in the cells of the contingency table should be proportionate to the marginal totals.

Thus, in the data for the first cell (first row, first column) the theoretical frequency would be obtained by distributing the total for the row (100) in proportion to the column totals or

$$100 \times \frac{118}{240} = 49.2$$

Generally, the theoretical frequency for each cell based on the null hypothesis is

$$\frac{n_r n_c}{N}$$

---

[1] The symbol H_0 is often used to designate the null hypothesis with the symbol H_1 referring to the original or alternative hypothesis.

where

> n_r = total of the number of observations in the row containing the specified cell
>
> n_c = total of the number of observations in the column containing the specified cell
>
> N = total number of observations in the entire table.

The theoretical frequencies for Table 13.1 are shown in parentheses in each cell of Table 13.2.

Table 13.2–Actual and Theoretical Frequencies

Reply	Elementary School	Secondary School	College	Total
Yes	66 (49.2)	22 (33.8)	12 (17.1)	100
No	32 (39.8)	36 (27.3)	13 (13.8)	81
No Answer	20 (29.0)	23 (19.9)	16 (10.1)	59
Total	118	81	41	240

The value of chi square is obtained from

$$\chi^2 = \Sigma \left[\frac{(f_o - f)^2}{f} \right]$$

where

> f = theoretical frequency
>
> f_o = observed frequency

The value $\dfrac{(f_o - f)^2}{f}$ is computed individually for each cell of the table and the values for all cells added to yield the value of χ^2. The method is shown in Table 13.3.

A table of chi square is consulted (see page 134)[1]. The probability level indicated in the column headings is the probability that a value of chi square as large as that indicated in the table might arise due to chance even if the classification of the columns and rows are unrelated.

---

[1] A more extended table of chi square is contained in Arkin, H. and Colton, R. R., Tables for Statisticians, Barnes and Noble, Inc., New York, 2nd ed., 1963.

Table 13.3—Calculation of Chi Square

Educational Level	Response	Observed Frequency (f_o)	Theoretical Frequency (f)	$(f_o - f)$	$(f_o - f)^2$	$\dfrac{(f_o - f)^2}{f}$
Elementary	Yes	66	49.2	16.8	282.24	5.74
Secondary	Yes	22	33.8	−11.8	139.24	4.12
College	Yes	12	17.1	− 5.1	26.01	1.52
Elementary	No	32	39.8	− 7.8	60.84	1.53
Secondary	No	36	27.3	8.7	75.69	2.77
College	No	13	13.8	− .8	.64	.05
Elementary	No Answer	20	29.0	− 9.0	81.00	2.79
Secondary	No Answer	23	19.9	3.1	9.61	4.83
College	No Answer	16	10.1	5.9	34.81	3.45
TOTAL		240	240.0	0.0		26.80

Calculated $\chi^2 = 26.80$

A suitably low significance level is selected (say .01). The table is consulted for this probability or significance level and for the appropriate degrees of freedom, calculated from

$$(r - 1)(c - 1)$$

where

    c = number of columns
    r = number of rows
In this problem there are

$$(3 - 1)(3 - 1) = 4 \text{ degrees of freedom}$$

The tabular value of chi square is 13.28 (see page 134)

$$\text{calculated } \chi^2 = 26.80$$
$$\text{tabular } \chi^2{}_{.01} = 13.28$$

Since the calculated value of chi square exceeds the tabular value, the probability that the difference between the observed and theoretical frequencies (from which the calculated chi square value was computed) is accidental is less than .01. It may be concluded that the null hypothesis has little likelihood of being true and thus may be considered false. If the null hypothesis is false, the original hypothesis is true. It may be concluded that the educational level of the respondents did affect their responses to the question.

Had the calculated value of chi square been less than the tabular value for the specified significance level, the conclusion would be that there was *no evidence* that the null hypothesis was either true or false or that the original hypothesis was true.

Limitations

1. The sample upon which the contingency table is based must be a probability (or random) sample.

2. The theoretical frequency in each cell of the contingency table must not be too small. A rule-of-thumb minimum generally agreed upon is 5 for any individual cell. If individual cells contain theoretical frequencies which are too small, it may be possible to raise the value by combining groups (columns or rows) until a sufficiently large theoretical frequency is secured. There is no limit on the magnitude of the observed frequencies in any cell.

3. In a 2 × 2 table (which has 1 degree of freedom) **Yates' correction for continuity** should be used. This correction is accomplished by computing chi-square as follows:

$$\chi^2 = \Sigma \left[\frac{(|f_0 - f| - 0.5)^2}{f} \right]$$

where

$|f_0 - f|$ indicates the absolute (signs ignored) difference between the observed and theoretical frequencies.

4. The chi square test cannot be applied to a contingency table which contains only relative frequencies (percentages).

ADDITIONAL BIBLIOGRAPHY

WALKER, H. M., & LEV, J., *Statistical Inference,* pp. 81–108, Henry Holt & Co., 1953.

YULE, G. U., & KENDALL, M. G., *An Introduction to the Theory of Statistics,* pp. 459–478, Hafner Publishing Co., 1950.

Chapter XIV
Theory of Sampling

THE SAMPLE
The analysis by statistical techniques of a mass of data is often made difficult or impossible when the mass of data is too great to be handled in its entirety. In these cases samples of the data subjected to the same techniques allow generalizations to be made about the larger mass of data from which the sample was drawn.

If the results are confined only to the cases studied they may be used for descriptive purposes. A number of problems, however, must be solved before the results can be generalized and applied to the larger number of cases not included in the sample.

Much time, energy, and money would be needed to make a comprehensive analysis of a great mass of statistical data. Consequently there is every incentive to resort to study of part or parts of the data. This process is known as **sampling**.

The Need for Sampling—Examples
In the physical sciences it is often impossible to obtain further data, as, for example, when an experiment cannot be performed more than a given number of times. Sampling must be resorted to in such a case. The results of a scientific experiment repeated ten times can be used as a generalization of the results which might logically be assumed to be obtainable if the experiment were performed an infinite number of times.

To obtain an average Intelligence Quotient for third grade public school students the use of any method other than sampling would mean the very expensive accumulation of an enormous mass of data.

Again, let us say it is desired to obtain the average price of bread in New York City. Obviously the cost factor and the time required for

141

a complete survey of the city's thousands of bakeshops and grocery stores would be prohibitive.

In the last instance cited if prices were obtained from chain stores only the sample would be prejudiced. To secure *representative* data it would be necessary to obtain sample prices from all the varied types of stores; from, in technical term, the entire **population**[1] or **universe**.

The limited sample is used to describe the larger population or group of data from which the sample was taken.[2]

METHODS

The items selected from the larger group of data for inclusion in the sample are called **sampling units**. The totality of observations from which these sampling units were drawn is called a population or universe. Upon occasion, potential sampling units may be missing from the population or otherwise not available when the sampling is performed. The sampling units actually available are called the *frame*. The frame is usually described as a listing but may constitute any specification of the sampling units available for sampling. If there is any disparity between the frame and the population, between the sampling units actually available for sampling and those comprising the population of interest, it must be remembered that the sample projection applies to the frame and not to the population.

When the measures computed from a sample are used to characterize the population, it is necessary to estimate the reliability of the measures, in other words the degree of error to which the generalization may be subject.

Samples may be drawn from the population in several different ways. However, to be susceptible to statistical evaluation the sample analyzed must be a random sample and is properly called a *probability* sample. When unrestricted random sampling is used all observations in the population are given an equal chance of inclusion in the sample.

However, it is often desirable to segregate a heterogeneous population into homogeneous subgroups and to draw from each subgroup at random. This process results in the **stratified sample**. A survey of the

---

[1] The population may be defined as the entire set of data from which the sample was drawn if all of it were available.

[2] A value, such as an average, computed from a sample is termed a "statistic." The equivalent value for the entire population is called the "parameter."

number of rooms in homes in a given city by sampling may be more effective if random samples are drawn from uniform areas of the city. Thus, a random sample of a low-income area may be combined with random samples taken from other income areas.

In a stratified random sampling operation, the opportunity for inclusion of each observation in the sample is constant for each stratum or segment of the population but may vary from stratum to stratum.

There are other variations on these basic sampling methods such as *cluster* sampling and *multi stage* sampling but these methods are beyond the scope of this work.

The **purposive** or **judgment sample** represents a deliberate selection of a sample manipulated by the statistician in such a fashion as to obtain a representative cross section of the population. The statistical techniques outlined in this chapter cannot be applied to judgment samples.

To constitute a probability sample and thus make available the techniques discussed in this chapter, the sampling units must be selected by the use of *random number* sampling or by *systematic* sampling.[1]   In random number sampling, the units to be included in the sample are usually selected by resort to tables of random numbers. The numbers listed in the table determine their selection. The units used are those with numbers listed in the table, starting at some randomly selected point.

In systematic sampling, every *n*th item is selected as one of the sampling units to be included in the sample, *starting with a randomly selected item* within the first sampling interval. Thus, if a sample of 100 sampling units is to be selected from a universe of 20,000, every 200th item starting with a randomly selected unit between 1 and 200 would be used.

MEASURES OF RELIABILITY AND SIGNIFICANCE— STANDARD ERRORS

In the bread price problem outlined above, let us assume that investigators are sent out to obtain a sample of prices for a given size and type of bread from 200 randomly selected stores. If the average prices for the samples were then arranged in the form of a frequency distri-

---

[1] For details on the proper methods for selecting a probability sample see Arkin, H., *Handbook of Sampling in Auditing and Accounting*, McGraw-Hill Book Co., New York, 1963, Chapter 3.

bution they would be found to tend toward a normal distribution. This is called the *sampling distribution of the average*.[1]

The average of the means of the sample prices drawn repeatedly at random from among the thousands of stores in New York City would undoubtedly result in a figure either the same as or very near the true mean of the underlying data. When the curve is idealized as though an infinite number of cases had been considered, the "true" mean for bread prices in New York can be obtained by computing the average of the distribution.[2]

From the hypothetical sampling distribution shown in Figure 14.1, it can be seen that the means of some of the samples were quite a distance from the true mean of the population. If the distance of the sample mean furthest away can be obtained, the greatest possible error will be known. It has already been shown that 99.7% of the cases will be included within a distance of 3 standard deviations from the mean (see page 43). If the value of the standard deviation of this distribution (of the means of the samples) can be obtained there are 99.7 chances out of 100 that no error (difference between sample mean and true mean) will occur larger than 3 times the value of the standard deviation.

The standard deviation of a sampling distribution of means, or of any other statistical measure computed from samples, is termed the **standard error of the mean** ($\sigma_{\overline{x}}$), or the standard error of the other statistical measure.

The error which will not be exceeded by 50% of the cases is known as the **probable error**.[3] It is equal to .6745 times the standard error.

---

[1] A sampling distribution is a tabulation of some specified statistical value from an infinite number of samples of a fixed size from a given population. The *central limit theorem* indicates that regardless of the nature of the distribution of individual values, the sampling distribution of the means will approach normality as the sample size increases. See Figure 14.1.

[2] By "true" mean here is meant the mean which would have been obtained if all of the observations in the population had been used in its computation, not just the values for the sample.

[3] Although sometimes used, the probable error is of comparatively little value. It can be interpreted as meaning that if another sample were drawn of the same number of items the chances are even that a discrepancy between the sample and true mean larger than the probable error would not exist.

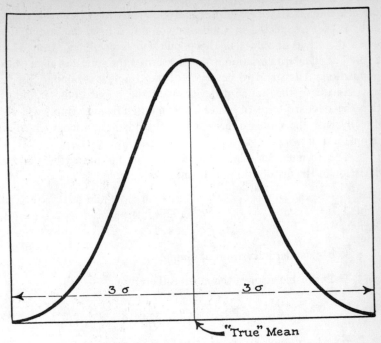

Fig. 14.1—Theoretical distribution of means of a large number of samples drawn at random from a given population.

Use of the probable error, notably by American statisticians, gives it a fictitious value far beyond its real worth as compared with the standard error.[1]

It is obvious that the greater the number of cases included in the sample the smaller the error to be expected, and the smaller the standard deviation (measure of sampling error) of the theoretical distribution of means (or other measure). In turn the standard error will vary inversely with the number of cases included in the sample.

When the range of the limits of variation in the population or

---

[1] R. A. Fisher (*Statistical Methods for Research Workers*, 1928, p. 46) points out that "the common use of the probable error is its only recommendation . . . when any critical test is required, the deviation must be expressed in terms of the standard error." It has also been pointed out that the standard error has been commonly used by European statisticians in preference to the probable error.

original data is large, for example from $1 to $1,000,000, a greater error is to be expected in a measure computed from the sample than when the range of values in the population is small; e.g., from $1 to $10. The standard deviation is the measure of the spread of data of the population. The standard deviation of the sample is commonly used as an estimate of the standard deviation of the population. It follows that the standard error of a measure computed from a sample will vary directly with the standard deviation of the population from which the sample was drawn.

The formula for the standard error of the mean (the standard deviation of the distribution of the means of samples) is[1]

$$\sigma_{\bar{x}} = \frac{\sigma}{\sqrt{n}}$$

where

σ = standard deviation of sample

The probable error of the mean is therefore

$$P.E._{x} = .6745 \frac{\sigma}{\sqrt{n}}$$

This enables the sample reliability to be evaluated. As an example, if the average price of bread is computed from one sample of prices obtained from 200 stores, and the following results were obtained,

$$\bar{X} = \$.30$$
$$\sigma = \$.05$$

the standard error of the mean would be

$$\sigma_x = \frac{\$.05}{\sqrt{200}} = \$.0035$$

As shown above there are 99.7 chances out of 100 that the mean computed from a random sample of 200 cases will not be further away than three standard errors of the mean from the true average price. It can now be assumed that the computed average of $.30 will not be more than $.0105 away from the true average (99.7 chances out of 100).

Thus, the sampler has a 99.7% assurance (*confidence level*) that

---

[1] See Technical Appendix IX for the derivation of this formula.

the sample result will not be farther from the true value or the value which would have been obtained if all stores had been included than $.0105 (the *sample reliability* or *precision*). The interval within which this value is contained ($.30 ± $.0105) is called the *confidence interval*, while the outside limits so established are called the *confidence limits*.

It is possible to select a sample size to provide given sample reliabilities at any selected confidence level. The choice of the confidence level and sample precision may be made by the sampler.

The probability of occurrence and the odds against the occurrence of an error as great as the given number of standard errors as given by Pearl[1] are shown in Table 14.1, p. 148.

The formula for the standard error of the mean, shown above, is for the mean of an unrestricted random sample. In a stratified random sampling operation the population is divided into segments or strata and an independent sampling is made for each stratum. The formula for the standard error of the mean of a stratified random sample is

$$\sigma_{\overline{X}} = \sqrt{\frac{\Sigma(N_s^2 \sigma_{\overline{X}_s}^2)}{N^2}}$$

where

$\sigma_{\overline{X}}$ = standard error of the mean of a stratified sample

N_s = number of observations in the population for each stratum

N = number of observations in the entire population

$\sigma_{\overline{X}_s}$ = standard error of the mean for sample from each stratum computed separately

The formulas discussed above apply to the means of samples drawn from infinitely large populations or of samples which are very small in comparison to the size of the underlying population. When the population is finite in size, particularly when the size of the sample is appreciable in proportion to the size of the universe, the standard error of the mean may be computed from the formula[2]

---

[1]Pearl, Raymond, *Medical Biometry and Statistics*, W. B. Saunders, Philadelphia, 1930

[2]A more precise definition of the finite population correction factor (f.p.c.) is

$$\sqrt{\frac{N-n}{N-1}}$$

However, for most populations the difference between these two versions is negligible and the above form is more generally used.

THEORY OF SAMPLING

Table 14.1 – The Probability of Occurrence of Statistical Deviations of Different Magnitudes Relative to the Standard Error

Number of standard errors	Probability of occurrence of a deviation as great as or greater than designated number of standard errors	Odds against the occurrence of a deviation as great as or greater than the designated number of standard errors
0.67449	50.00%	1.00 to 1
0.7	48.39	1.07 to 1
0.8	42.37	1.36 to 1
0.9	36.81	1.72 to 1
1.0	31.73	2.15 to 1
1.1	27.13	2.69 to 1
1.2	23.01	3.35 to 1
1.3	19.36	4.17 to 1
1.4	16.15	5.19 to 1
1.5	13.36	6.48 to 1
1.6	10.96	8.12 to 1
1.7	8.91	10.22 to 1
1.8	7.19	12.92 to 1
1.9	5.74	16.41 to 1
2.0	4.55	20.98 to 1
2.1	3.57	26.99 to 1
2.2	2.78	34.96 to 1
2.3	2.14	45.62 to 1
2.4	1.64	60.00 to 1
2.5	1.24	79.52 to 1
2.6	.932	106.3 to 1
2.7	.693	143.2 to 1
2.8	.511	194.7 to 1
2.9	.373	267.0 to 1
3.0	.270	369.4 to 1
3.1	.194	515.7 to 1
3.2	.137	726.7 to 1
3.3	.0967	1,033 to 1
3.4	.0674	1,483 to 1
3.5	.0465	2,149 to 1
3.6	.0318	3,142 to 1
3.7	.0216	4,637 to 1
3.8	.0145	6,915 to 1
3.9	.00962	10,390 to 1
4.0	.00634	15,770 to 1
5.0	.0000573	1,744,000 to 1
6.0	.00000020	500,000,000 to 1
7.0	.00000000026	400,000,000,000 to 1

$$\sigma_{\overline{x}_F} = \sigma_{\overline{X}} \sqrt{1 - \frac{n}{N}}$$

or

$$\sigma_{\overline{x}_F} = \frac{\sigma}{\sqrt{n}} \sqrt{1 - \frac{n}{N}}$$

where

> n = number of items in sample
> N = number of items in population
> $\sigma_{\overline{X}}$ = the standard error of the mean of a random sample as above

The sample multiplier $\left(\sqrt{1 - \dfrac{n}{N}} \right)$ may be used to correct other standard error formulae for the effect of a finite population. However, it is to be noted that when the sample is only a small proportion of the population (say 1% or 2%), the correction is insignificant.

In a similar fashion the standard errors of other statistical measures may be computed. The formulae for some of these standard errors are shown below.

STANDARD ERROR FORMULAE

(Infinite Populations)

Measure	*Formula*
Mean ($\overline{X}$)	$\sigma_{\overline{X}} = \dfrac{\sigma}{\sqrt{n}}$
Sum (ΣX)	$\sigma_{\Sigma X} = \sigma \sqrt{n}$
Median	$\sigma_{mdn} = 1.2533 \dfrac{\sigma}{\sqrt{n}}$
Standard Deviation (for sample from normally distributed population)	$\sigma_\sigma = \dfrac{\sigma}{\sqrt{2n}}$
Mean Deviation	$\sigma_{MD} = .6028 \dfrac{\sigma}{\sqrt{n}}$
Coefficient of Variation (V)	$\sigma_v = \dfrac{V}{\sqrt{2n}} \sqrt{1 + 2(V)^2}$
Coefficient of Correlation (r)	$\sigma_r = \dfrac{1 - r^2}{\sqrt{n}}$
Coefficient of Rank Correlation (ρ) (n over 25)	$\sigma_\rho = \dfrac{1}{\sqrt{n-1}}$

Multiple Correlation Coefficient
$(R_{1.23...n})$

$$\sigma_{R_{1.23}\,...\,n} = \frac{1 - R^2_{1.23...n}}{\sqrt{n}}$$

Partial Correlation Coefficient $(r_{12.34\,...\,n})$

$$\sigma_{r_{12.34}\,...\,n} = \frac{\sqrt{1 - r^2_{12.34\,...\,n}}}{\sqrt{n}}$$

Proportion (%)[1]

$$\sigma_{\%} = \sqrt{\frac{pq}{n}}$$

Coefficient of Regression (b_{yx})

$$\sigma_b = \frac{S_y}{\sigma_x \sqrt{n}}$$

Regression Equation Y Intercept (a)

$$\sigma_a = \frac{S_y}{\sqrt{n}}$$

Coefficient of Multiple Regression

$$\sigma_a = \frac{S_{1.23}}{\sqrt{n}}$$

$$\sigma_{b\,12.3} = \frac{S_{1.23}}{\sigma_2 \sqrt{n}\,(1 - r^2_{23})}$$

This list represents merely a small number of the standard error formulae. For a more complete listing see the list of formulas at the rear of this book and Dunlap, J. W. and Kurtz, A. K., *Handbook of Statistical Nomographs, Tables and Formulae.*

SIGNIFICANCE OF THE DIFFERENCE BETWEEN TWO MEANS

Very frequently it is desirable to test the means of two samples to determine whether there is any significant difference between them, or whether the difference, if any, is merely due to chance.

---

[1] See Technical Appendix XII for proof.

In scientific fields it is customary to use a "control" when carrying out an experiment. This control, to which the new technique being tested is not applied, is used as a basis for comparison. The measured results for the "control" group are compared with those from the experimental group to determine whether they differ significantly.

As an instance, if a new technique for teaching spelling is subjected to an experimental test two groups of pupils are used. The new technique is applied to group 1, and group 2 is used solely as a control. Both groups are comprised of *randomly* selected students. The results (grades on examinations, etc.) are then tested to determine the significance of the difference between the mean grades for the two groups.

If two samples of any data are drawn from a given population undoubtedly there will be a difference between the means of the samples, a difference due solely to chance, because of variations in the selection of items.

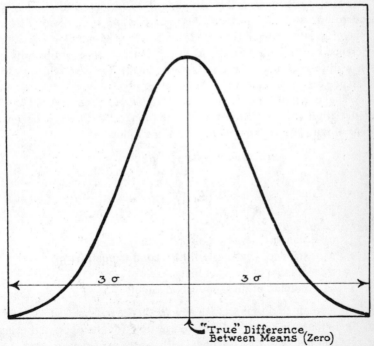

Fig. 14.2 – Theoretical distribution of difference between means of a large number of pairs of samples drawn at random from a given population.

If a very large number of these pairs of samples are drawn from the population, and if the differences between the means of the pairs are arranged in the form of a frequency distribution, the resulting distribution will be normal. The true difference is in reality zero, since all these pairs of samples are drawn from the same population and only *chance or accidental* differences arise between the samples. If all the differences determined from a very large or more exactly an infinite number of pairs are averaged (signs retained), the true difference (zero) will result.

The situation may then be shown by a normal curve representing a distribution of the differences between the means of an infinite number of pairs of samples. The mean of this distribution would then be zero, the true difference, and positive and negative differences due to chance would arise about this mean. See Figure 14.2, p. 151.

It is obvious from the information previously given about the normal curve that, in 99.7 cases out of 100, no difference larger than 3 standard deviations of this sampling distribution of differences (3 standard errors of the difference between the two means) would arise. If, therefore, the actual difference is larger than 3 standard errors of the difference between the means (or, in other words, if the probability of such a difference due to chance is very small) it can be said that the difference is significant and not due to chance.

The standard error of the difference between two means (standard deviation of the theoretical distribution of differences between means of samples) can be obtained from the following[1]

$$\sigma_D = \sqrt{\sigma_{\overline{X}_1}^{-2} + \sigma_{\overline{X}_2}^{-2}}$$
$$= \sqrt{\frac{\sigma_1^2}{n_1} + \frac{\sigma_2^2}{n_2}}$$

where

σ_1 = standard deviation of first sample
σ_2 = standard deviation of the second sample
n_1 = number of items in first sample
n_2 = number of items in second sample

As a result of a time study a new method was suggested for a certain operation in a factory. The average time for the operation using the old method was 17.5 seconds with a standard deviation of 1.5

[1]This is based on the assumption that the two samples from which the means were computed are uncorrelated.

seconds in 50 trials. After learning the new method a sample of work-men were timed with a resulting average time for 50 trials of 15 seconds and a standard deviation of 1.2 seconds. It is now possible by means of the technique outlined above to determine whether the difference in average time of 2.5 seconds is significant or merely due to chance.

$$\sigma_D = \sqrt{\frac{(1.5)^2}{50} + \frac{(1.2)^2}{50}} = .27 \text{ seconds}$$

A difference as large as .81 seconds (3 standard errors of the dif-ference between the two means) might conceivably (.3 chances in 100) arise due to chance. Since the actual difference (2.5 seconds) is much larger than this amount (.81 seconds) it is extremely unlikely that it arose due to chance.

This minimum difference necessary to demonstrate a statistically significant difference is sometimes called the least significant difference (LSD). Its formula for the difference between two means from an in-finite population becomes

$$\text{LSD} = t \sqrt{\sigma_{\overline{X}_1}^2 + \sigma_{\overline{X}_2}^2}$$

where

$t =$ the number of standard errors required to reach a given confidence level based on the normal deviates (see page 122)

Thus, if a value of t (say 1.96) is used to secure a given probabil-ity or confidence level (in this case 95%), it is apparent that although there is *no* difference between the two averages tested, there is never-theless an .05 (1.00 − .95) probability that the observed difference may exceed the LSD only due to sampling variability. This is a risk involved in any sampling test of significance; it is called a Type I error and is usually represented by the Greek letter α.

Conversely, even when the two arithmetic means are actually sig-nificantly different, there is a probability that due to the vagaries of sampling, the observed difference is less than the LSD, and it will not be possible to detect the significance of the difference. This is called a Type II or β error.

To avoid the possibility of contradictory results, the right way to describe a test of the statistical significance is to state that the observed difference is statistically significant or alternatively that there is *no evidence* that the difference is statistically significant.

When the statement is made that the difference is statistically significant, it is meant that there is only a small (or perhaps negligible) probability that the observed difference is an accident of sampling.

Another approach is to divide the observed difference between the two means by the standard error of the difference (σ_D) resulting in a value t where

$$t = \frac{\overline{X}_1 - \overline{X}_2}{\sigma_D}$$

This value of t is then compared to the value for the number of standard errors (standard deviations) required to reach the desired p (confidence level). For instance, the tabular value of t for the 99.7% confidence level has been previously established as 3. If the value of t as computed from the above formula exceeds 3, then the difference may be said to be *statistically significant.* This method is often referred to as the *t test.* It yields the same result as the method of the least significant difference.

The significance of the difference between any two statistical measures computed from two samples may be obtained from the following formulae.

If the two samples are correlated

$$\sigma_D = \sqrt{\sigma_{\theta_1}^2 + \sigma_{\theta_2}^2 - 2r_{12}\sigma_{\theta_1}\sigma_{\theta_2}}$$

where

σ_{θ_1} is the standard error of any statistic θ computed from sample #1

σ_{θ_2} is the standard error of any statistic computed from sample #2

r_{12} is the coefficient of correlation between samples #1 and #2

If the samples are uncorrelated the formula becomes

$$\sigma_D = \sqrt{\sigma_{\theta_1}^2 + \sigma_{\theta_2}^2}$$

SIGNIFICANCE OF THE DIFFERENCE BETWEEN PROPORTIONS

If two random samples are drawn and indicate that a given characteristic is in a certain proportion, the difference between the two proportions can be tested to determine whether it is significant or arises out of a sampling fluctuation by use of the formula

$$\sigma_{D\%} = \sqrt{pq\left(\frac{1}{n_1} + \frac{1}{n_2}\right)}$$

where

      p is the total percentage of occurrence
      $q = 1 - p$
      n_1 = number in first sample
      n_2 = number in second sample

In a study of the effectiveness of slogans it was found that 75.7% of the males questioned recognized a certain slogan while 66.3% of the females questioned recognized the slogan. The above formula may be applied to determine whether there was a significant difference in the percentage of recognition by the two sexes.

Table 14.2—Results of Recognition Test of a Slogan, as Given to 374 College Students

Sex	Number Recognizing	Percent Recognizing	Number Questioned
Male	209	75.7%	276
Female	65	66.3%	98
Total	274	73.3%	374

Source: Glick, S., Commercial Slogans, Thesis, College of City of New York, 1935.

$$p = 73.3\%$$
$$q = 26.7\%$$

$$\sigma_{D\%} = \sqrt{pq\left(\frac{1}{n_1} + \frac{1}{n_2}\right)} = \sqrt{(.733)(.267)\left(\frac{1}{276} + \frac{1}{98}\right)}$$

$$= .052 = 5.2\%$$

Since the actual difference between the two proportions (75.7% - 66.3% = 9.4%) is 1.81 times the standard error of the difference, there are a little over 7 chances in 100 that the difference is a chance difference due to sampling.

STANDARD ERROR OF MEASUREMENTS

A certain degree of variation must be expected when physical measurements are made. If a distance is measured repeatedly or if a quantity is weighed several times, the results will show a degree of variation.

If the average of the several measurements is taken as the true measurement it must be remembered that this average is a measurement obtained from a sample. It is therefore subject to a sampling error which may be computed. If a measurement is made 10 times it constitutes a sample of 10 measurements drawn from a universe of an infinite number of measurements which may be made.

It has been shown previously that the error of such a mean can be computed through the use of the standard error or the probable error of the mean.

For large samples $(n > 30)$

$$\sigma_{\overline{X}} = \frac{\sigma}{\sqrt{n}}$$

For small samples $(n < 30)$[1]

$$S_{\overline{X}} = \frac{s}{\sqrt{n}}$$

To combine measurements to obtain areas, volumes, etc., the standard error of the resulting value must be obtained.

1. When the individual measurements are combined by addition the standard error of the resulting value may be obtained from[2]

$$\sigma^2_{\overline{X}_1 + \overline{X}_2 + \overline{X}_3 \ldots, + \overline{X}_n} = \sigma^2_{\overline{X}_1} + \sigma^2_{\overline{X}_2} + \sigma^2_{\overline{X}_3} + \ldots + \sigma^2_{\overline{X}_n}$$

To obtain the distance between two points the distance was measured in two separate sections, with the following results, as an average of a number of measurements

Distance #1 = 500 yds.
Distance #2 = 600 yds.

The standard error of the measurements of the first distance $\left(\sigma_{\overline{X}} = \frac{\sigma}{\sqrt{n}} \right)$ was 2 yards; that of the second, 2.5 yards.

---

[1] But see page 160.
[2] These relationships are true only if the items are mutually independent, for when they are not the relationship is

$$\sigma_{\overline{X}_1 + \overline{X}_2} = \sqrt{\sigma^2_{\overline{X}_1} + \sigma^2_{\overline{X}_2} + 2r_{12} \, \sigma_{\overline{X}_1} \, \sigma_{\overline{X}_2}}$$

The standard error of the entire distance, 500 yards + 600 yards, is calculated.

$$\sigma^2_{\overline{X}_1+\overline{X}_2} = 4 + 6.25 = 10.25$$

$$\sigma_{\overline{X}_1+\overline{X}_2} = 3.20 \text{ yards}$$

2. If the measurement is raised to any power (n) the standard error of the resulting value may be obtained from[1]

$$\frac{\sigma_{\overline{X}^n}}{\overline{X}^n} = n\left(\frac{\sigma_{\overline{X}}}{\overline{X}}\right)$$

The measurements of one side of a square result in an average length of 10 feet with a standard error of .05 feet. The standard error of the area can be obtained as follows

$$\text{Area} = L^2 = 10^2 = 100 \text{ square feet}$$

$$\frac{\sigma_{\overline{X}^n}}{100} = 2\left(\frac{.05}{10}\right) = .01$$

$$\sigma_{\overline{X}^n} = 1 \text{ square foot}$$

3. The standard error of the product of a series of means the standard errors of which are known, is obtained from[2]

$$\left(\frac{\sigma_{\overline{X}_1 \ldots \overline{X}_2 \ldots \overline{X}_n}}{\overline{X}_1 . \overline{X}_2 \ldots \overline{X}_n}\right)^2 = \left(\frac{\sigma_{\overline{X}_1}}{\overline{X}_1}\right)^2 + \left(\frac{\sigma_{\overline{X}_2}}{\overline{X}_2}\right)^2 + \ldots + \left(\frac{\sigma_{\overline{X}_n}}{\overline{X}_n}\right)^2$$

4. The standard error of a quotient can be obtained from[2]

$$\left(\frac{\dfrac{\sigma_{\overline{X}_1}}{\overline{X}_2}}{\dfrac{\overline{X}_1}{\overline{X}_2}}\right)^2 = \left(\frac{\sigma_{\overline{X}_1}}{\overline{X}_1}\right)^2 + \left(\frac{\sigma_{\overline{X}_2}}{\overline{X}_2}\right)^2$$

The standard error of the volume of a cylindrical tank obtained from a series of measurements of its radius and height is obtained as follows:

$$\text{radius} = 10 \text{ feet (r)}$$
$$\text{height} = 20 \text{ feet (h)}$$

---

[1] Providing $\sigma_{\overline{X}}$ is small compared with $\overline{X}$, which is ordinarily the case, and that the data are uncorrelated.

[2] Provided the data are uncorrelated.

$$V = \pi r^2 h = 3.14159(10)^2 (20) = 6283.18$$

and

$$\sigma_r = .1 \text{ feet}$$
$$\sigma_h = .2 \text{ feet}$$

The standard error of r^2 can be obtained from

$$\frac{\sigma_{\overline{X}^n}}{\overline{X}^n} = n\frac{\sigma_{\overline{X}}}{\overline{X}} = 2\frac{\sigma_r}{r} = 2\frac{.1}{10}$$

and the standard error of the volume from

$$\left(\frac{\sigma_{\overline{X}_1 \cdot \overline{X}_2}}{\overline{X}_1 \cdot \overline{X}_2}\right)^2 = \left(\frac{\sigma_{\overline{X}_1}}{\overline{X}_1}\right)^2 + \left(\frac{\sigma_{\overline{X}_2}}{\overline{X}_2}\right)^2$$

or in this case where $V = \pi r^2 h$

$$\left(\frac{\sigma_V}{V}\right)^2 = \left(\frac{\sigma_V}{\pi r^2 h}\right)^2 = \left(2\frac{\sigma_r}{r}\right)^2 + \left(\frac{\sigma_h}{h}\right)^2$$

$$\left(\frac{\sigma_V}{6283.18}\right)^2 = \left(2\frac{.1}{20}\right)^2 = \left(\frac{.2}{20}\right)^2$$

$$\sigma_V = 140.5 \text{ cubic feet}$$

SIGNIFICANCE OF COEFFICIENT OF CORRELATION

When a coefficient of correlation is computed it is necessary to determine whether or not the correlation indicated signifies a real association between the two series considered, or whether the indicated relationship has arisen from the accidental selection of values in the samples.

Although no real association exists between the two series constituting the sample it is possible to obtain a definite value for r when computed from a sample drawn from the universe. It has already been noted that a difference between the means of two samples may make its appearance even when both samples are drawn from the same population. In the same way the value for r may be due to sampling fluctuations.

If the coefficient of correlation is computed for each of a large number of samples of paired values, a frequency distribution (sampling distribution) of the resulting coefficients will be normal (if the true association is zero). Through application of the standard deviation (standard error of the coefficient of correlation) it can be foretold that no value of r greater than three times its standard error will arise due to chance 99.7 times out of 100. If, therefore, the computed r is more than three times σ_r, 99.7 times out of 100 it is significant.

To determine the error likely to arise due to sampling, the standard error of the coefficient of correlation may be used in the same fashion as the standard error of the mean.

In 50 chances out of 100 the difference between the observed and the actual r will not be larger than $.6745\ \sigma_r$ (the probable error of r), and 99.7 chances out of 100 this difference will not be larger than $3\ \sigma_r$.

The formula for the standard error of the coefficient of correlation is

$$\sigma_r = \frac{1 - r^2}{\sqrt{n - 2}}$$

However, when the coefficient of correlation for the underlying population approaches 100%, the sampling distribution cannot be normal or symmetrical, since the possibilities of extremes in one direction are limited by the maximum obtainable value for r of 1.00, while the range of possible values of r in the opposite direction is still great. Here the value of r may be converted for tests of reliability and significance into a more useful value of z.[1]

$$z = \frac{1}{2}\left[(\log_e(1 + r) - \log_e(1 - r)\right]$$

The sampling distribution of this value approaches normality and is symmetrical.

Its standard error is

$$\sigma_z = \frac{1}{\sqrt{n - 3}}$$

---

[1] Tables for converting r to z and z to r are available to facilitate the computation; see Arkin H. and Colton, R. R., *Tables for Statisticians*, pp. 127–128, Barnes and Noble Inc., New York, 1963.

Assume that a coefficient of correlation has been computed from a sample of thirty ($n = 30$) pairs of observations resulting in a coefficient of correlation (r) of +.60. By reference to a conversion table, it is found that when $r = .60$, $z = .6931$.

The standard error of z is computed as

$$\sigma_z = \frac{1}{\sqrt{n-3}} = \frac{1}{\sqrt{30-3}} = .1925$$

Using a $3\sigma_z$ confidence level (probability 99.7%) it is seen that the expected range of sampling variation for this set of conditions is

$$z \pm 3\sigma_z = .6931 \pm .5775$$
$$= .1156 \text{ to } 1.2706$$

Using a table to convert these values of z to r (or performing the formula computation) it is found that when $z = .12$ (the lower limit above) then $r = .12$ while when z is 1.27 (the upper limit) r is .85. The 3σ confidence limits of r may then be stated as .12 to .85.

SMALL SAMPLES—STANDARD ERROR OF MEAN

Due to serious errors which arise where the number of items constituting the sample is small (generally less than 30) the standard errors outlined above can no longer be used for certain measures. If, for instance, the sample is small, a new standard error is computed for the arithmetic mean.[1]

$$s^2 = \frac{\Sigma(x^2)}{n-1} = \frac{n\sigma^2}{n-1}$$

$$S_{\bar{x}} = \frac{s}{\sqrt{n}}$$

But for small samples the usual numbers of multiples of standard errors taken to include a given percentage of the cases are no longer valid. The multiples to be used for various percentages of probability of occurrence of deviation not greater than a given size are shown in

---

[1]If σ is used, rather than s, this formula may be written $S_{\bar{x}} = \frac{\sigma}{\sqrt{n-1}}$.

Table 14.3—t Table

n'	50%	95%	99%
1	1.000	12.706	63.657
2	.816	4.303	9.925
3	.765	3.182	5.841
4	.741	2.776	4.604
5	.727	2.571	4.032
6	.718	2.447	3.707
7	.711	2.365	3.499
8	.706	2.306	3.355
9	.703	2.262	3.250
10	.700	2.228	3.169
11	.697	2.201	3.106
12	.695	2.179	3.055
13	.694	2.160	3.012
14	.692	2.145	2.977
15	.691	2.131	2.947
16	.690	2.120	2.921
17	.689	2.110	2.898
18	.688	2.101	2.878
19	.688	2.093	2.861
20	.687	2.086	2.845
21	.686	2.080	2.831
22	.686	2.074	2.819
23	.685	2.069	2.807
24	.685	2.064	2.797
25	.684	2.060	2.787
26	.684	2.056	2.779
27	.684	2.052	2.771
28	.683	2.048	2.763
29	.683	2.045	2.756
30	.683	2.042	2.750

Table 14.3.[1]   (The n' in the table represents the degrees of freedom. For the standard error of the mean this is $n - 1$.)

Other Standard Errors for Small Samples

The standard errors for three other statistical measures when computed for small samples are shown below. The results obtained from these formulae may be applied in the same manner as the standard er-

---

[1] This table is generally referred to as the t table and the value in the body of the table as t. It is to be used only when the standard deviation of the universe is estimated by using the sample standard deviation in the standard error formula.

rors for large samples (see pages 141–154), using the appropriate multiples of the standard error obtained from the table above.

Measure	Standard Error (For small samples)	Value of n' in table of multiples
Difference between[1] two means	$s^2 = \dfrac{\Sigma(x_1{}^2) + \Sigma(x_2{}^2)}{(n_1 - 1) + (n_2 - 1)}$ $S_D = \dfrac{s}{\sqrt{\dfrac{n_1 n_2}{n_1 + n_2}}}$	$n' = n_1 + n_2 - 2$
Median	$S_{MDN} = 1.2533 \dfrac{S}{\sqrt{n}}$	$n' = n - 1$

ADDITIONAL BIBLIOGRAPHY

ARKIN, HERBERT, *Handbook of Sampling for Auditing and Accounting*, pp. 27–143. McGraw-Hill Book Co., New York, 1963.

COCHRAN, WILLIAM G., *Sampling Techniques*, 3nd ed., pp. 2–152, John Wiley & Sons, Inc., New York, 1963.

FERBER, ROBERT, *Statistical Techniques in Market Research,* pp. 65–226, McGraw-Hill Book Co., New York, 1949.

KISH, LESLIE, *Survey Sampling,* John Wiley & Sons, Inc., New York, 1965.

SMITH, J. G., and DUNCAN, A. J., *Sampling Statistics and Applications*, pp. 153–307, McGraw-Hill Book Co., New York, 1945.

WALKER, H. M., & LEV, J., *Statistical Inference*, pp. 1–179, Henry Holt & Co., New York, 1953.

---

[1] x_1 is the deviation of the actual values from the mean of all X_1 values.

Chapter XV

Analysis of Variance

PURPOSE

The purpose of the analysis of variance technique is to provide a method for testing the statistical significance of the differences between the means of *several* samples. The method provides an indication as to whether the observed differences among the means of the samples may or may not be ascribed to *sampling fluctuations*.

Methods for testing the significance of the difference between two means have been described in Chapter XIV. The analysis of variance technique can be used for two means but extends the test to problems involving more than two means.

ONE-FACTOR ANALYSIS OF VARIANCE—PRINCIPLES

Given several groups of measurements, the "null" hypothesis that they all represent random samples from the same population is tested. This null hypothesis is

$$\overline{X}_1 = \overline{X}_2 = \overline{X}_3 = \ldots = \overline{X}$$

where

$\overline{X}_1, \overline{X}_2, \overline{X}_3$, etc., are the arithmetic means of groups of observations

$\overline{X}$ = arithmetic mean of population

To perform the test, two estimates of the population variances, obtained in different ways from the samples, are compared with each other.

It was shown in Chapter XIV that the means of samples drawn from a population (the sampling distribution) have a standard deviation (standard error) of

$$\sigma_{\overline{X}} = \frac{\sigma}{\sqrt{n}}$$

or a variance of

$$\sigma_{\overline{X}}^2 = \frac{\sigma^2}{n}$$

where

$\sigma_{\overline{X}}^2$ = variance of sample means
σ = standard deviation of values in population
n = number of samples

Thus,

$$\sigma^2 = n\sigma_{\overline{X}}^2$$

Given several sample groups of observations and their means, $\sigma_{\overline{X}}^2$ may also be computed empirically as follows, assuming that the groups are random samples from the same population (as in the null hypothesis)

$$\sigma_{\overline{X}}^2 = \frac{\Sigma(\overline{X}_i - \overline{X})^2}{n_r - 1}$$

where

$\overline{X}_i$ = mean of each group or sample
$\overline{X}$ = overall mean of all groups
n_r = number of sample means

An estimate of the population variance can be obtained from

$$\sigma^2 = n\sigma_{\overline{X}}^2 \text{ or}$$

$$\sigma^2 = n\sigma_{\overline{X}}^2 = n_i \left(\frac{\Sigma(\overline{X}_i - \overline{X})^2}{n_r - 1} \right)$$

If the groups of samples contain unequal numbers of observations this becomes

$$\sigma^2 = \Sigma \left[n_i \left(\frac{\Sigma(\overline{X}_i - X)^2}{n_r - 1} \right) \right]$$

where

n_i = number of observations in each sample (assuming equal size sample groups)
n_r = number of sample means

Since this variance is based on the differences among the several sample means, it is referred to as the **among (or between) variance.**

Another estimate of the population variance is obtained from the variation of the observations within each sample using the pooled estimate technique described for small samples (see pp. 160–162).

$$S^2 = \frac{\Sigma(X - \overline{X}_1)^2 + \Sigma(X - \overline{X}_2)^2 + \ldots \Sigma(X - \overline{X}_n)^2}{(n_1 - 1) + (n_2 - 1) + \ldots (n_n - 1)}$$

$$= \frac{\Sigma(x_1^2) + \Sigma(x_2^2) + \ldots \Sigma(x_n^2)}{(n_1 - 1) + (n_2 - 1) + \ldots (n_n - 1)}$$

$$= \Sigma \left[\frac{\Sigma(X - \overline{X}_i)^2}{n_i - 1} \right]$$

Since this estimate is based on the variation of the values within each sample, it is called the **within variance.**[1]

If the null hypothesis is true, these two estimates of the population variance should differ only by an amount equal to that which might arise from sampling fluctuations. If the variance estimated from the means of the groups (among variance) is significantly greater than that estimated from the variations within the group (within variance), it may be said that the differences among the group means must be greater than that ascribable to sampling fluctuations and the groups are not from the same population.

The comparison of the two variance estimates is accomplished through the *F test* where *F* is the ratio between the two variances.

$$F = \frac{\sigma^2_{\text{Among}}}{\sigma^2_{\text{Within}}}$$

Theoretically, the two variance estimates should be equal if the null hypothesis is true and *F* would then equal 1.0. Because of the vagaries of sampling, however, the *F* ratio will generally not be equal to 1.0 precisely even when the groups are all drawn at random from the same population.

Reference to a table of *F* provides the largest value of *F* that might arise due to sampling fluctuations at a given significance or probability level (in most tables 0.05 or 0.01). If the observed *F* ratio exceeds the tabular value for the selected significance level and appro-

---

[1] The within variance is also called the residual or error variance.

166   ANALYSIS OF VARIANCE

priate degrees of freedom it may be said that the group means differ
significantly.

METHOD
ANOVA Table

The computations are generally accomplished in a form known as
an ANOVA (analysis of variance) table. The calculation consists of determining the sum of squares of the deviations and dividing by the appropriate degrees of freedom to obtain the estimates of the variances.
The ANOVA table takes the following general form.[1]

Source of Variation	Sum of Squares of Deviations	Degrees of Freedom	Variance
Among groups	$\Sigma\,[n_i(\overline{X}_i - \overline{X})^2]$	$c - 1$	$\Sigma\,[n_i(\overline{X}_i - \overline{X})^2]/c - 1$
Within groups	$\Sigma\,[\Sigma(X - \overline{X}_i)^2]$	$\Sigma(n_i - 1)$	$\Sigma\,[\Sigma(X - \overline{X}_i)^2]/\Sigma(n_i - 1)$
Total	$\Sigma(X - \overline{X})^2$	$n - 1$	

where

X = each observation
$\overline{X}_i$ = arithmetic mean of each group
$\overline{X}$ = overall arithmetic mean
n_i = number of observations in each group
n = number of observations in all groups combined
c = number of groups

Computational Method

The value for the sums of squares may be more readily computed
from the following equations.[2]

Total sum of squares (of deviations)

$$SS_{Total} = \Sigma(X^2) - \frac{T^2}{n}$$

Among sum of squares

$$SS_{Among} = \Sigma\left(\frac{T_i{}^2}{n_i}\right) - \frac{T^2}{n}$$

---

[1] Since $S^2 = \dfrac{\Sigma(x^2)}{n-1} = \dfrac{\Sigma(X - \overline{X})^2}{n-1}$ then $(n-1)S^2 = \Sigma(X - \overline{X})^2$

[2] See Technical Appendix XIII for proof.

Within sum of squares

$$SS_{\text{Within}} = \Sigma(X^2) - \Sigma\left(\frac{T_i^2}{n_i}\right)$$

where

X = each observation
T = Total value of all observations
T_i = Total value of all observations in each group
n = Total number of observations in all groups
n_i = Number of observations in each group

An example of the calculation may be drawn from the data in Table 15.1. Samples of parts from 3 manufacturers were tested for thickness of plating. The table is used to produce the following equations.

Table 15.1 – Zinc Plating Thickness (10^{-5} Inches) for Parts from 3 Manufacturers

	A	B	C	
	40	25	27	
	38	32	24	
	30	13	20	
	47	35	13	
T	155	105	84	344
$\Sigma(X^2)$	6153	3043	1874	11070

Source: Hicks, C. R., The Analysis of Covariance, Industrial Quality Control, Vol. XII, No. 6, Dec. 1965, p. 282.

Total sum of squares

$$SS_{\text{Total}} = \Sigma(X^2) - \frac{T^2}{n} = 11070 - \frac{(344)^2}{12}$$

$$= 1208.67$$

Among sum of squares

$$SS_{\text{Among}} = \Sigma\left(\frac{T_i^2}{n_i}\right) - \frac{T^2}{n}$$

$$= \frac{(155)^2}{4} + \frac{(105)^2}{4} + \frac{(84)^2}{4} - \frac{(344)^2}{12}$$

$$= 665.17$$

Within sum of squares

$$SS_{\text{Within}} = \Sigma(X^2) - \Sigma\left(\frac{T_i^2}{n_i}\right)$$

$$= 11,070 - \left(\frac{(155)^2}{4} + \frac{(105)^2}{4} + \frac{(84)^2}{4}\right)$$

$$= 543.50$$

The ANOVA table is then completed (Table 15.2).

Table 15.2–ANOVA Table

Source of Variation	Sum of Squares of Deviations	Degrees of Freedom	Variance
Among Manufacturers	665.17	2	332.59
Within Manufacturers	543.50	9	60.39
	1208.67	11	109.88*

*This value is ordinarily not computed but is shown here for discussion purposes.

The degrees of freedom for the total are $n - 1$ ($12 - 1 = 11$). For the among, the degrees of freedom are $c - 1$ (number of groups $- 1$) while for the within they are $\Sigma(n - 1)$ or $(4 - 1 + 4 - 1 + 4 - 1)$.

The variances are secured by dividing the sum of squares of deviations on each line by the appropriate degrees of freedom.

The F ratio is computed by dividing the among manufacturers variance by that for the within manufacturer variance

$$F = \frac{332.59}{60.39} = 5.51$$

The F table gives values of F at the .05 level for 2 degrees of freedom for the larger variance and 9 degrees for the smaller variance.

$$F_{.05} = 4.26$$

This implies that there is more than a 95% assurance that the plating thickness of the parts obtained from the manufacturers are different.

It is to be noted that the columns for sums of squares and degrees of freedom in the ANOVA table both total correctly but that the variance column is not additive. Thus, since the computation of the variance for the total line results in a value not used in the F computation nor serving as a check value, it is usually not computed.

Assumptions

1. The observations were obtained by probability sampling methods.

2. The within (error) variations are normally distributed.

3. The within (error) variances for several samples are for a population or populations with a common value for the variance[1]

$$\sigma_1^2 = \sigma_2^2 = \sigma_3^2 = \sigma_n^2 = \sigma^2$$

where σ^2 is a population variance.

4. The various effects are additive. The differences among groups tend to be by constant amounts rather than ratios.

5. The within (error) variances are independent of the effects.

MULTIFACTOR ANALYSIS OF VARIANCE

The analysis of variance technique may be used when there is more than one factor to be considered. For several factors, a cross-tabulation of values may be prepared.

The application of the analysis of variance technique to multifactor problems enables the simultaneous evaluation of the significance of each of the factors involved and of the effect of all combinations of factors (interactions). The details of such methods are beyond the scope of this book.

ADDITIONAL BIBLIOGRAPHY

COWDEN, D. J., *Statistical Methods in Quality Control*, pp. 128–138, Prentice-Hall, Inc., Englewood Cliffs, N.J., 1957.

DIXON, W. J., & MASSY, F. J., *Introduction to Statistical Analysis*, pp. 139–168, McGraw-Hill Book Co., New York, 1957, 2nd ed.

STEEL, R. G. D., & TORRIE, J. H., *Principles and Procedures of Statistics*, pp. 99–128, McGraw-Hill Book Co., New York, 1960.

---

[1] Bartlett's test is often used to test for homogeneity of variance.

Chapter XVI
Index Numbers

The **index number** is a statistical device for measuring changes in groups of data.

The method may be employed in the measurement of many general conditions; examples are employment, prices, group health, and academic grades. Data descriptive of these general conditions fluctuate widely; but such data nevertheless exhibit definite and measureable general tendencies.

In order to measure the changes in the large number of constantly varying items in the data it is necessary to resort to some relative[1] averaging device that will serve as a yardstick. This is a comparative measurement usually relative to some reference or base period. The index number is such a yardstick.

The index number measures fluctuations during intervals of time, group differences of geographical position or degree, etc. It is possible to obtain an index number showing the relative sales possibilities for a given product in different territories; the academic standing of a group of college students as compared with other groups of students; or to ascertain the relative credit position of a single corporation as compared with many others in the same industry.

For simplicity the discussion below is largely confined to index numbers of prices.

---

[1]Index numbers are not always relative (percentage) in form. Occasionally they are expressed in terms of absolute (actual) values.

INDEX NUMBER CONSTRUCTION PROBLEMS

1. The purpose of the index number has a definite bearing upon the choice of the data and the method followed.

2. Careful selection of the number and types of items to be used is necessary so that the index fluctuations will be truly representative of the fluctuations in the series.

3. After determination of the proper method of data collection the available sources must be found and the data actually collected.

4. The problems of selecting the base period and the best method of computation must be solved.

5. The degree of relative importance of each constituent item to the purpose of the index must be determined. This designation of the relative importance of each item is known as *weighting*.

Number and Kinds of Commodities

Index numbers must be constructed from samples or limited portions of the types of prices considered; two rules should be observed in selecting the commodities.

1. The sample used should be representative (see page 142). The items selected should be chosen for their representative quality rather than because of the ease of securing quotations.

2. A sufficient number of items should be used. Dr. Irving Fisher points out that index numbers of prices are seldom of much value "unless they consist of more than 20 commodities and 50 is a much better number." He also shows that "after 50 the improvement obtained from increasing the number of commodities is gradual and it is doubtful if the gain from increasing the number beyond 200 is ordinarily worth the extra trouble and expense."[1]

The Base Period

The base period is assigned the value of 100% and is thereby arbitrarily established as a reference period. Index numbers are then computed relative to the base period.

In the selection of the base period the following should be considered.

1. The base period should not be too far in the past; this in order that a comparison of the price level relative to the base period will be of definite present value.

---

[1] Fisher, Irving, *The Making of Index Numbers*.

2. Comparison is generally made to a normal period; therefore the base period should not be extreme.

Shifting the Base

For comparative purposes the base of an index number series is sometimes shifted from one period to another. The shift is made by dividing each number in the series by the index number indicated for the new base year, and multiplying the result by 100.[1]

In the following illustration the base year of the index series, 1966, is shifted to 1968 by dividing each index number by the index value for 1968 (150.0) and multiplying by 100.

1966	1967	1968	1969
100.0	110.1	150.0	125.3
66.7	73.4	100.0	83.5

SELECTION OF METHOD OF COMPUTATION

Irving Fisher gives over 150 different formulae for the construction of index numbers.[2] These formulae, however, are largely variations of a limited number of main types.

Some of the major groups of methods of constructing index numbers may be classified.

1. The Unweighted (Simple) Method
   a. The aggregate of actual prices.
   b. The average of relative prices.
2. The Weighted Method
   a. The weighted aggregate of actual prices.
   b. The weighted average of relative prices.

Simple Aggregate of Actual Prices

The index number constructed by the simple aggregate method is a comparison of the sum of the prices for the commodities considered to the sum of the prices for the same commodities in the base period.[3]

$$\frac{\Sigma p_n}{\Sigma p_o}$$

(Index number formula 1)

---

[1] All types of index numbers cannot have their base shifted in this manner (see discussion below for types which can). In order to use a new base period certain types must be completely reconstructed.

[2] Fisher, Irving, *The Making of Index Numbers.*

[3] The index may be expressed as a sum of money rather than a relative or percentage figure.

where

$$\Sigma p_n = \text{sum of prices of commodities of any given period.}$$
$$\Sigma p_o = \text{sum of prices of commodities in base period.}$$

However, the index number computed by the simple aggregate method is subject to a serious defect in that those commodities which have large figure quotations will dominate the index. For instance, if there should be a decrease of 10% in the price of pig iron while all other commodities rose 10%, with the indication of an increase in the price level, the predominating influence of the pig iron quotation will nevertheless cause the index to fall.

	Number of Commodities	1962	19—	
Pig Iron	1	$66.4000	$59.7600	Decrease of 10%
All other commodities	5	1.8417	2.0299	Increase of 10%
Total	6	$68.2417	$61.7899	
Index		100%	90.5%	

The difficulty indicated above cannot be avoided by reducing all commodities to a common unit such as a pound—as done in the Bradstreet index. Such procedure would only give rise to new inequalities. Applied to the problem in Table 16.1 pig iron would be $.0332 a pound while copper would be $.3083 a pound (1962).

Table 16.1—Computation of Index of Wholesale Metal Prices By Unweighted Aggregate of Actuals Method (1962 Used as Base Year)

Metal	Unit	Prices In Dollars		
		1962	1963	1964
Pig Iron	Ton	$66.4000	$66.2500	$63.1100
Copper	Pound	.3080	.3083	.3218
Aluminum	Pound	.2388	.2262	.2372
Lead	Pound	.0943	.1094	.1342
Zinc	Pound	.1163	.1201	.1357
Silver	Ounce	1.0840	1.2790	1.2930
Total		$68.2417	$68.2930	$65.2319
Index		100%	100.07%	95.59%

Average of Relative Prices

A method which avoids the price inequalities shown above involves the conversion of the price figures into *relatives*. A price relative is a statement of the price of a commodity as a percent of its price in the base period. Expressed in formula form

$$\frac{p_n}{p_o}$$

where

p_n = price in the given period
p_o = price in the base period

The relative for each commodity in the base period is 100%. The relative for the period under consideration is averaged to obtain the index number. The arithmetic mean, the median, or the geometric mean may be used for averaging.

The formula below demonstrates the computation of an index number by the average of relative prices method, using the arithmetic mean. Its use can be seen in Table 16.2.

$$\frac{\sum\left(\dfrac{p_n}{p_o}\right)}{N} \qquad \text{(Index number formula 2)}$$

Table 16.2–Computation of Index of Wholesale Metal Prices by Unweighted Arithmetic "Mean of Relatives" Method (1962 Used as Base Year)

Metal	Unit	1962		1963		1964	
		Price in Dollars	Relative Price	Price in Dollars	Relative Price	Price in Dollars	Relative Price
Pig Iron	Ton	$66.4000	100%	$66.2500	99.8%	$63.1100	95.0%
Copper	Lb.	.3083	100	.3083	100.0	.3218	104.4
Aluminum	Lb.	.2388	100	.2262	94.7	.2372	99.3
Lead	Lb.	.0943	100	.1094	116.0	.1342	142.3
Zinc	Lb.	.1163	100	.1201	103.3	.1357	116.7
Silver	Oz.	1.0840	100	1.2790	118.0	1.2930	119.3
Totals			600%		631.8%		677.0%
Index			100%		105.3%		112.8%

Source: Statistics. Standard and Poors. 1967.

ADVANTAGES AND DISADVANTAGES OF VARIOUS AVERAGES IN INDEX NUMBER CONSTRUCTION
Arithmetic Mean

Advantages
1. The mean is relatively easy to compute.
2. Due to long and common usage the arithmetic mean is commonly understood.

3. If a weighted average is taken the means of subgroups can be averaged to obtain the means of values. (A weighted average may be necessary if there are a varying number of items in the various groups).

Disadvantages

1. The mean is greatly affected by extremes (see pp. 23-24).

2. Increases are given a greater emphasis than decreases. For instance, if commodity A should rise from $1 to $2, an increase of 100 percent, while commodity B fell from $2 to $1, a decrease of 50 percent, an index of the price level of these two commodities (if the arithmetic mean of relatives is used) will show an increased instead of an unchanged price level.

Commodity	1966 Price	1966 Relative	1968 Price	1968 Relative
A	$1	100%	$2	200%
B	$2	100	$1	50
Total		200%		250%
Index		100%		125%

3. The base of the index number computed by the average of relatives method cannot be shifted by the above described method.

Median

Advantages

1. Unlike the arithmetic mean the median will not overemphasize increases.

2. The median is less affected by extremes than the arithmetic mean (see pp. 23-24).

3. It is easy to compute. The relatives are arranged according to size and the middle one is selected as the median.

Disadvantages

1. The median cannot be treated algebraically; i.e. the medians of subgroups cannot be averaged to obtain the median of all the data.

2. Its value is erratic when the number of items is small.

3. The index constructed by this method cannot be shifted to a new base by the short method.

Geometric Mean[1]

Advantages

1. The geometric mean does not overemphasize increases; rather, it gives equal importance to equal ratios of change.

---

[1] For an explanation of the geometric mean see pp. 29-30.

In the problem illustrating the disadvantage of the arithmetic mean the correct index number can be secured with the geometric mean.

Commodity	Price	1926 Relative	Price	1928 Relative
A	$1	100%	$2	200%
B	$2	100	$1	50
Geometric Mean		100%		100%

2. The base of an index number constructed by this technique can be shifted by the short method.

Disadvantages

1. The calculation of the geometric mean is laborious.
2. It is an unfamiliar form of average.

THE WEIGHTING OF INDEX NUMBERS

It is often desirable to assign a varying degree of importance to the items composing the index numbers. If this action is not taken each commodity will be given a weight or importance which depends on its price, or on some other extraneous factor, rather than a proportionate weight depending upon its real importance.

This objection may be eliminated by introducing a deliberate system of weights. To measure the weight or importance of the items composing a price index the quantity of each commodity produced may be used.

The Weighted Average

A weighted average (arithmetic mean) may be obtained:

1. By multiplying each item by its corresponding weight.
2. By totaling the results obtained.
3. By dividing by the sum of the weights.

$$\text{Weighted average} = \frac{\Sigma(\text{items} \times \text{weights})}{\Sigma(\text{weights})}$$

Thus if it is found that in a particular section there are two quotations on the price of bread, say 6¢ in chain stores selling 10,000 loaves and 8¢ in independent bakers selling 1000 loaves, a weighted average of the prices may be determined as follows

	Price	Quantity Sold	Price Times Quantity
Chain Store	$.06	10,000	600
Bakery	.08	1,000	80
		11,000	680
		680 ÷ 11,000 =	$.062

Weighted Aggregate of Actual Prices

A weighted aggregate[1] of actual prices may now be computed by using as a weight the quantity of each commodity produced. The quantities produced in some fixed period, such as the base year, may be used as weights. The index is obtained by comparing the weighted aggregate (total) for the given year to that for the base year.

$$\frac{\Sigma(p_n q_o)}{\Sigma(p_o q_o)} \qquad \text{(Index number formula 3a)}$$

where

p_n = Price given year
p_o = Price base year
q_o = Quantity base year

In Table 16.3 (p. 178) the index for 1963 will be

$$\frac{\Sigma(p_1 q_o)}{\Sigma(p_o q_o)} = \frac{\$6,414,127.29}{\$6,459,464.94} = 99.30\%$$

and for 1964

$$\frac{\Sigma(p_2 q_o)}{\Sigma(p_o q_o)} = \frac{\$6,309,683.79}{\$6,459,464.94} = 97.68\%$$

However, since conditions change, the quantity of the commodities produced in any one fixed period will not be a good measure of their relative importance for all other periods. To meet this objection a set of weights which change every year may be used. Thus the quantity produced in each given year may be used as a weight when constructing the index for that particular period. The formula will then read

$$\frac{\Sigma(p_n q_n)}{\Sigma(p_o q_n)} \qquad \text{(Index number formula 3b)}$$

For 1963 in Table 16.4 (p. 179)

$$\frac{\Sigma(p_1 q_1)}{\Sigma(p_o q_1)} = \frac{\$6,964,020.41}{\$6,898,242.90} = 100.95\%$$

---

[1] The *weighted average* (arithmetic mean) is obtained by dividing the sum of the various items multiplied by their respective weights by the sum of the weights. The *weighted aggregate* or sum is obtained by securing the sum of the various items times their corresponding weights *without* dividing by the sum of the weights.

Table 16.3—Computation of Index of Wholesale Prices of Metals in the United States by Weighted Aggregate of Actuals Method Using Base Year Weights (1962 Used as Base Year)

Metal	Unit	1962			1963		1964	
		Price in dollars p_0	Production (Thousands) q_0	Price times quantity $p_0 q_0$	Price in dollars p_1	Price Times quantity $p_1 q_0$	Price in dollars p_2	Price times quantity $p_2 q_0$
Pig Iron	Ton	$66.4000	67,595	$4,488,308.00	$66.2500	$4,478,168.75	$63.1100	$4,265,920.45
Copper	Pound	.3083	2,456,800	757,431.44	.3083	757,431.44	.3218	790,598.24
Aluminum	Pound	.2388	4,236,000	1,011,556.80	.2262	958,183.20	.2372	1,004,779.20
Lead	Pound	.0943	474,000	44,698.20	.1094	51,855.60	.1342	63,610.80
Zinc	Pound	.1163	1,011,000	117,579.30	.1201	121,421.10	.1357	137,192.70
Silver	Ounce	1.0840	36,800	39,891.20	1.2790	47,067.20	1.2930	47,582.40
				$6,459,464.94		$6,414,127.29		$6,309,683.79

Table 16.4–Computation of Index of Wholesale Prices of Metals in the United States by Weighted Aggregate of Actuals Method Using Given Year Weights (1962 Used as Base Year)

Metal	Unit	1962 Price in dollars p_0	1963 Price in dollars p_1	1963 Production (Thousands) q_1	1963 $p_1 q_1$	1963 $p_0 q_1$	1964 Price in dollars p_2	1964 Production (Thousands) q_2	1964 $p_2 q_2$	1964 $p_0 q_2$
Pig Iron	Ton	$66.4000	$66.2500	73,715	$4,883,618.75	$4,894,676.00	$63.1100	87,932	$5,549,388.52	5,838,684.80
Copper	Pound	.3083	.3083	2,426,400	748,059.12	748,059.12	.3218	2,501,600	805,014.88	771,243.28
Aluminum	Pound	.2388	.2262	4,626,000	1,046,401.20	1,046,401.20	.2372	5,106,000	1,211,143.20	1,219,312.80
Lead	Pound	.0943	.1094	506,800	55,443.92	47,791.24	.1342	566,600	76,037.72	53,430.38
Zinc	Pound	.1163	.1201	1,058,600	127,137.86	123,115.18	.1357	1,144,800	155,349.36	133,140.24
Silver	Ounce	1.0840	1.2790	35,240	45,071.96	38,200.16	1.2930	36,130	46,716.09	39,164.92
					$6,964,020.41	$6,898,242.90			$7,843,649.77	$8,054,976.20

and for 1964

$$\frac{\Sigma(p_2 q_2)}{\Sigma(p_o q_2)} = \frac{\$7,943,649.77}{\$8,054,976.20} = 97.38\%$$

WEIGHTED AVERAGE OF RELATIVES

An index number may be constructed by securing a weighted average of the relative prices for the period under consideration. Quantities of production, however, cannot be used as weights since each quantity is expressed in different units (tons, pounds, ounces, bushels, etc.). The column of figures resulting from the multiplication of the price relatives by these weights would be expressed in different units and could not be totaled. It becomes necessary to use weights expressed in common units. The most usual common unit is the dollar. The money value rather than the quantity of production may be used as a weight.

With base period weights and using a weighted arithmetic mean the formula will be

$$\frac{\sum \left[\dfrac{p_n}{p_o} \times (p_o q_o) \right]}{\Sigma(p_o q_o)} \qquad \text{(Index number formula 4a)}$$

for $p_o q_o$ equals value of production in the base period (the price times the quantity).

Through cancellation the formula reduces to

$$\frac{\Sigma(p_n q_o)}{\Sigma(p_o q_o)}$$

or the same as the weighted aggregate using base year weights (Formula 3a).

If given year weights are used a new formula is evolved.

$$\frac{\sum \left[\dfrac{p_n}{p_o} \times (p_n q_n) \right]}{\Sigma(p_n q_n)} \qquad \text{(Index number formula 4b)}$$

For 1963 in Table 16.5

$$\frac{\sum \left[\dfrac{p_1}{p_o} \times p_1 q_1 \right]}{\Sigma(p_1 q_1)} = \frac{\$6,861,685.84}{\$6,964,020.41} = 98.53\%$$

Table 16.5—Computation of Index of Wholesale Prices of Metals in the United States by Weighted Average of Relative Methods Using Given Year Weights (1962 Used as Base Year)

Metal	Unit	1962		1963					1964				
		Price in dollars p_0	Price relative $\frac{p_0}{p_0}$	Price in dollars p_1	Price relative $\frac{p_1}{p_0}$	Production (Thousands) q_1	$p_1 q_1$	Relative Times Weight $\frac{p_1}{p_0} \times p_1 q_1$	Price in dollars p_2	Price relative $\frac{p_2}{p_0}$	Production (Thousands) q_2	$p_2 q_2$	Relative Times Weight $\frac{p_2}{p_0} \times p_2 q_2$
Pig Iron	Ton	$66.4000	100%	$66.2100	99.8	73,715	$4,883,618.75	4,873,851.51	$63.1100	95.0%	87,932	5,549,388.52	$5,271,919.09
Copper	Pound	.3083	100	.3083	100.0	2,426,400	748,059.12	748,059.12	.3218	104.4	2,501,600	805,014.88	840,435.53
Aluminum	Pound	.2388	100	.2262	94.7	4,626,400	1,046,401.20	990,941.94	.2372	99.3	5,106,000	1,211,143.20	1,202,665.20
Lead	Pound	.0943	100	.1094	116.0	506,800	55,443.92	64,314.95	.1342	142.3	566,600	76,037.72	108,201.68
Zinc	Pound	.1163	100	.1201	103.3	1,058,600	127,137.86	131,333.41	.1357	116.7	1,144,800	155,349.36	181,292.70
Silver	Ounce	1.0840	100	1.2790	118.0	35,240	48,071.96	53,184.91	1.2930	119.3	36,130	46,716.09	55,732.30
							$6,964,020.41	6,861,685.84				7,843,649.77	7,660,246.50

and for 1964

$$\frac{\sum\left[\dfrac{p_2}{p_o} \times p_2 q_2\right]}{\Sigma(p_2 q_2)} = \frac{\$7,660,246.50}{\$7,843,649.77} = 97.66\%$$

THE IDEAL INDEX NUMBER

Irving Fisher has developed an index number which meets the requirements of certain tests (see below). His formula is a "cross" or geometric average of two formulae which are subject to opposite error. It is the geometric average of the aggregate of actuals weighted by base year quantities (Formula 3a) and the aggregate with given year weights.

$$\sqrt{\text{Formula 3a} \times \text{Formula 3b}}$$

$$\sqrt{\frac{\Sigma(p_n q_o)}{\Sigma(p_o q_o)} \times \frac{\Sigma(p_n q_n)}{\Sigma(p_o q_n)}} \qquad \text{(Index number formula 5)}$$

INDEX NUMBER TESTS
Time Reversal Test

Index numbers computed by using different periods as bases should not be inconsistent. For example, if an index number with a base at 1966 (1966 = 1.00) should give rise to an index of 2.00 for 1968, reconstructing the index with a base of 1968 the index for 1966 should be .50, the reciprocal[1] of 2.00.

	1966	1968
Index A	1.00	2.00
B	.50	1.00

Cross-multiplying the index numbers as indicated by the arrows should give a value of 1.00, since these numbers are reciprocal.

Factor Reversal Test

The change in the price multiplied by the change in the quantity should be equal to the change in the value of the commodities produced.

The index of prices can be obtained by any of the methods; for example, the aggregate of actuals weighted by base or fixed year weights (Formula 3a) may be used for the purpose.

---

[1] A reciprocal of a number is 1 divided by that number.

$$\frac{\Sigma(p_n q_o)}{\Sigma(p_o q_o)}$$

An index of the quantity of production can be secured by reversing the positions of the price figures (p) with the quantity figures (q).

$$\frac{\Sigma(q_n p_o)}{\Sigma(q_o p_o)}$$

An index of the value of production can be obtained by comparing the value of production in the given period (V_n) to the value of production in the base period (V_o).

Therefore

$$\frac{\Sigma(p_n q_o)}{\Sigma(p_o q_o)} \times \frac{\Sigma(q_n p_o)}{\Sigma(q_o p_o)} \text{ should equal } \frac{V_n}{V_o}$$

where V_n, the value of production in the given period, may be obtained by multiplying the price in the base period by the quantity produced or $\Sigma(p_n q_n)$ and for the base period $\Sigma(p_o q_o)$.

The test will then read

$$\frac{\Sigma(p_n q_o)}{\Sigma(p_o q_o)} \times \frac{\Sigma(q_n p_o)}{\Sigma(q_o p_o)} \text{ should equal } \frac{\Sigma(p_n q_n)}{\Sigma(p_o q_o)}$$

QUANTITY INDEX NUMBERS

Index number technique can be applied to measurement of changes in quantity groups as well as of price changes. Index numbers of this type are used to measure changes in business activity, industrial production, commodity stocks, etc.

The methods of construction are the same. The simplest form is the simple aggregate type

$$\frac{\Sigma q_n}{\Sigma q_o}$$

where

Σq_n is the sum of the quantities in any given period
Σq_o is the sum of the quantities in the base period

Since this form of index involves the sums of series, the various quantities must all be in the same units (tons, bushels, etc.) to make the summation possible.

When the units are different for the various items in the series and an unweighted index number is desired the average of relatives may be used. If the arithmetic mean is used as the average the formula is

$$\frac{\sum\left(\dfrac{q_n}{q_o}\right)}{N}$$

It is generally desirable however, to weight the index numbers in order to assign various degrees of importance to the several items composing the index number. Either the price of the commodity or some arbitrary weight may be used for this purpose.

The weighted aggregate form for use in measuring quantity changes is

$$\frac{\Sigma(q_n p_o)}{\Sigma(q_o p_o)} \text{ with base year weights}$$

or

$$\frac{\Sigma(q_n p_n)}{\Sigma(q_o p_n)} \text{ with given year weights}$$

Unless the units are the same for all items only the prices can be used as weights and not arbitrary weights since if the latter are used no summation will be possible.

The weighted average of relatives may be used where the units are different and it is desired to use arbitrary weights

$$\frac{\sum\left(\dfrac{q_n}{q_o} \times \text{wt}\right)}{\Sigma(\text{wt})}$$

where wt = weight

The "ideal index" can also be converted to quantity form

$$\sqrt{\frac{\Sigma(q_n p_o)}{\Sigma(q_o p_o)} \times \frac{\Sigma(q_n p_n)}{\Sigma(q_o p_n)}}$$

ADDITIONAL BIBLIOGRAPHY

FISHER, IRVING, *The Making of Index Numbers*, Houghton Mifflin Company, Boston, 1927.

MITCHELL, WESLEY C., *Index Numbers of Wholesale Prices in the United States and Foreign Countries*, United States Bureau of Labor Statistics, Bulletin No. 173, Washington, D. C., 1921.

MUDGETT, B. D., *Index Numbers*, John Wiley & Sons, New York, 1951.

YULE, G. U., & KENDALL, M. G., *An Introduction to the Theory of Statistics*, pp. 590–609, Hafner Publishing Co., New York, 1950.

Chapter XVII

Moments of a Frequency Distribution

MOMENTS

A frequency distribution can be more accurately analyzed if certain constants or **moments** of the distribution are computed. Moments are used for computing measures which are descriptive of the distribution, and for the determination of the appropriate curve to be used in smoothing the distribution mathematically (see page 124).

I. The first moment of a frequency distribution as measured about any arbitrary origin is[1]

$$v_1 = \frac{\Sigma f(d)}{N}$$

II. The second moment (about an arbitrary origin) is

$$v_2 = \frac{\Sigma f(d^2)}{N}$$

III. The third moment (about an arbitrary origin) is

$$v_3 = \frac{\Sigma f(d^3)}{N}$$

IV. The fourth moment (about an arbitrary origin) is

$$v_4 = \frac{\Sigma f(d^4)}{N}$$

---

[1] The symbol v is the Greek small letter nu.

The most important moments are those which are measured with the mean as the origin.[1]

$$\mu_1 = \frac{\Sigma f(x)}{N} = 0$$

$$\mu_2 = \frac{\Sigma f(x^2)}{N}$$

$$\mu_3 = \frac{\Sigma f(x^3)}{N}$$

$$\mu_4 = \frac{\Sigma f(x^4)}{N}$$

where x represents the deviation of the actual value from the mean.

The sum of the deviations about the mean is zero and, therefore, the first moment will equal zero. The other moments about the mean can be obtained readily from

$$\mu_2 = \nu_2 - \nu_1^2$$

$$\mu_3 = \nu_3 - 3\nu_1\nu_2 + 2\nu_1^3$$

$$\mu_4 = \nu_4 - 4\nu_1\nu_3 + 6\nu_1^2\nu_2 - 3\nu_1^4$$

SHEPPARD'S CORRECTIONS FOR GROUPING[2]

The computation of the moments from a frequency distribution involves the assumption that the values may be dealt with as though they were all located at the midpoint of the class interval. This assumption is subject to a certain error, allowance for which can be made by use of the corrections shown below.

I. Corrected first moment

$$\mu_1' = 0$$

II. Corrected second moment

$$\mu_2' = \mu_2 - \frac{1}{12}$$

III. Corrected third moment

$$\mu_3' = \mu_3$$

---

[1] The symbol μ is the Greek small letter mu.

[2] The corrections apply only when (a) the distribution is continuous (see page 6), and when (b) the distribution tapers off gradually in both directions.

IV. Corrected fourth moment

$$\mu_4' = \mu_4 - \frac{1}{2}\mu_2 + \frac{7}{240}$$

For convenience, the moments calculated by the methods outlined above are generally computed in terms of class intervals rather than in original units. To convert the moments back to the original units the following relationships are used.

$$\mu_2' \text{ (in original units)} = C^2 \ \mu_2' \text{ (in class interval units)}$$

$$\mu_3' \text{ (in original units)} = C^3 \ \mu_3' \text{ (in class interval units)}$$

$$\mu_4' \text{ (in original units)} = C^4 \ \mu_4' \text{ (in class interval units)}$$

where C = size of class interval groupings.

CALCULATION OF MOMENTS

The computation of the moments is illustrated in Table 17.1.

$$\nu_1 = \frac{\Sigma f(d')}{N} = \frac{-2}{600} = -.0033$$

$$\nu_2 = \frac{\Sigma f(d'^2)}{N} = \frac{2718}{600} = 4.5300$$

$$\nu_3 = \frac{\Sigma f(d'^3)}{N} = \frac{10}{600} = .0167$$

$$\nu_4 = \frac{\Sigma f(d'^4)}{N} = \frac{32502}{600} = 54.1700$$

$\mu_1 = 0$

$\mu_2 = \nu_2 - \nu_1^2 = 4.5300 - (-.0033)^2 = 4.52999$

$\mu_3 = \nu_3 - 3\nu_1\nu_2 + 2\nu_1^3 = .0167 - 3\,(-.0033)\,(4.5300) + 2(-.0033)^3$

$$= .06155$$

$\mu_4 = \nu_4 - 4\nu_1\nu_3 + 6\nu_1^2\nu_2 - 3\nu_1^4 = 54.1700 - 4(-.0033)\,(.0167)$

$$+ 6\,(-.0033)^2\,(4.5300) - 3\,(-.0033)^4 = 54.170269$$

$\mu_1' = 0$

$\mu_2' = \mu_2 - \frac{1}{12} = 4.52999 - .08333 = 4.44666$

$\mu_3' = \mu_3 = .06155$

$\mu_4' = \mu_4 - \frac{1}{2}\mu_2 + \frac{7}{240} = 54.170269 - 2.264995 + .029167 = 51.93444$

Table 17.1—Calculation of Moments: Variation of Thickness in 600 Brass Washers Manufactured by the ABC Co.*

Thickness (In Inches)	Number of Washers (f)	Deviation From Arbitrary Origin in Class Intervals d'	(5) f(d'²)	(6) f(d'³)	(7) f(d'⁴)
.0180–.01839	6	−5	− 30	150	− 750	3750
.0184–.01879	30	−4	−120	480	−1920	7680
.0188–.01919	42	−3	−126	378	−1134	3402
.0192–.01959	66	−2	−132	264	− 528	1056
.0196–.01999	94	−1	− 94	94	− 94	94
.0200–.02039	120	0	0	0	0	0
.0204–.02079	102	1	102	102	102	102
.0208–.02119	60	2	120	240	480	960
.0212–.02159	54	3	162	486	1458	4370
.0216–.02199	14	4	56	224	896	3584
.0220–.02239	12	5	60	300	1500	7500
	600		− 2	2718	10	32502

*Hypothetical data based on a smaller distribution given by W. A. Shewhart, *Economic Control of Quality of Manufactured Product*.

CURVE TYPE CRITERIA

The curve type that best describes the distribution may be identified from criteria calculated on the basis of the values of the moment. The criteria may be computed as follows.[1]

$$\beta_1 = \frac{\mu_3^2}{\mu_2^3}$$

$$\beta_2 = \frac{\mu_4}{\mu_2^2} = \frac{\mu_4}{\sigma^4}$$

$$\kappa = \frac{\beta_1 (\beta_2 + 3)^2}{4(4\beta_2 - 3\beta_1)(2\beta_2 - 3\beta_1 - 6)}$$

By using these criteria the type of Pearson curve best describing the distribution may be identified.[2]

---

[1] The symbol β is the Greek small letter beta and κ is Greek small letter kappa.

[2] See Pearson, Karl, *Tables for Statisticians and Biometricians*, pp. lx-lxv and page 123, section on generalization of curves.

Other similar curve criteria used are[1]

$$\alpha_3 = \frac{\mu_3}{\sigma^3} = \sqrt{\beta_1}$$

and

$$\alpha_4 = \frac{\mu_4}{\sigma^4} = \beta_2$$

KURTOSIS

The kurtosis of a frequency distribution is its peakedness.

If the curve has a higher degree of kurtosis than the normal curve ($\beta > 3$) the curve may be said to be **leptokurtic**. If β is less than 3, the curve is more flat-topped than the normal curve and is said to be **platykurtic**.

The measure of kurtosis is sometimes given as

$$\beta_2 - 3$$

or

$$a_4 - 3$$

Where the result is

1. zero, the curve is mesokurtic;
2. a positive value, the curve is leptokurtic;
3. a negative value, the curve is platykurtic.

The calculation of the β_2 value for the distribution of Table 17.1 is

$$\beta_2 = \frac{51.93444}{(4.44666)^2} = 2.63$$

OTHER MEASURES OF SKEWNESS

A more exact determination of skewness can be computed from a_3.

The value of a_3 will be zero for a normal curve. For a right skewed distribution a_3 will have a positive value, for a left skewed distribution, a negative.

Another formula for skewness is[2]

$$\chi = \frac{\sqrt{\beta_1}(\beta_2 + 3)}{2(5\beta_2 - 6\beta_1 - 9)}$$

---

[1] The symbol α is the Greek small letter alpha.
[2] Suggested by Karl Pearson. χ is not to be confused with χ^2 used in testing goodness of fit. (χ is the Greek small letter chi.)

This value χ (the measure of skewness) can be used to locate the mode more accurately than the methods outlined previously.

$$\text{mode} = \overline{X} - \chi\sigma$$

ADDITIONAL BIBLIOGRAPHY

ELDERTON, W. P., *Frequency Curves and Correlation*, C. and E. Layton, London, 1929.

YULE, G. U., & KENDALL, M. G., *An Introduction to the Theory of Statistics*, pp. 151-168, Hafner Publishing Co., 1950.

Chapter XVIII

Collection of Data

Data may be obtained from primary original sources, i.e., by interview or mailed questionnaire; or from secondary sources, i.e., from data compiled by other individuals or agencies.

PRIMARY SOURCES
Interview Method
Advantages
1. A higher degree of accuracy is attained through the acquisition of material direct from the source.
2. Material is often obtained that cannot be through the questionnaire alone.
3. There is opportunity personally to check information acquired.
4. The "no response" proportion is usually minimized.
Disadvantages
1. Only relatively small samples can be gathered because of the cost.
2. The subjective factor is involved in recording by interviewer.
Mail Questionnaire Method
Characteristics
1. The questions should be easily understood.
2. If possible they should be arranged in logical sequence.
3. The answers should consist of yes or no, check or blank space, or numerical indication where possible.
4. The questionnaire should be concise.
5. It should be in the most convenient, answerable form.
6. It should be constructed so as to facilitate the tabulation of data.

Advantages

1. A large area may be easily and quickly covered.

2. The method of assembling data is relatively inexpensive.

Disadvantages

1. Frequently questions cannot be answered without a supplementary explanation.

2. In many cases the results are unreliable due to a large "non response."

SECONDARY SOURCES

Advantages

1. The data are already compiled, thereby saving time and expense.

2. The responsibility for accuracy may be shifted.

Disadvantages

1. The data obtained by the primary agent cannot be verified.

2. The statistical technique used may not be obtainable and therefore the accuracy of the results may not be verifiable.

3. Subjective compiling and interpretation may have influenced the result shown.

4. The purpose of the study may have prejudiced the choice of material and technique adopted.

5. A representative sample may not have been taken.

ADDITIONAL BIBLIOGRAPHY

BOYD, H. W., & WESTFALL, R., *Marketing Research*, pp. 244–353, Richard D. Irwin, Inc., Homewood, Ill., 1964.

PARTEN, MILDRED, *Surveys, Polls and Samples*, pp. 48–218, Harper & Bros., New York, 1950.

Chapter XIX
Statistical Tables

A statistical table is a systematic arrangement of numerical data presented in columns and rows for purpose of comparison.

Statistical tables, classified according to purpose, are of two types, general purpose (primary) tables and special purpose (derived or text) tables.

GENERAL PURPOSE TABLES
Functions
1. The primary function of the general purpose table is to present original data in tabular form for reference purposes.

2. It serves as a source of information where original data are needed.

3. It is used in the construction of special purpose tables.

Characteristics
1. The general purpose table presents varied information on the same subject.

2. It should contain absolute, not percentage, figures.

3. Information should be presented in such form that it can easily be used for reference.

4. Actual figures, not round numbers, should be included.

SPECIAL PURPOSE TABLES
Functions
1. The primary function of the special purpose table is to present data so as to emphasize specific relationships.

2. It is used to emphasize a particular phase of the general information contained in a general purpose table.

3. It permits the presentation of selected materials in simple form.

Characteristics

1. Round numbers may be used at times.

2. The selected material in a special purpose table is presented in a small space to facilitate interpretation.

TITLE →

BOXHEAD →

STUB →

SOURCE →

COLUMN CAPTIONS ←

UNITS ←

FOOTNOTE ←

Table 19.1-Iron Ore Production and Prices in the United States 1953–1964

Year	Production (Millions of Gross Tons)	Prices* (Dollars per Long Ton)
1953	117.99	9.66
1954	78.09	9.90
1955	103.00	10.10
1956	97.88	10.85
1957	106.15	11.40
1958	67.71	11.45
1959	60.28	11.45
1960	88.78	11.45
1961	71.33	11.45
1962	71.83	10.85
1963	73.60	10.65
1964	81.34	10.56

*Average price per long ton, basic f.o.b. valley furnaces.

Source: American Metal Market, Metal Statistics, New York, 1968 edition.

RULES FOR TABLE CONSTRUCTION

Practice varies, but the generally accepted rules for the construction of a statistical table are as follows.[1]

1. **Title.** The title should be self-explanatory and should indicate, in the following order,

    a. the nature of the data presented

    b. the locality covered

    c. the time period included.

The title is placed above the table. The lettering is usually larger in the title than in any other section.

---

[1]Exceptions to the generally accepted procedure of table construction are usually justified by the particular purpose of a specific table.

2. Source. The source of the material should always be indicated on a table, except when original data has been obtained, since it is used

    a. to indicate the authority for the data

    b. as a means of verification

    c. as a reference for additional data.

The source is placed below the table at the left.

3. Footnotes. A footnote is used to further explain a figure in the table etc. It is placed immediately beneath the table, above the source. The footnote should be indicated by symbols as *, #, etc., or by a letter of the alphabet, never by a number, since the latter might be interpreted as part of the table.

4. Arrangement of data. Careful arrangement of items in a table facilitates reading of the table, analysis and comparison of data and permits emphasis on selected groups of data. Items may be arranged:

    a. **alphabetically.** This is the most frequently used arrangement for general purpose tables.

    b. **chronologically**—according to time of occurrence in comparing subjects over a period of time. Dates should move from the earliest to the latest date from the top of the stub to the bottom or in the boxhead from the left to the right of the table.[1]

    c. **geographically**—according to location in the customary classification, for example: country, state, county, etc., or Maine, New Hampshire, Vermont, etc. This arrangement is generally confined to reference (general purpose) tables.

    d. **by magnitude**—according to size. The largest number is placed at the top of the column and the others arranged in order of size. The row captions correspond to their values. When the row captions are numerical, as class intervals in the frequency distribution, they are arranged by size, the smallest number at the top, the largest at the bottom. With columns, the smallest is placed at the left, the largest at the right.

    e. **by customary classification.** There is a customary arrangement for many types of data which do not follow any serial

---

[1] An exception to this rule occurs when the latest figures are of primary interest as when the figures are published for the first time. In this instance the latest figure may be listed before the others, and then separated from them by a double or heavy line.

arrangement. For instance the classification men, women, and children, is rarely listed in the order women, children, and men.

5. Columns. When there are a number of columns in a table they may be numbered or lettered for reference purposes.

6. Column Captions. The heading of each column is known as the column caption. It should be concise. A miscellaneous column is placed at the right end of the table.

7. Stub. The heading of a row is known as the row caption. The section of the table containing row headings is designated as the stub. Items in the stub should be grouped, as months grouped by quarters, to facilitate interpretation of the data.

8. Totals. The totals of columns should be placed at the bottom of the columns, while row totals should be placed at the extreme right.[1]

9. Units of Measurement. These should be included in the box-head under the column captions.

10. Rulings. Lines may be ruled on a table as follows.

    a. A horizontal line is placed below the title, and below the body of the table.

    b. Columns are separated by single lines. In typewritten columns these lines are not essential but are useful.

    c. The stub and boxhead are separated from the figures by double or heavy lines, especially in non printed tables.

    d. Totals are separated from the other figures in a column by a single line.

11. Emphasis. A double line, heavy line, italics and light and bold face type contrasts are all used for emphasis on tables.

ADDITIONAL BIBLIOGRAPHY

LUCK, D. J., WALES, H. G., & TAYLOR, D. A., *Marketing Research*, pp. 312–316, Prentice-Hall, Inc., Englewood Cliffs, N. J., 1961.
PARTEN, M., *Surveys, Polls and Samples*, pp. 477–483, Harper & Bros., New York, 1950.

---

[1] The United States Census Bureau places totals at column headings and on the extreme left. This practice is explained by the Department to be due to the major interest in totals (see footnote 1 above).

Chapter XX

Graphic Presentation

A graph is a method of presenting statistical data in visual form.

TYPES OF GRAPHS

There are many varieties of graphs. The use of a particular type is dependent upon the data and upon the purpose for which the graph is constructed.

Graphs may be divided into the following types.

I. **Line** or **Curve Graphs**
   a. Arithmetic ruling
   b. Semi-logarithmic or logarithmic ruling
   c. Other special rulings
   Special Types of Line Graphs
   a. Silhouette chart
   b. Band chart
   c. High Low Chart
   d. Histogram
II. **Bar charts**
III. **Area Diagrams**
IV. **Solid Diagrams**
V. **Statistical Maps**

RULES FOR CONSTRUCTION OF GRAPHS[1]

1. Every graph must have a clear and concise title which is generally placed at the top center of the graph.[2] As a rule the title includes

---

[1] For a more complete discussion of the technique of graphic presentation the reader is referred to Arkin, H., and Colton R., *Graphs*.

[2] Graphs in printed form generally have the title placed below the graph; see graphs in this text.

information about

    a. the nature of the data

    b. the geographical location

    c. the period covered.

These elements of the title customarily appear in the order given above.

    2. Coordinate lines should be held to a minimum and curve lines emphasized so that the curves stand out sharply against the background.

    3. The source of the data should be indicated just under the graph at the lower left.

    4. Footnotes, if any, should be placed under and to the right of the graph.

    5. If the graph is to be readily understood the curve lines, segments and other details should be as few in number as possible.

    6. Each scale must have a scale caption indicating the units used.

        a. The X axis scale caption should be centered directly beneath the X axis.

        b. The Y axis scale caption should be placed at the top of the Y axis.

    7. The zero point should be indicated on the scale (Y axis), otherwise a misleading comparison may result. The necessity of indicating the zero point is seen by comparison of the peaks in the two graphs in Figure 20.1.[1]

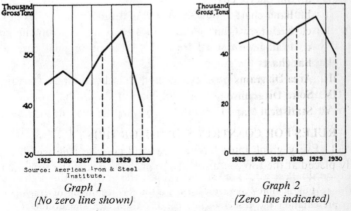

Graph 1
(No zero line shown)

Graph 2
(Zero line indicated)

Fig. 20.1. Steel ingot production in the United States.

---

[1] An exception to this rule occurs when the graph is in percentage form. In this case the 100% line is emphasized.

Inclusion of the zero point (Y axis) in Graph 2 shows an entirely different ratio in the heights of the points at 1928 and 1930 (5 to 4) to that of Graph 1 (with the misleading suggestion of 2 to 1).

If, however, lack of space makes it inconvenient to use the zero point line a scale break may be inserted to indicate its omission. Two types of scale breaks are shown in Figure 20.2.

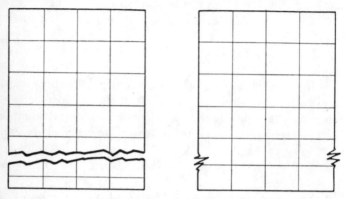

Fig. 20.2. Types of scale breaks.

9. The scales of values should be placed along the X and Y axis, thus giving a general indication of the size of the variations occurring in the graph.

It is unnecessary to indicate fine gradations on the scales of value since it is not intended that actual values be read off from the graph. Actual values can be obtained from the table of original data which usually accompanies the graph.

10. If a space on the X axis is used to indicate time intervals, the point representing the value for each period should be plotted at the midpoint for the period. If desired, however, the periods may be made to coincide with, and the points may then be plotted on, given coordinate lines.

11. On the Y axis the scale of values should run from zero (or from the smallest value) on the bottom of the graph to the highest value at the top. On the X axis the values should run from lowest on the left to highest on the right.

The various elements composing the graph are shown in Figure 20.3.

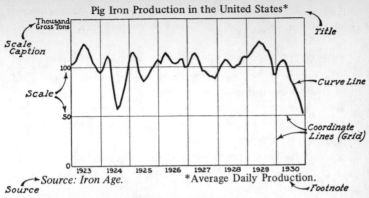

Fig. 20.3. Elements of the graph.

LINE OR CURVE GRAPHS

The line (or curve) graph is distinguished by the fact that the variations in the data are indicated by means of a line or curve (see Figure 20.3).

This type of graph is constructed by plotting points whose positions are determined by their respective values on the X and Y scales. The points are connected by straight lines.

Line graphs may be classified according to the type of scale ruling used.

    a. Arithmetic ruling

    b. Logarithmic rulings

    c. Other rulings[1]

ARITHMETIC RULINGS

Arithmetically ruled paper has equal distances between the co-ordinate lines. Equal quantities will then have equal distances. Thus the distance between 1 and 3 on the background ruling will be the same as that between 8 and 10.

An arithmetic progression will plot as a straight line on arithmetic paper since there are constant differences between the successive values in this type of series.

Since equal amounts are assigned equal distances, equal changes indicate identical absolute differences.

---

[1] Various other rulings are available but are beyond the scope of this discussion.

The line or curve type of graph is the most commonly used form of graphic presentation.

LOGARITHMIC AND SEMI-LOGARITHMIC RULINGS[1]

When it is desired to compare percentage rather than absolute changes a somewhat different form of ruling is used.

It can be shown[2] that where there is a constant percentage change between two pairs of figures the differences between the logarithms of the figures will be equal.

Thus if the logarithms of the values rather than the original figures are plotted constant differences (rises or falls) will then equal constant percentage changes.

Numbers	Logarithms	
2	0.30103	
4	0.60206	
Difference	0.30103	100% increase

Numbers	Logarithms	
5	0.69897	
10	1.00000	
Difference	0.30103	100% increase

Since, however, a great deal of time and effort is required to convert the original data into the form of logarithms, a more convenient procedure is to arrange the scale so that the logarithms may be plotted directly by reference to a special scale.

The logarithms in the longer procedure may be plotted on the arithmetic scale in the usual fashion. Thus, if it is desired to plot the value 2 its logarithm is determined (0.30103) and this value plotted. If, however, a scale is prepared in advance with the position 0.30103 marked 2, then the data may be plotted without previously determining the logarithms of the values.

The relation between a simple arithmetic scale and a scale corresponding to it but prepared for plotting logarithms is shown in Figure 20.4.

If a logarithmic ruling is used on both the X and Y axis the paper is known as logarithmic, if used only on one axis it is semi-logarithmic.

---

[1] Semi-logarithmic paper is also known as ratio paper.
[2] See any text on elementary mathematics.

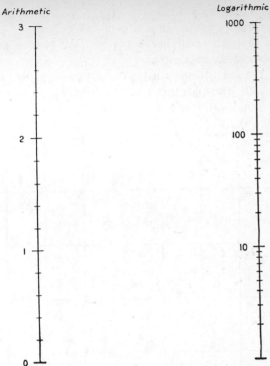

Fig. 20.4. Arithmetic and logarithmic scales.

Since time is generally placed on the X axis an arithmetic ruling is used on this axis in semi-logarithmic paper while the logarithmic ruling is retained on the Y axis.

Characteristics of Logarithmic and Semi-Logarithmic Charts

1. There is no zero or base line.

2. Semi-logarithmic charts have an arithmetic scale on the horizontal axis. Logarithmic charts are ruled logarithmically on both scales.

3. When plotted on logarithmic paper geometric progressions in both the X and Y values form a straight line, since the logarithms of a geometric progression form an arithmetic progression. On semi-logarithmic paper a geometric progression in the Y values will also form a straight line.

4. Equal rises or falls indicate equal percentage changes.

5. Equal slopes on a logarithmic chart denote equal rates of change.

204

GRAPHIC PRESENTATION

Use of Logarithmic Charts
1. To compare proportional rates of change.
2. To show the relationship between the two or more series which differ widely in amount.
   a. The unsatisfactory nature of arithmetic paper as compared with semi-logarithmic paper for this purpose can be seen from Figures 20.5 and 20.6.

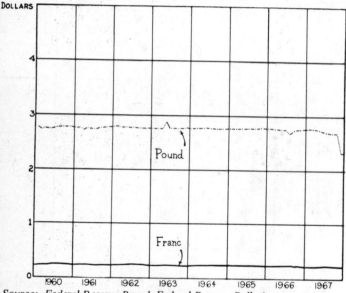

Source: Federal Reserve Board, Federal Reserve Bulletin.

Fig. 20.5 Market rates on the franc and the pound sterling, 1960–1967. (Plotted on arithmetic paper.)

SPECIAL TYPES OF LINE GRAPHS
1. **Silhouette charts** are line graphs showing the positive and negative deviations from a zero or base line with the area between the zero or base line and the curve filled in (see Figure 20.7, p. 206).

Silhouette charts are constructed by plotting points indicating the actual deviations from the base line. The points are then connected and the area between the curve and the base line filled in.

2. **Band charts,** another form of line graph, show variations in the component parts as well as the total.

The chart is prepared by first plotting the variation in the largest

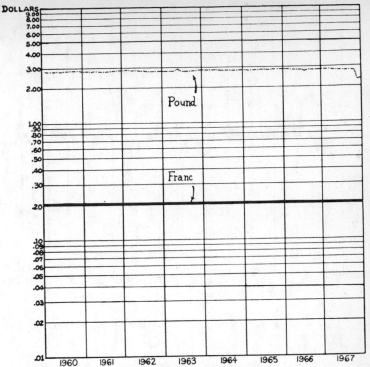

Source: Federal Reserve Board, Federal Reserve Bulletin.

Fig. 20.6 Market rates on the franc and the pound sterling, 1960–1967. (Plotted on semi-logarithmic paper.)

component part of the total. This segment may then be shaded in or cross-hatched. The next component part is then added to this first segment and the result plotted. This cumulative process is then continued until all of the component parts have been included. The variations in the top line will then represent variations in the total while the variations in the width of any segment will indicate the variations in that particular component part. Figure 20.8, p. 207, illustrates this type of graph.

3. **High-low graphs,** a form of line graph, present not only the changes occurring over a period of time but the fluctuations occurring within each period (as day, week, month, etc.) as well, indicating the high and low values.

The high-low chart is constructed by first plotting the lowest

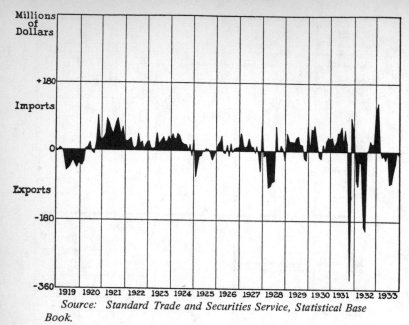

Source: *Standard Trade and Securities Service, Statistical Base Book.*

Fig. 20.7. *Gold movements from the United States, 1919–1933. (Silhouette chart.)*

value for a period and then the highest value for the same period. This procedure is continued until the end of the time covered by the graph. The low point and the high point for each period are connected by means of heavy lines. These lines, since they are closely spaced, tend to take the appearance of an irregular band.

4. **The histogram,** also known as a rectangular frequency polygon, is constructed from a frequency distribution in the following manner. Rectangles are erected using as the width the size of the class interval and as the height the frequency in each class interval. For an example see Figure 1.2, page 5.

BAR CHARTS

Bar charts visually contrast quantities by a comparison of bars of varying length but uniform width.

Bar charts may be subdivided into four types, namely:
1. Absolute
   a. Simple
   b. Subdivided

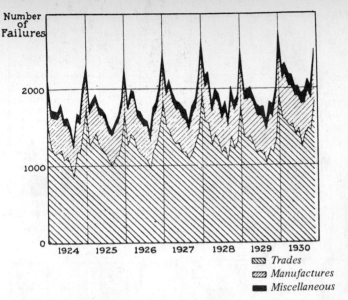

Trades
Manufactures
Miscellaneous

Source: *Standard Trade and Securities Service, Statistical Base Book.*

Fig. 20.8. *Commercial failures in the United States, by types, 1924–1930. (Band chart.)*

2. Percent
   a Simple
   b. Subdivided

Simple Absolute Bar Charts

Rectangular bars of the same width are erected from the same base line to proportionate lengths based on absolute or actual data. The bars may be set up either horizontally or vertically; however, when the scale involves time the vertical type of bar is indicated. See Graph A, Figure 20.9, p. 208.

Subdivided Absolute Bar Charts

The bars here are subdivided according to the size of each component. The components of each bar are arranged in similar order with the largest subdivisions at the base. The figures may vary to such an extent that the largest subdivision may not be the same, but the order of arrangement should remain fixed. Subdivided charts are cumulative in that each subdivision in plotting is added to the total of the subdivisions below it (see Graph B, Figure 20.9).

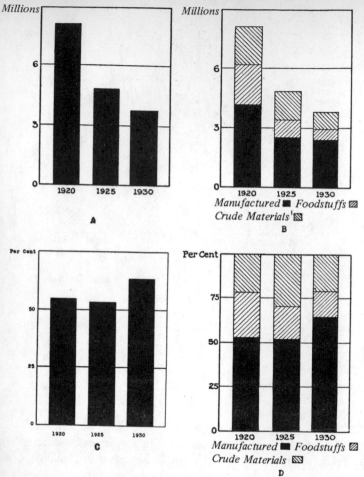

Source: United States Department of Commerce.

Fig. 20.9. Exports from the United States, 1920, 1925, and 1930. Shown as various forms of bar charts. (Chart C is for manufactured exports as percent of total.)

Simple Percentage Bar Charts

The bars are constructed in a similar fashion to the method used in simple absolute bar charts except that lengths of the bars represent percentage values (see Graph C, Figure 20.9).

Subdivided Percentage Bar Charts

Rectangular bars of the same width and the same length are constructed on the same base. The length represents 100 per cent. Each bar is divided into segments, the size of each segment being dependent on the percentage of the total figure which each subdivision represents.[1]

The subdivisions of each bar are arranged in the same order of presentation with the largest percentage at the base (see Graph D, Figure 20.9).

A special type of subdivided percentage bar chart is that which makes use of a single bar. The single bar is used when interest is centered on the component parts of a single total. The entire length of the bar represents 100 percent and each segment is represented in order of size from left to right.

PICTORIAL BAR CHARTS

Bar charts may be constructed in pictorial form. Pictures of different heights may be used for comparative purposes. Thus, to represent the number of persons in the armed forces, rows of soldiers with each representing a given number of men may be used. Each row corresponds in length to the values involved (see Figure 20.10).

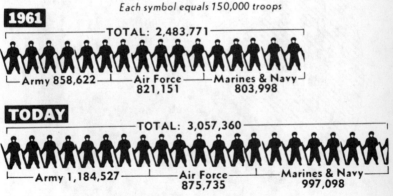

Each symbol equals 150,000 troops

1961

TOTAL: 2,483,771

Army 858,622 — Air Force 821,151 — Marines & Navy 803,998

TODAY

TOTAL: 3,057,360

Army 1,184,527 — Air Force 875,735 — Marines & Navy 997,098

Source: N. Y. Times, July 17, 1966.

Fig. 20.10. Growth of U. S. Armed Forces. (Pictorial bar chart.)

LOSS AND GAIN BAR CHARTS

This type of bar chart is constructed by having the bars extended from a zero line. If the bar chart is constructed horizontally the bars representing losses extend to the left, profits to the right. If the bar

---

[1]In subdivided bar charts the number of subdivisions should be as few as possible.

chart is constructed vertically the bars above the horizontal normal line represent profits, the bars below the normal line represent losses.

AREA DIAGRAMS

The area diagram contrasts quantities by comparing figures with varying areas. Area diagrams may use many techniques, the simplest making use of geometric figures (such as circles and squares). Area diagrams are of two main types. In the first type total areas of different sizes may be contrasted by varying the sizes of the figures. In the second type subdivisions of a single area may be compared.

The most usual type of area diagram is the **pie chart** (see Figure 20.11).

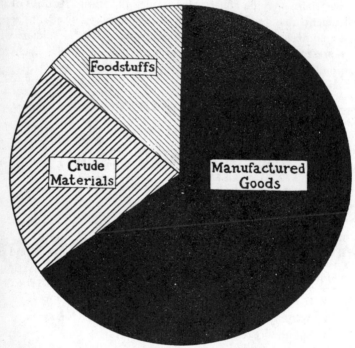

Source: United States Department of Commerce.

Fig. 20.11. Exports from the United States, by Economic Classes, 1930.

Pie Chart

A pie diagram is a chart of circular shape broken into subdivisions. The size of the section indicates the proportion of each component part to the whole.

To construct it, the whole circle is considered to represent 100 percent. Each circle is divided into 360°, and each percentage point equals 3.6°.

Characteristics

1. The arrangement of the size of sectors is generally clockwise according to size.

2. A uniform arrangement of sectors must be made to compare charts.

3. Whenever possible wording and percentages should be placed horizontally on the sector.

4. If shading, cross-hatching, colors, etc. are used in place of working on sectors, a legend should be constructed to indicate their their meaning.

5. The effectiveness of the pie chart is enhanced by cross-hatching, colors, shades, etc.

6. A pie chart should have a minimum of sectors.

7. The pie chart is difficult to construct accurately.

8. It is difficult to estimate visually with any degree of accuracy the proportionate size of the sectors of a pie chart where percentages are not indicated.

SOLID DIAGRAMS

Solid diagrams consist of geometric forms (cubes, spheres, cylinders, etc.) or irregular figures constructed to illustrate comparisons of magnitudes through comparison of volumes at the figures (see Figure 20.12, p. 212). The volumes of the figures in a solid diagram are compared and not the heights or lengths of the figures.

The solid diagram makes accurate comparisons difficult and for this reason it should not be used if some other method of illustration is possible.

MAP GRAPH

The map graph presents in pictorial form the facts in a geographic distribution.[1]

Map graphs are of five major types:
1. Shaded
2. Cross-hatched
3. Dotted

---

[1] A geographic distribution is known also as a "spatial" distribution.

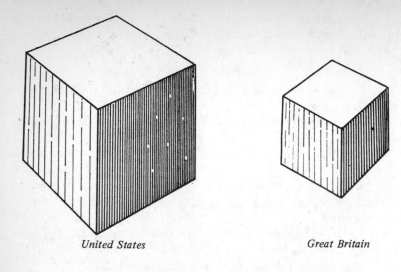

United States   *Great Britain*

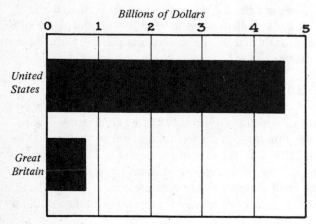

Fig. 20.12. Gold Stock of the United States and Great Britain. Shown in solid diagram and bar chart form.

4. Colored
5. Pin (Tacks, pins, or flags)

Shaded Maps

The proportionate quantities for particular areas may be indicated by using various degrees of shading ranging from solid black to white.

Cross-Hatched Maps

Cross-hatching may be used to indicate varying quantities by varying the proportions of black and white space.

Dotted Maps

Dots (circular areas) on maps are used primarily in three ways

a. Dots of similar size are placed on a map the primary purpose of which is to indicate the *density* of the numbers, etc., in an area by varying the number of these dots.

b. Dots of proportional sizes are placed on a map to indicate the total number of sizes in an area.

c. Dots of a fixed size each with the same assigned value may be varied in number to indicate the various quantities for each area.

Care must be taken in denoting relative sizes since comparisons must be made by varying the area of the dot.

Colored Maps

Colored maps are constructed by using various colors to indicate variations in sizes, etc.

A variety of colors should not be used to indicate relative values since an individual color is of no greater value than any other color to the observer.

Various degrees of the same color may be used to illustrate relative positions of different areas. The difficulty with using a single color scheme is that there is a limited number of shades which can be satisfactorily used.

Map-Tack System

Maps using tacks, flags, etc., are used for a large number of purposes in indicating relative sizes and also densities in a geographic area.

The heads of the pins or tacks may be of various colors, sizes, and shapes and thus extend their flexibility for the indication of sizes, locations, routes, etc.

ADDITIONAL BIBLIOGRAPHY

SCHMID, C. F., *Handbook of Graphic Presentation,* Ronald Press, New York, 1954.
SPEAR, M. E., *Charting Statistics,* McGraw-Hill Book Co., New York, 1952.
Department of Army, *Standards of Statistical Presentation,* Department of Army Pamphlet 325–10, U. S. Government Printing Office, April, 1966.

Chapter XXI

Special Techniques in Education, Psychology, Biology, and Industry

SPECIAL TECHNIQUES IN EDUCATION AND PSYCHOLOGY
Standard Scores[1]

In order to compare the results of two or more tests given to a number of students the tests should be of equal difficulty. If they are unequal in difficulty the average grades for each examination may differ widely, as will the resulting dispersion of the grades. The difficulty in the comparison of grades can be seen in the following distribution of the results for two examinations.

| Student | Examination | |
Number	Number 1	Number 2
1	100	100
2	99	85
3	99	70
4	98	68
5	97	65
6	96	60
7	90	60
8	90	60
9	87	50
10	80	45

The grades or scores may be standardized by converting each grade into a deviation from its respective arithmetic mean and dividing

---

[1] Compare Kelley, T. L., *Statistical Methods.*

by the standard deviation to allow for differing average attainment and dispersion of grade.

$$z = \frac{x}{\sigma}$$

where

z = standard score

x = deviation of given score from mean

σ = standard deviation of original grade

The grades on both examinations will average to the same value, zero (since $\Sigma(x) = 0$) and since each grade has been made relative to its standard deviation[1] the standard scores on both tests will have the same degree of dispersion.

If a pupil is consistent the standard scores attained on two tests measuring different traits, such as silent reading and arithmetic, should be equal ($z_1 = z_2$).

A large divergence in the value of these standard scores indicates a definite idiosyncrasy in the pupil considered.

The Coefficient of Reliability

The reliability of any test is measured by the similarity of results attained when the same test is given a number of times. If the reliability of the test or other measuring instrument is perfect, exactly similar results (allowing for chance variation) should be obtained when the test is given twice or the coefficient of correlation between the two sets of scores will be 1.00.

The coefficient of correlation of the scores secured from two applications of the same measuring instrument is thus a coefficient of reliability.

However it is usually not practical to give the same examination or the same form of examination twice. In place of this it is preferable to use the coefficient of correlation between the scores attained by dividing the questions into two parts. For this purpose the odd numbered questions (1, 3, 5, 7, etc.) may be used as one set and the even numbered questions (2, 4, 6, 8, etc.) as the other. The distribution of questions between the two sets must be random or at least the two sets of questions must be of equal difficulty.

The increase in the reliability of a test secured by lengthening it

---

[1] If it is desired to have a fixed average score of 50 in each test the formula $z = 50 + \frac{x}{\sigma} 10$ may be used.

or repeating it a number of times in the same form may be computed from the Spearman-Brown formula

$$r_n = \frac{nr_{1I}}{1 + (n - 1)r_{1I}}$$

where

r_n is the increased reliability coefficient resulting from either increasing the length of the test n times or repeating it n times

r_{1I} is the coefficient of reliability for the original test

The Intelligence Quotient

The scores on various intelligence tests given an individual will increase with his age. It is necessary, therefore, to relate the *mental age* as indicated by the intelligence tests to his *chronological age*.

$$I.Q. = \frac{MA}{CA} 100$$

where

$I.Q.$ = Intelligence Quotient
MA = Mental age
CA = Chronological age.

For statistical purposes the *normal* intelligence quotient for any age will then be 100.

In order to calculate the value of the intelligence quotient the chronological age is increased until it reaches a maximum at which it is retained. Otis[1] argues that the maximum age should be 18 rather than the generally used maximum of 16 years.

Subject Quotients and Ratios

Pupil accomplishment in a particular subject varies according to chronological age. To obtain a true picture of relative ability in a given subject it is necessary to compare "subject age" to chronological age.

$$\text{Arithmetic quotients} = \frac{\text{arithmetic age}}{\text{chronological age}}$$

$$\text{Reading quotients} = \frac{\text{reading age}}{\text{chronological age}}$$

$$\text{Any subject quotients} = \frac{\text{any subject age}}{\text{chronological age}}$$

---

[1] Otis, Arthur S., *Statistical Method in Educational Measurement.*

The average of a pupil's subject quotients is known as his **educational quotient**.

If the mental age is used in the quotients outlined above the result is known as a **subject ratio**, etc.

$$\text{Subject ratio} = \frac{\text{Subject Age}}{\text{Mental Age}}$$

The **accomplishment ratio** may then be obtained by averaging the subject ratios.

SPECIAL TECHNIQUES IN BIOLOGY
Index of Abmodality

A deviation from average or type is of little significance unless the deviation is related to the customary dispersion of the data. A deviation of two inches from the average height of a man of a certain age is of little import unless compared to the ordinary or usual dispersion of male heights of the same age. The same would be true for a deviation of an inch from the average length of a squirrel of a specified age, and most biological deviations.

In order to take the dispersion of the data into consideration the deviation from the mean may be related to the standard deviation of the data.[1]

$$\frac{x}{\sigma}$$

This measure is known to biologists as the **index of abmodality**.

The index of abmodality for an essentially normal distribution indicates the number of standard deviations the given value is from the mean. Thus if the index attains a certain value it may be further interpreted in light of the previous discussion on the normal curve (see chapter XII).

It is known that a deviation larger than 3 standard deviations from the mean (in any one direction plus or minus) will occur less than 2 times in 1000. This knowledge is obtained from the demonstrable fact that the area within 3 standard deviations from the mean will include 49.87% of the cases (on one side only of the mean), and therefore only .15% of the cases will be larger than the value of the index of abmodality if it equals 3. Any given value of the index of abmodality may thus be interpreted with the aid of the normal curve area table.

---

[1]In the field of education this is known as the standard score (see page 215).

Coefficient of Heredity

When the coefficient of correlation is applied to the measurement of the association between a specific characteristic of a parent and the same characteristic of an offspring it is known as the **coefficient of heredity**.

The coefficient of heredity between fathers and offspring is assigned the symbol r_1; between mothers and offspring r_2.

Coefficient of Assortative Mating

When the coefficient of correlation is used to measure the association between a specified characteristic of fathers and the same characteristic of mothers it is known as the **coefficient of assortative mating**. The symbol assigned to this coefficient is r_3.

Variability of Offspring

The variability (standard deviation) of a group of offspring from particular parents may be determined from the following formula

$$\sigma_{3.12} = \sigma_3 \sqrt{1 - \frac{2r_1^2}{1 + r_3}}$$

where

$\sigma_{3.12}$ = standard deviation of an array of offspring
σ_3 = standard deviation of offspring in general
r_1 = coefficient of heredity between offspring and parents, assuming parents to be equipotent ($r_1 = r_2$)
r_3 = coefficient of assortative mating

Abmodality of Offspring

The average abmodality of a group of offspring from parents of fixed characteristics may be computed from the formula

$$h_3 = \frac{r_1 \sigma_3}{(1 + r_3)\sigma_1} \left(h_1 + \frac{\sigma_1}{\sigma_2} h_2 \right)$$

where

h_3 = deviation of mean of given characteristic of offspring from mean of characteristic of all offspring
h_1 = deviation (abmodality) of father
h_2 = deviation of mother
σ_1 = standard deviation of characteristic in fathers in general
σ_2 = standard deviation of characteristic of mothers in general
σ_3 = standard deviation of offspring in general
r_1 = coefficient of heredity (assuming $r_1 = r_2$)
r_3 = coefficient of assortative mating

Vital Statistics (Demography)

Data relative to deaths, births, and sickness are significant only if considered in relation to the size and kind of population from which they were drawn. Thus the fact that 2,000 deaths were recorded in a year in a particular city is of no significance unless the population of the city is known. If the 2,000 deaths were recorded in New York City with a population of approximately 7,800,000 an entirely different significance would be attached to such a record than if it were recorded in a city of 25,000 population.

A **rate** is an expression of the number of times a specific kind of event occurs in a given population in relation to the total number in the population exposed to the possibility of its occurrence. This may be expressed in the form of a formula as

$$\text{rate} = \frac{a}{a + b}$$

where

a = number of times event appears in the population

b = number of times event does not appear in the population

The resulting value is in decimal form but is generally multiplied by 100, 1,000, 100,000 or 1,000,000 to give the result as percent (per 100), per 1000, per 100,000 or per million.

A **ratio** expresses the relation of occurrence of a given kind of event to the occurrence of other events or of one kind of data to another. In formula form this is

$$\text{ratio} = \frac{a}{c}$$

where

a = number of times event occurs

b = number of times another event occurs

Vital statistics make use of birth, death, and morbidity rates. These rates are important also in medical and actuarial statistics. Birth, death, and morbidity rates may be classified as follows.[1]

A. Mortality rates (Death rates)
   1. Observed
      a. Crude death rates
      b. Specific death rates

---

[1] This classification is after Pearl, Raymond in *Medical Biometry and Statistics.*

    2. Theoretic
        a. Standardized death rates
        b. Corrected death rates
  B. Natality rates (Birth rates)
    1. Observed
        a. Crude birth rates
        b. Specific birth rates
    2. Theoretic
        a. Standardized birth rates
        b. Corrected birth rates
  C. Morbidity Rates
    1. Observed
        a. Crude
        b. Specific

Crude Death, Birth, and Morbidity Rates

The crude death, birth, or morbidity rates are merely the total number of deaths, births, or cases of sickness divided by the total population.

$$\text{Crude death rate} \quad = \frac{D}{P}$$

$$\text{Crude birth rate} \quad = \frac{B}{P}$$

$$\text{Crude morbidity rate} = \frac{M}{P}$$

where

D = number of deaths
B = number of births
M = number of persons sick

Specific Death Rate

Although these rates must be specified as to time and place they are crude in that they do not include specifications as to age or sex. When such specifications are made the rate is known as the **specific** death, birth, or morbidity rate.

$$\text{Specific death, birth, or morbidity rate} = \frac{D' \text{ or } B' \text{ or } M'}{P'}$$

where

D' = deaths in a specified class of population
B' = births in a specified class of population
M' = number of persons sick in a specified class of population

P' = total number of persons in the specified population group
The specific death rates at various ages will of course vary greatly.
If it is assumed that 100,000 persons are born at the same instant
a hypothetical table showing the number of survivors, the number

Table 21.1–Stationary Life Table Population of 1,000,000 Persons
Number Living in Each Yearly Interval of Age

Age interval	Persons per million in current age interval	Age interval	Persons per million in current age interval	Age interval	Persons per million in current age interval
0– 1	17,841	35–36	14,146	70– 71	6,373
1– 2	16,916	36–37	14,031	71– 72	5,979
2– 3	16,612	37–38	13,912	72– 73	5,597
3– 4	16,448	38–39	13,791	73– 74	5,178
4– 5	16,338	39–40	13,667	74– 75	4,776
5– 6	16,255	40–41	13,540	75– 76	4,375
6– 7	16,186	41–42	13,411	76– 77	3,978
7– 8	16,127	42–43	13,278	77– 78	3,589
8– 9	16,078	43–44	13,141	78– 79	3,210
9–10	16,036	44–45	13,000	79– 80	2,843
10–11	15,998	45–46	12,854	80– 81	2,490
11–12	15,962	46–47	12,702	81– 82	2,152
12–13	15,927	47–48	12,545	82– 83	1,835
13–14	15,890	48–49	12,383	83– 84	1,546
14–15	15,851	49–50	12,216	84– 85	1,287
15–16	15,808	50–51	12,045	85– 86	1,058
16–17	15,761	51–52	11,867	86– 87	859
17–18	15,708	52–53	11,683	87– 88	687
18–19	15,650	53–54	11,489	88– 89	541
19–20	15,586	54–55	11,284	89– 90	418
20–21	15,516	55–56	11,067	90– 91	318
21–22	15,441	56–57	10,836	91– 92	236
22–23	15,363	57–58	10,592	92– 93	172
23–24	15,282	58–59	10,336	93– 94	123
24–25	15,200	59–60	10,069	94– 95	86
25–26	15,117	60–61	9,791	95– 96	59
26–27	15,032	61–62	9,501	96– 97	39
27–28	14,946	62–63	9,199	97– 98	26
28–29	14,857	63–64	8,884	98– 99	17
29–30	14,765	64–65	8,556	99–100	10
30–31	14,671	65–66	8,217	100–101	6
31–32	14,573	66–67	7,868	101–102	4
32–33	14,472	67–68	7,508	102–103	2
33–34	14,367	68–69	7,139	103–104	1
34–35	14,259	69–70	6,760	104–105	1

Source: Pearl, Raymond, Medical Biometry and Statistics, p. 259.

dying, the rate of mortality, and the stationary population at each age interval can be constructed. A table of this kind is known as a life table.

From the life table a stationary life table may be prepared showing the number of persons per million of each yearly interval of age. Table 21.1, p. 221, is such a table.

Standardized Death Rates

Due to various factors such as immigration, type of community, etc. the actual age distribution in one location may differ greatly from that in another community making it impossible to compare the crude death rates for all ages for the two localities directly. An allowance must be made for the difference. This is accomplished by means of standardized and corrected death rates.

The standardized death rate is obtained by applying the specific death rate obtained from the general population or a life table to the actual age distribution of the given population. The rate obtained is the rate that would exist if the hypothetical specific rate existed with the actual distribution of age.

$$\text{Standardized death rate} = \frac{\Sigma(pq)}{\Sigma(q)}$$

where

p = actual population for each age
q = specific death rate from life table

A comparison of this rate to the death rate of the standard population (from the life table) gives a correction factor.

$$\text{Correction factor} = \frac{R}{R'},$$

where

R = death rate from life table
R' = standardized death rate

Multiplying the crude death rate by this correction factor will make an allowance for the different age distributions.

Corrected Death Rates

The corrected death rate is obtained by using the specific death rates of the locality and hypothetical age distribution of the life table. This computation places the rates on a strictly comparable basis insofar as the age distribution is concerned.

$$\text{Corrected death rate} = \frac{\Sigma(p'q')}{\Sigma(p')}$$

where

p' = population in age group from life table

q' = actual specific death rates

Corrections have thus been made for varying age distributions. In a similar fashion corrections may be made for other differences such as sex, race, etc. Rates other than death rates may be corrected in the same way.

SPECIAL TECHNIQUES IN INDUSTRIAL PRODUCTION
Quality Control

The quality of a given product may be defined as its conformity to given standards or specifications. A manufactured product exhibits a certain amount of variation in its conformity to specifications, no matter how carefully guarded the process may be, due to innumerable chance causes. When only these chance, uncontrollable factors are the cause of variation the quality is said to be controlled. As soon as some controllable cause causes a variation from specifications, control is lost and the product ceases to conform to standard.

The problem of quality control is to detect the existence of these assignable causes and thus to maintain quality at the optimum level.

The two large areas of quality control are

1. Process control
2. Acceptance sampling

Process Control

Process control applies statistical methods to continuing surveillance of the manufacturing process to maintain quality standards by detection of assignable causes of quality variation. This method is sometimes called the control chart technique.

Process control involves continuous inspection of samples from production lines. There are two types, one testing attributes, one variables.

Attribute testing classifies the product into categories such as defective and nondefective.

Variables tests involve measurements on a full continuous scale (such as pounds, inches), and the actual numerical variation is recorded.

The principle of the control chart technique is that the quality measurements obtained from samples from production will vary. When such variations are beyond predictable sampling variation limits, an assignable cause of quality variations must exist.

Attributes Process Control

1. Select samples of size n from production, at random.[1]

2. Determine the number of defective items in the sample. This may be expressed as percent defective if so desired.

3. Compute the limits of sampling variation, using either

    a. for number of defects

$$\sigma = \sqrt{npq}$$

    b. for percent defective

$$\sigma_\% = \sqrt{\frac{pq}{n}}$$

where p is the percent defective established on past data or as the desired minimum attainable level. Multiply by the appropriate number of standard errors (t) to reach the confidence level desired. For instance for the 99.7% level use 3σ, for the 99% level use 2.58σ, etc.

4. Prepare a graphic control chart (see Figure 21.1) by drawing a horizontal line extending from the vertical quality scale axis at the number of defectives or percent defective established as previous average performance or the level of the desired performance. A pair of horizontal lines (control limits) are drawn on both sides at the distance

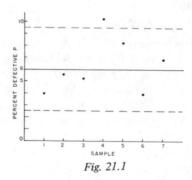

Fig. 21.1

---

[1] The size of the sample used will affect the sensitivity of the method in detecting departures from quality standards. Sample sizes used are the result of such determinations together with practical considerations such as the cost of testing, etc.

$t\sigma$ from the central line. The number or percent defective for each sample are then plotted. The result is a control chart.

5. If any of the plotted points fall outside of the limit lines, the process is said to be out of control.

Variables Process Control

Control of Quality Level

1. Select samples of size n from production at random and inspect.

2. Compute an average (arithmetic mean, median, midrange, etc.) for each set of sample measurements.[1]

3. Compute the appropriate standard error of the average used

$$\sigma_{\bar{x}} = \frac{\sigma}{\sqrt{n}} \;,\; \sigma_{mdn} = 1.2533 \frac{\sigma}{\sqrt{n}} \;,\; \text{etc.}$$

4. Prepare a graphic control chart (see Figure 21.1) by drawing a horizontal line extending from the vertical quality scale at the average value. A pair of horizontal lines (control limits) are drawn on either side of this central line at a distance t times the standard error.

5. Plot the averages obtained from the sample average values. If any of the plotted points fall outside of the established control limits, the process is out of control.

Control of Quality Uniformity

1. Select samples of size n from production at random and inspect.

2. Compute a measure of variability for each set of sample measurements such as the standard deviation, range, etc.

3. Compute the standard errors of the measure of variability used

$$\sigma_{\sigma} = \frac{\sigma}{\sqrt{2n}}$$

If the range is used, resort is had to special tables to secure the limits of its sampling variation.[2]

4. Prepare a graphic control chart by drawing a horizontal line from the vertical quality variability scale at a point established by past performance and continue as above.

---

[1] The midrange is a value half way between the two most extreme values.

[2] See Grant, E., *Statistical Quality Control,* McGraw-Hill Book Co., New York, 1952, 2nd Edition, Chapter VI.

5. Plot the measure computed for each sample. If any of the plotted points fall beyond the control limits the process is out of control with respect to quality uniformity.

Special Methods

The computation of the standard errors used to establish control limits requires the best possible estimate of the standard deviation of the basic population of items.

Since the standard deviations are usually computed from a series of samples the method used is to compute the average standard deviation (σ) for a series of such samples and to correct this average by the ratio between that value and that of the population as predicted by statistical theory. The factor used for this correction, c_2, varies in accordance with the size of the samples.

The estimate is obtained by dividing the average standard deviation by c_2

$$\sigma' = \frac{\overline{\sigma}}{c_2}$$

where

σ' is the estimated population standard deviation

$\overline{\sigma}$ is the average standard deviation of a number of samples of size n

Values of c_2 for some values of n are[1]

n	c_2
2	.5642
3	.7236
4	.7979
5	.8407
6	.8686

The standard deviation of the population may be estimated from the average range of the samples by dividing the average range of a series of samples by a factor, d_2. This is most useful for small sample sizes.

$$\sigma' = \overline{R}/d_2$$

Values of d_2 for some values of n are[1]

n	d_2
2	1.128
3	1.693

---

[1] See Grant, *op. cit.*, for completed table.

4	2.059
5	2.326
6	2.534

Acceptance Sampling

The aim of acceptance sampling is to assure acceptance of lots of a desirable quality level and the rejection of lots of an undesirable quality level on the basis of samples.

The specification for sampling is called a sampling plan. Sampling can be accomplished on the basis of attributes (percent defective) or variables. Variables plans are beyond the scope of this book.

The attributes sampling plan is specified by

N — the lot size from which the sample is drawn.

n — the sample size to be drawn from the lot.

c — the maximum number of defectives in the sample which will permit acceptance of the lot.

It is possible to compute mathematically the probability that a lot of given percent defective will be accepted under a given sampling plan. When computed for all possible percent defectives in lots submitted to the test, these results may be plotted with all possible percent defectives (P) for incoming lots on the X axis and the probabilities of acceptance (Pa) on the Y axis. The resulting curve is called an operating characteristic (OC) curve. The OC curve for the plan $N = 800$, $n = 38$, and $c = 1$ is shown in Figure 21.2.

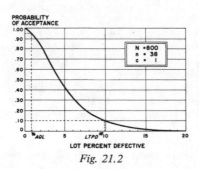

Fig. 21.2

On the basis of this curve, certain criteria may be determined.

1. Lot tolerance percent defective, $LTPD$ or P_t, is the percent defective for an inspected lot which for the specified plan will give a small probability (usually 10%) of being accepted. See Figure 21.2.

2. Acceptable quality level is the percent defective for an inspected lot which for the specified plan will give a high probability of acceptance (usually 95%). See Figure 21.2.

Sampling plan tables which list various plans and specify the *LTPD* or *AQL* for each plan have been published.

Types of Acceptance Sampling Plans

1. Single Sample, where one sample is drawn from the lot and a decision reached as to acceptance or rejection.

2. Double Sampling, where the first sample decides whether to accept, reject, or draw a second sample to reach a decision.

3. Multiple or Sequential Sampling, where the first sample decides whether to accept, reject, or draw a second sample. The second sample decides whether to accept, reject, or draw a third sample, and so on for a fixed series of samples.

ADDITIONAL BIBLIOGRAPHY

Techniques in Education and Psychology
GUILFORD, J. P., *Fundamental Statistics in Psychology and Education,* McGraw-Hill Book Co., New York, 1956, 2nd ed.
PEATMAN, J. J., *Introduction to Applied Statistics,* Harper & Row, New York, 1963.
RAY, W. S., *Statistics in Psychological Research,* The Macmillan Co., New York, 1962.
Techniques in Biology
DAVENPORT, C. B., *Statistical Method with Special Reference to Biological Variations,* 3rd rev. ed., John Wiley & Sons, New York, 1904.
FINNEY, D. J., *Statistical Method in Biological Assay,* Hafner Publishing Co., New York, 1952.
STEEL, R. G., & TORRIE, J. H., *Principles and Procedures of Statistics with Special Reference to the Biological Sciences,* McGraw-Hill Book Co., New York, 1960.
Techniques in Vital Statistics
GLOVER, J. W., *United States Life Tables, 1890, 1901, 1910, and 1901–1910,* Bureau of Census, Washington, 1921.
PEARL, RAYMOND, *Medical Biometry and Statistics,* W. B. Saunders, Philadelphia, 1930.
Techniques in Production Control
COWDEN, D. J., *Statistical Methods in Quality Control,* Prentice-Hall, Englewood Cliffs, N. J., 1957.
DUNCAN, A. J., *Quality Control and Industrial Statistics,* Richard D. Irwin, Inc., Homewood, Ill., 3rd ed. 1959.
GRANT, E. L., *Statistical Quality Control,* McGraw-Hill Book Co., New York, 3rd ed.
SHEWHART, W. A., *Economic Control of Quality of Manufactured Product,* D. Van Nostrand Co., New York, 1931.

List of Formulae

The following reference list includes most of the formulae contained in this volume and many others dealt with in more extended texts.

Averages

Arithmetic mean

    from ungrouped data

$$\overline{X} = \frac{\Sigma(X)}{N}$$

    from grouped data

      A. Long method

$$\overline{X} = \frac{\Sigma(f \times M.P.)}{N}$$

      B. Short (unit deviation) method

$$\overline{X} = \overline{Z} + \frac{\Sigma(fd)}{N}$$

      C. Short (group deviation) method

$$\overline{X} = \overline{Z} + \frac{\Sigma(fd')}{N} C$$

Median

    from grouped data

$$\text{Median} = L_{mc} + \frac{i}{f} C$$

Mode

from grouped data

A. Moments of force method

$$\text{mode} = L_{mo} + \frac{f_a}{f_a + f_b} C$$

B. Empirical method

$$\text{mode} = \text{mean} - 3\,(\text{mean} - \text{median})$$

C. Skewness method

$$\text{mode} = \overline{X} - \chi\sigma$$

Geometric mean

from ungrouped data

$$G_m = \sqrt[n]{X_1 \cdot X_2 \cdot X_3 \cdots X_n}$$

$$\log G_m = \frac{\log X_1 + \log X_2 + \log X_3 + \ldots + \log X_n}{N}$$

from grouped data

$$\log G_m = \frac{\Sigma(f \log M.P.)}{N}$$

Quadratic mean

$$Q_m = \sqrt{\frac{\Sigma(X^2)}{N}}$$

Harmonic mean

$$\frac{1}{H_m} = \frac{\dfrac{1}{X_1} + \dfrac{1}{X_2} + \dfrac{1}{X_3} + \ldots + \dfrac{1}{X_n}}{N}$$

Measures of Dispersion

Mean deviation

from ungrouped data

$$MD = \frac{\Sigma|x|}{N} \text{ or } \frac{\Sigma|d|}{N}$$

from grouped data

$$MD' = \frac{\Sigma |fd'|}{N} + \frac{(N_S - N_L)c}{N}$$

$$MD = MD' + C$$

$$MD = \frac{\Sigma |fd| + (N_S - N_L)c}{N} C$$

$$MD' = \frac{\Sigma (fd)}{N} + \frac{(N_a + N_b)c + f_m(.25 + c^2)}{N}$$

Standard deviation

$$\sigma = \sqrt{\frac{\Sigma(x^2)}{N}}$$

when computed from sample

$$s = \sqrt{\frac{\Sigma(x^2)}{n-1}}$$

from ungrouped data

$$\sigma = \sqrt{\frac{\Sigma(X^2)}{N} - \left(\frac{\Sigma X}{N}\right)^2}$$

from grouped data
Short (unit deviation) method

$$\sigma = \sqrt{\frac{\Sigma f(d^2)}{N} - \left(\frac{\Sigma fd}{N}\right)^2}$$

Short (group deviation) method

$$\sigma = C\sqrt{\frac{\Sigma f(d'^2)}{N} - \left(\frac{\Sigma fd'}{N}\right)^2}$$

Correction of standard deviation for grouping

$$\sigma^2 = (\sigma'^2 - 1/12)\, C^2$$

Charlier check for computation of standard deviation

$$\Sigma f(d' + 1)^2 = \Sigma fd'^2 + 2\,\Sigma fd' + N$$

Variance of a sum or difference (correlated data)

$$\sigma^2{}_{x_1 \pm x_2} = \sigma_{x_1}{}^2 + \sigma_{x_2}{}^2 \pm 2r\sigma_{x_1}\sigma_{x_2}$$

Variance of a sum or difference (uncorrelated data)

$$\sigma^2_{x_1 \pm x_2} = \sigma_{x_1}{}^2 + \sigma_{x_2}{}^2$$

Variance of a product of 2 variables
where σ_{x_1} and σ_{x_2} are small compared to $\overline{X}_1$ and $\overline{X}_2$
and X_1 and X_2 are independent

$$\sigma_{x_1 x_2}{}^2 = \overline{X}_1 \sigma_{x_2}{}^2 + \overline{X}_2 \sigma_{x_1}{}^2$$

Variance of a quotient of 2 variables
where σ_{x_1} and σ_{x_2} are small compared to $\overline{X}_1$ and $\overline{X}_2$

$$\frac{\sigma_{x_2}{}^2}{x_1} = \frac{\overline{X}_1 \sigma_{x_2}{}^2 + \overline{X}_2 \sigma_{x_1}{}^2}{\overline{X}_1{}^4}$$

Measures of Skewness and Kurtosis

Coefficient of skewness

$$S_K = \frac{\text{mean} - \text{mode}}{\sigma}$$

$$S_K = \frac{3\,(\text{mean} - \text{median})}{\sigma}$$

$$S_K = \frac{(Q_3 - \text{median}) - (\text{median} - Q_1)}{QD}$$

Other measures of skewness

$$\alpha_3 = \frac{\mu_3}{\sigma_3} = \sqrt{\beta_1}$$

$$\beta_1 = \frac{\mu_3{}^2}{\mu_2{}^3}$$

$$\chi = \frac{\sqrt{\beta_1}\,(\beta_2 + 3)}{2(5\beta_2 - 6\beta_1 - 9)}$$

Kurtosis

$$\beta_2 - 3$$

$$a_4 - 3$$

Moments

about arbitrary origin

The first moment

$$\nu_1 = \frac{\Sigma(fd)}{N}$$

The second moment

$$\nu_2 = \frac{\Sigma f(d^2)}{N}$$

The third moment

$$\nu_3 = \frac{\Sigma f(d^3)}{N}$$

The fourth moment

$$\nu_4 = \frac{\Sigma f(d^4)}{N}$$

about arithmetic mean

$$\mu_1 = \frac{\Sigma f(x)}{N} = 0$$

$$\mu_2 = \frac{\Sigma f(x^2)}{N}$$

$$\mu_3 = \frac{\Sigma f(x^3)}{N}$$

$$\mu_4 = \frac{\Sigma f(x^4)}{N}$$

$$\mu_2 = \nu_2 - \nu_1^2$$

$$\mu_3 = \nu_3 - 3\nu_1\nu_2 + 2\nu_1^3$$

$$\mu_4 = \nu_4 - 4\nu_1\nu_3 + 6\nu_1^2\nu_2 - 3\nu_1^4$$

Sheppard's corrections for grouping

Corrected first moment (in class intervals)

$$\mu_1' = 0$$

Corrected second moment (in class intervals)

$$\mu_2' = \mu_2 - 1/12$$

Corrected third moment (in class intervals)

$$\mu_3{}' = \mu_3$$

IV. Corrected fourth moment (in class intervals)

$$\mu_4{}' = \mu_4 - \tfrac{1}{2}\mu_2 + 7/240$$

$\mu_2{}'$ (in original units) $= C^2 \mu_2{}'$ (in class interval units)

$\mu_3{}'$ (in original units) $= C^3 \mu_3{}'$ (in class interval units)

$\mu_4{}^1$ (in original units) $= C^4 \mu_4{}'$ (in class interval units)

Frequency Distribution

Variance of weighted sum

$$\sigma_{ws}^2 = \Sigma W_i{}^2 \sigma_i{}^2 + 2\Sigma r_{ij} W_i \sigma_i W_j \sigma_j$$

Quartile deviation (Semi-interquartile range)

$$QD = \frac{Q_3 - Q_1}{2}$$

Coefficient of variation

$$V = \frac{\sigma}{\overline{\overline{X}}}$$

$$V = \frac{MD}{\text{median (or mean)}}$$

$$V_Q = \frac{\dfrac{Q_3 - Q_1}{2}}{\dfrac{Q_3 + Q_1}{2}} = \frac{Q_3 - Q_1}{Q_3 + Q_1}$$

Tchebycheff inequality

$$1 - \frac{1}{T^2}$$

Camp-Meidel inequality

$$1 - \frac{1}{2.25\,T^2}$$

Bartlett's test of homogeneity of variances

$$B^2 = 2.3026 \left[(\log \bar{s}^2)(n - k) - \Sigma(n_i - 1)(\log s_1^2) \right]$$

Curve criterion

$$k = \frac{\beta_1 (\beta_2 + 3)^2}{4(4\beta_2 - 3\beta_1)(2\beta_2 - 3\beta_1 - 6)}$$

Time Series Analysis

Straight line

$$Y = a + bX$$

Normal equations for straight line

$$\text{(I)} \quad \Sigma(Y) = Na + b\Sigma(X)$$

$$\text{(II)} \ \Sigma(XY) = a\Sigma(X) + b\Sigma(X^2)$$

Simplified normal equations for straight line (origin at midpoint of data)

$$\text{(I)} \quad \Sigma(Y) = Na$$

$$\text{(II)} \ \Sigma(XY) = b\Sigma(X^2)$$

Potential equations

$$Y = a + bX + cX^2 \text{ (second degree)}$$

$$Y = a + bX + cX^2 + dX^3 \text{ (third degree)}$$

$$Y = a + bX + cX^2 + dX^3 + eX^4 \text{ (fourth degree)}$$

$$Y = a + bX + cX^2 + dX^3 + eX^4 \ldots \text{etc.}$$

Normal equations for a second degree potential

$$\text{(I)} \quad \Sigma(Y) = Na + b\Sigma(X) + c\Sigma(X^2)$$

$$\text{(II)} \ \ \Sigma(XY) = a\Sigma(X) + b\Sigma(X^2) + c\Sigma(X^3)$$

$$\text{(III)} \ \Sigma(X^2 Y) = a\Sigma(X^2) + b\Sigma(X^3) + c\Sigma(X^4)$$

Exponential Equations

$$Y = a\, b^x$$

$$\log Y = \log a + X \log b$$

$$Y = a\, X^b$$

$$\log Y = \log a + b \log X$$

Other Non-Linear Curves

Hyperbola

$$Y = \frac{1}{a + bX}$$

Gompertz curve

$$Y = a\, b^{c^x}$$

$$\log Y = \log a + C^x \log b$$

Pearl-Reed curve

$$Y = \frac{k}{1 + e^{a + bX}}$$

Linear Correlation

Coefficient of correlation

$$r = \sqrt{1 - \frac{S_y{}^2}{\sigma_y{}^2}}$$

Product moment method

$$r = \frac{p}{\sigma_x \sigma_y}$$

$$p = \frac{\Sigma(XY)}{N} - \left(\frac{\Sigma X}{N}\right)\left(\frac{\Sigma Y}{N}\right)$$

Standard error of estimate

$$S_y = \sqrt{\frac{\Sigma(d^2)}{N}}$$

$$S_y = \sigma_y \sqrt{1 - r^2}$$

Line of regression

$$y = r\frac{\sigma_y}{\sigma_x}\, x$$

$$Y - \bar{Y} = r\frac{\sigma_y}{\sigma_x}(X - \bar{X})$$

Coefficient of determination

$$r^2 = 1 - \frac{S_y^2}{\sigma_y^2}$$

Coefficient of alienation

$$k = \sqrt{\frac{S_y^2}{\sigma_y^2}} = \sqrt{1 - r^2}$$

Correction of S_y and r for number of cases

$$\overline{S}_y^2 = S_y^2 \frac{N - 1}{N - 2}$$

$$\overline{r}^2 = 1 - (1 - r^2) \frac{N - 1}{N - 2}$$

Spearman's correlation from ranks

$$\rho = 1 - \frac{6\Sigma(D^2)}{N(N^2 - 1)}$$

Relation between coefficient of correlation from ranks (ρ) and r

$$r = 2 \sin\left(\frac{\pi}{6} \rho\right)$$

"Spearman's footrule"

$$R = 1 - \frac{6\Sigma G}{N^2 - 1}$$

Non-Linear Correlation

Index of correlation

$$\rho = \sqrt{1 - \frac{S_y^2}{\sigma_y^2}}$$

$$\rho^2 = \frac{a\Sigma(Y) + b\Sigma(XY) + c\Sigma(X^2 Y) + \ldots - Nc_y^2}{\Sigma(Y^2) - Nc_y^2}$$

Standard error of estimate (non-linear correlation)

$$S_y^2 = \frac{\Sigma(Y)^2 - a\Sigma(X) - b\Sigma(XY) - c\Sigma(X^2 Y) - \ldots}{N}$$

Correlation ratio

$$\eta = \sqrt{1 - \frac{\sigma_{ay}^2}{\sigma_y^2}}$$

Correction of correlation ratio for grouping

$$\eta'^2 = \frac{\eta^2 - \frac{(\kappa - 3)}{N}}{1 - \frac{(\kappa - 3)}{N}}$$

Test for linearity of regression

$$\zeta = \eta^2 - r^2$$

Other Correlation Methods

Coefficient of multiple correlation

$$R_{1\cdot234} = \sqrt{1 - \frac{S^2_{1\cdot234}}{\sigma_1^2}}$$

$$R^2_{1\cdot234} = \frac{b_{12\cdot34}p_{12} + b_{13\cdot24}\,p_{13} + b_{14\cdot23}\,p_{14}}{\sigma_1^2}$$

Multiple correlation regression (linear)

$$X_1 = a + b_{12\cdot34}X_2 + b_{13\cdot24}X_3 + b_{14\cdot23}X_4$$

Standard error of estimate for multiple correlation

$$S_{1\cdot234} = \sqrt{\frac{\Sigma d^2}{N}}$$

$$S^2_{1\cdot234} = \sigma_1^2 - b_{12\cdot34}p_{12} - b_{13\cdot24}\,p_{13} - b_{14\cdot23}\,p_{14}$$

Non-linear multiple correlation regression

$$X_1 = a + f(X_2) + f(X_3) + f(X_4) + \ldots$$

Coefficient of partial correlation

$$r_{13\cdot24} = \sqrt{b_{13\cdot24} \cdot b_{31\cdot24}}$$

$$r_{12\cdot3} = \frac{r_{12} - r_{13}r_{23}}{\sqrt{1 - r_{13}^2}\,\sqrt{1 - r_{23}^2}}$$

$$r_{14 \cdot 23} = 1 - \frac{(1 - R_{1 \cdot 234}^2)}{(1 - R_{1 \cdot 23}^2)}$$

Coefficient of part correlation

$$_{12}r_{34}^2 = \frac{b_{12 \cdot 34}^2 \, \sigma_2^2}{b_{12 \cdot 34}^2 \, \sigma_2^2 + \sigma_1^2 \, (1 - R_{1 \cdot 234}^2)}$$

Beta coefficients

$$\beta_{12 \cdot 34} = b_{12 \cdot 34} \frac{\sigma_2}{\sigma_1}$$

$$\beta_{13 \cdot 24} = b_{13 \cdot 24} \frac{\sigma_3}{\sigma_1}$$

$$\beta_{14 \cdot 23} = b_{14 \cdot 23} \frac{\sigma_4}{\sigma_1}$$

$$\beta_{12 \cdot 3} = \frac{r_{12} - r_{13} r_{23}}{1 - r_{23}^2}$$

$$\beta_{13 \cdot 2} = \frac{r_{13} - r_{12} r_{23}}{1 - r_{23}^2}$$

Mean square contingency

$$\phi^2 = \frac{\chi^2}{N}$$

Coefficient of contingency

$$CC = \sqrt{\frac{\chi^2}{N + \chi^2}} = \sqrt{\frac{\phi^2}{1 + \phi^2}}$$

Probability

Probability of success

$$p = \frac{a}{n}$$

Probability of failure

$$q = \frac{b}{n}$$

Binomial distribution

$$\Pr(r) = \frac{n!}{r!(n - r)!} \, p^r q^{n-r}$$

Arithmetic mean of bionomial distribution

$$\overline{X} = np$$

Standard deviation of binomial distribution

$$\sigma = \sqrt{npq}$$

In relative form

$$\sigma_\% = \sqrt{\frac{pq}{n}}$$

Poisson distribution

$$\Pr(r) = \frac{e^{-\overline{X}}\,\overline{X}^r}{r!}$$

Variance of a Poisson distribution

$$\sigma^2 = np = \sqrt{\overline{X}}$$

Lexis ratio

$$L = \frac{\sigma}{\sigma_B}$$

Charlier coefficient of disturbancy

$$\rho = \frac{\sqrt{\sigma^2 - \sigma_B^2}}{np}$$

Normal curve

$$Y = Y_0\, e^{\frac{-x^2}{2\sigma^2}}$$

$$y = \frac{N}{\sigma\sqrt{2\pi}} e^{\frac{-x^2}{2\sigma^2}}$$

Maximum ordinate of normal curve

$$Y_0 = \frac{N}{\sigma\sqrt{2\pi}} = \frac{N}{2.506628\,\sigma}$$

Chi square test

$$\chi^2 = \Sigma\left(\frac{(f_o - f)^2}{f}\right)$$

Chi square with Yates' correction

$$\chi^2 = \Sigma \left[\frac{(|f_o - f| - 0.5)^2}{f} \right]$$

Theory of Sampling

Standard error of arithmetic mean

$$\sigma_{\bar{x}} = \frac{\sigma}{\sqrt{n}}$$

Standard error of sum (ΣX)

$$\sigma_{\Sigma X} = \sigma \sqrt{n}$$

Standard error of median

$$\sigma_{mdn} = 1.2533 \frac{\sigma}{\sqrt{n}}$$

Finite population correction factor

$$\sqrt{\frac{N - n}{N - 1}}$$

or approximately

$$\sqrt{1 - \frac{n}{N}}$$

Standard error of standard deviation for a sample drawn from a normally distributed universe

$$\sigma_\sigma = \frac{\sigma}{\sqrt{2n}}$$

Standard error of standard deviation for a sample from any universe, normal or non-normal

$$\sigma_\sigma = \sqrt{\frac{\mu_4 - \mu_2^2}{4\mu_2 \cdot n}}$$

Standard error of mean deviation

$$\sigma_{MD} = .6028 \frac{\sigma}{\sqrt{n}}$$

Standard error of coefficient of variation

$$\sigma_v = \frac{V}{\sqrt{2n}} \sqrt{1 + \frac{2(V)^2}{(10)^4}}$$

Standard error of a percentage

$$\sigma_\% = \sqrt{\frac{pq}{n}}$$

Standard error of coefficient of correlation

$$\sigma_r = \frac{1 - r^2}{\sqrt{n}}$$

Standard error of coefficient of rank correlation (ρ) (n over 25)

$$\sigma_\rho = \frac{1}{\sqrt{n - 1}}$$

Standard error of coefficient of rank correlation (Spearman's)

$$\sigma_\rho = \frac{1 - \rho^2}{\sqrt{n}} (1 + .086\rho^2 + .013\rho^4 + .002\rho^6)$$

Standard error of coefficient of multiple correlation

$$\sigma_{R_{12 \cdot 3 \ldots n}} = \frac{1 - R_{1 \cdot 23 \ldots n}^2}{\sqrt{n}}$$

Standard error of coefficient of partial correlation

$$\sigma_{r_{12 \cdot 34 \ldots n}} = \frac{1 - r_{12 \cdot 34 \ldots n}^2}{\sqrt{n}}$$

Standard error of the difference between two means

$$\sigma_D = \sqrt{\sigma_{\bar{X}_1}^2 + \sigma_{\bar{X}_2}^2}$$

$$= \sqrt{\frac{\sigma_1^2}{n_1} + \frac{\sigma_2^2}{n_2}}$$

z transformation

$$z = \frac{1}{2}[(\log_e (1 + r) - \log_e (1 - r)]$$

$$\sigma_z = \frac{1}{\sqrt{n - 3}}$$

Standard error of second moment about mean

$$\sigma_{\mu_2} = \sqrt{\frac{\mu_4 - \mu_2^2}{n}}$$

$$\sigma_{\mu_2} = \sigma^2 \sqrt{\frac{2}{n}}$$

Standard error of third moment about mean

$$\sigma_{\mu_3} = \sqrt{\frac{\mu_6 - \mu_3^2}{n}}$$

$$\sigma_{\mu_3} = \sigma_3 \sqrt{\frac{6}{n}}$$

Standard error of fourth moment about mean

$$\sigma_{\mu_4} = \sqrt{\frac{\mu_8 - \mu_4^2}{n}}$$

$$\sigma_{\mu_4} = \sigma^4 \sqrt{\frac{96}{n}}$$

Standard error of β_2

$$\sigma_{\beta_2} = \sqrt{\frac{24}{n}}$$

Standard error of coefficient of skewness as measured by $\dfrac{\overline{X} - Mode}{\sigma}$

$$\sigma_{SK} = \sqrt{\frac{3}{2n}}$$

Standard error of semi-interquartile range

$$\sigma_Q = .7867 \frac{\sigma}{\sqrt{n}}$$

Standard error of difference between two mutually independent standard deviations

$$\sigma_{\sigma_1 - \sigma_2} = \sqrt{\sigma_{\sigma_1}^2 + \sigma_{\sigma_2}^2}$$

Standard error of coefficient of regression

$$\sigma_b = \frac{Sy}{\sigma_x \sqrt{n}}$$

Standard error of coefficient of multiple regression

$$\sigma_a = \frac{S_{1 \cdot 23}}{\sqrt{n}}$$

$$\sigma_{b12.3} = \frac{S_{1 \cdot 23}}{\sigma_2 \sqrt{n} \ (1 - r_{23}^2)}$$

Standard error of Y intercept of regression equation

$$\sigma_a = \frac{Sy}{\sqrt{n}}$$

Standard error of total

$$\sigma_T = N \frac{\sigma}{\sqrt{n}}$$

Standard error of correlation ratio

$$\sigma_\eta = \frac{1 - \eta^2}{\sqrt{n}}$$

Standard error of arithmetic mean for small samples

$$S_{\overline{X}} = \frac{s}{\sqrt{n}}$$

where

$$s^2 = \frac{\Sigma(x^2)}{n - 1} = \frac{n\sigma^2}{n - 1}$$

Standard error of difference between two arithmetic means for small samples

$$S_D = \frac{s}{\sqrt{\dfrac{n_1 \ n_2}{n_1 + n_2}}}$$

$$s^2 = \frac{\Sigma(x_1^2) + \Sigma(x_2^2)}{n_1 + n_2 - 2}$$

Standard error of the difference between two proportions

$$\sigma_{D\%} = \sqrt{pq \left(\frac{1}{n_1} + \frac{1}{n_2} \right)}$$

Standard error of the sum of a series of means with given standard errors (when samples used are mutually independent)

$$\sigma_{\bar{X}_1 + \bar{X}_2 + .. + \bar{X}n}{}^2 = \sigma_{\bar{X}_1}{}^2 + \sigma_{\bar{X}_2}{}^2 + \cdots + \sigma_{\bar{X}_n}{}^2$$

Standard error of a mean raised to a power

$$\frac{\sigma_{\bar{X}n}}{\bar{X}^n} = n\left(\frac{\sigma_{\bar{X}}}{\bar{X}}\right)$$

Standard error of a product of a series of means
(mutually independent)

$$\left(\frac{\sigma_{\bar{X}_1 \cdot \bar{X}_2 \ldots \bar{X}n}}{\bar{X}_1 \cdot \bar{X}_2 \ldots \bar{X}_n}\right)^2 = \left(\frac{\sigma_{\bar{X}_1}}{\bar{X}_1}\right)^2 + \left(\frac{\sigma_{\bar{X}_2}}{\bar{X}_2}\right)^2 = \left(\frac{\sigma_{\bar{X}_3}}{\bar{X}_3}\right)^2 \ldots \left(\frac{\sigma_{\bar{X}n}}{\bar{X}_n}\right)^2$$

Standard error of a quotient of two means
(mutually independent)

$$\left(\frac{\sigma_{\frac{\bar{X}_1}{\bar{X}_2}}}{\frac{\bar{X}_1}{\bar{X}_2}}\right)^2 = \left(\frac{\sigma_{\bar{X}_1}}{\bar{X}_1}\right)^2 + \left(\frac{\sigma_{\bar{X}_2}}{\bar{X}_2}\right)^2$$

Standard error of a sum (not mutually independent)

$$\sigma_{A+B} = \sqrt{\sigma_A^2 + \sigma_B^2 + 2r_{AB}\sigma_A \sigma_B}$$

Standard error of a difference (not mutually independent)

$$\sigma_{A-B} = \sqrt{\sigma_A^2 + \sigma_B^2 - 2r_{AB}\sigma_A \sigma_B}$$

Standard error of biserial r

$$\sigma_{\text{biserial } r} = \frac{\sqrt{\dfrac{pq}{n}} - r^2_{\text{biserial } r}}{\sqrt{n}}$$

Standard error of β coefficient

$$\sigma_{\beta 12 \cdot 34 \ldots n}^2 = \frac{1 - R_{1 \cdot 234 \ldots n}^2}{(1 - R_{2 \cdot 34 \ldots n}^2)(n - m)}$$

Index Numbers

Simple aggregate of actual prices

$$\frac{\Sigma p_n}{\Sigma p_o}$$

Average of relative prices (arithmetic mean)

$$\frac{\Sigma \left(\dfrac{p_n}{p_o} \right)}{N}$$

Weighted aggregate of actual prices
    a. base year weights

$$\frac{\Sigma(p_n q_o)}{\Sigma(p_o q_o)}$$

    b. given year weights

$$\frac{\Sigma(p_n q_n)}{\Sigma(p_o q_n)}$$

Weighted average of relative prices
    a. arithmetic mean (base year weights)

$$\frac{\Sigma \left[\dfrac{p_n}{p_o} \times (p_o q_o) \right]}{\Sigma (p_o q_o)}$$

(given year weights)

$$\frac{\Sigma \left[\dfrac{p_n}{p_o} \times (p_n q_n) \right]}{\Sigma (p_n q_n)}$$

    b. geometric mean (logarithmic form—given year weights)

$$\log \text{ index} = \frac{\log \left[\dfrac{p'_n}{p'_o} (p'n q'n) \right] + \log \left[\dfrac{p_n''}{p_o''} (pn\, qn) \right] + \text{etc.}}{\Sigma(pn\, qn)}$$

Ideal index number formula

$$\sqrt{\frac{\Sigma (p_n q_o)}{\Sigma (p_o q_o)} \times \frac{\Sigma (p_n q_n)}{\Sigma (p_o q_n)}}$$

Weighted average
Arithmetic mean

$$\frac{\Sigma(\text{items} \times \text{weights})}{\Sigma(\text{weights})} = \frac{\Sigma(I \times W)}{\Sigma(W)}$$

Geometric mean

$$\Sigma w \sqrt{I_1{}^w . I_2{}^w . I_3{}^w \ldots I_n{}^w}$$

Logarithm of weighted geometric mean

$$\log \text{weighted } G_m = \frac{\Sigma(w \log I)}{\Sigma(w)}$$

Education and Psychology

Standard score

$$z = \frac{x}{\sigma}$$

Conversion of standard scores (average score of 50)

$$z = 50 + \frac{x}{\sigma} 10$$

Coefficient of correlation from standard scores

$$r_{12} = \frac{\Sigma(z_1 z_2)}{N}$$

Coefficient of reliability resulting from test given n times

$$r_n = \frac{n r_{1I}}{1 + (n-1) r_{1I}}$$

Intelligence quotient

$$\text{I. Q.} = \frac{MA}{CA}$$

Subject quotients
Arithmetic quotient

$$\frac{\text{arithmetic age}}{\text{chronological age}}$$

Reading quotient

$$\frac{\text{reading age}}{\text{chronological age}}$$

Any subject quotient

$$\frac{\text{any subject age}}{\text{chronological age}}$$

Subject ratio

$$\frac{\text{subject age}}{\text{mental age}}$$

Biology

Index of abmodality

$$\frac{x}{\sigma}$$

Variability of offspring

$$\sigma_{3.12} = \sigma_3 \sqrt{1 - \frac{2r^2_1}{1 + r_3}}$$

assuming $r_1 = r_2$

Abmodality of offspring

$$h_3 = \frac{r_1 \sigma_3}{(1 + r_3)\sigma_1} \left(h_1 + \frac{\sigma_1}{\sigma_2} h_2 \right)$$

assuming $r_1 = r_2$

Miscellaneous Formulae

Sum of first n natural numbers

$$\Sigma n = \frac{n(n + 1)}{2}$$

Sum of squares of first n natural numbers

$$\Sigma(n^2) = \frac{2n^3 + 3n^2 + n}{6}$$

Sum of cubes of first n natural numbers

$$\Sigma(n^3) = \frac{n^2(n + 1)^2}{4}$$

Sum of 4th powers of first n natural numbers

$$\Sigma(n^4) = \frac{n(n + 1)\,(2n + 1)\,(3n^2 + 3n - 1)}{30}$$

Number of combinations of n things taken r at a time

$$C_r^n = \frac{n!}{r!(n - r)!}$$

Number of permutations of n things taken r at a time

$$nPr = \frac{n!}{(n - r)!}$$

List of Symbols

a	Y intercept
a	number of ways in which a favorable outcome can appear
b	coefficient of slope (or regression)
b	possible number of unfavorable results
$b_{12 \cdot 34}$	coefficient of net regression of X_2 on X_1 excluding X_3 and X_4
C	size of class interval
CA	chronological age
CC	coefficient of contingency
c	difference between arbitrary origin and mean or median
c	constant in trend or regression equation
D	differences in rank
d	deviation of an individual value or midpoint of a class interval from an arbitrary or guessed mean (or average—other than arithmetic mean)
d	deviation from line of trend ($Y - Yc$)
d	constant in trend or regression equation
d'	d in class interval units
e	a constant $= 2.71828$
f	frequency
f_a	frequency of class interval above modal group
f_b	frequency of class interval below modal group
G	positive differences in rank
G_m	geometric mean

H_m	harmonic mean
h_1	deviation (abmodality) of father
h_2	deviation (abmodality) of mother
h_3	deviation of mean of given characteristic of offspring from mean of characteristic of all offspring
I	item
I. Q.	intelligence quotient
k	coefficient of non-determination
L_{me}	lower limit of class interval containing median
L_{mo}	lower limit of modal group
LSD	least significant difference
M.A.	mental age
MD	mean deviation
MD'	mean deviation in class intervals
M.P.	midpoint
N or n	number of cases
N_L	number of cases overstated or too large
N_S	number of cases understated or too small
$P.E._{\bar{x}}$	probable error of mean
$P.E._\theta$	probable error of any statistical measure (θ) such as $P.E._{mdn}, P.E._r$, etc.
p	product moment
p	probability of success
p	confidence level
p_n	price of a commodity in period n
p_o	price of a commodity in base period
p_o'	price of first commodity in base period
p_o''	price of second commodity in base period
Q_1	first quartile
Q_3	third quartile
QD	quartile deviation (semi-inter-quartile range)
Q_m	quadratic mean
q	probability of failure
q_n	quantity of a commodity produced or consumed in period n
q_o	quantity of a commodity produced or consumed in base period
q_o'	quantity of first commodity produced or consumed in base period

$q_o{}''$	quantity of second commodity produced or consumed in base period
R	coefficient of correlation computed by "Spearman's footrule" method
$R_{1.234}$	coefficient of multiple correlation between X_1 and X_2, X_3, X_4
r or r_{12} or r_{xy}	coefficient of correlation
$\bar{r}$	coefficient of correlation corrected for number of cases
r_1	coefficient of heredity (fathers and offspring)
r_2	coefficient of heredity (mothers and offspring)
r_3	coefficient of assortative mating (fathers and mothers)
r_{11}	coefficient of reliability
$r_{12.3}$	coefficient of partial correlation between X_1 and X_2 excluding X_3
$_{12}r_{34}$	coefficient of partial correlation between X_1 and X_2 excluding X_3 and X_4
$S_{1.234}$	standard error of estimate measured about regression surface $X_1 = f(X_2) + f(X_3) + f(X_4)$
S_y	standard error of estimate
$\bar{S}_y$	standard error of estimate corrected for number of cases
W or Wt	weight
X	an individual value
$\bar{X}$	arithmetic mean
x	deviation of individual value from its arithmetic mean
Y	an individual value
$\bar{Y}$	arithmetic mean of Y values
Y_c	computed Y value as determined from line of trend or regression
y	deviation of individual Y value from its arithmetic mean
$\bar{Z}$	an arbitrarily selected value—guessed mean
z	standard score
z'	residual-difference between actual value and theoretical line of regression value
z_1	standard score on first test
z_2	standard score on second test

a_3	measure of skewness
β_1	curve criterion
β_2	curve criterion (measure of kurtosis)
$\beta_{12\cdot3}, \beta_{12\cdot34},$ $\beta_{13\cdot24}, \beta_{24\cdot23}$	beta coefficients
ζ	test for linearity of regression
η	correlation ratio
θ	any statistic
κ	curve criterion
κ	number of arrays in correlation table
$\mu_1{'}\,\mu_2{'}\,\mu_3{'}\,\mu_4{'}$	moments about arithmetic mean corrected for grouping (Sheppard's corrections)
$\mu_2\,\mu_3\,\mu_4$	moments about arithmetic mean
$\nu_1\,\nu_2\,\nu_3$	moments about arbitrary origin
π	a constant = 3.141593
ρ	index of correlation (measured on basis or curvilinear line of regression)
ρ	coefficient of correlation from ranks
Σ	sum of
σ	standard deviation
σ'	standard deviation corrected for grouping error
σ_{xy}	standard deviation of values about means of respective columns in correlation table
$\sigma_{\overline{X}}$	standard error of arithmetic mean; the standard error of any statistical measure (σ_{mdn}, σ_r, σ_σ, etc.) is similarly written
$\sigma_{3\cdot12}$	standard deviation of an array of offspring
σ_3	standard deviation of offspring in general
ϕ^2	mean squared contingency
χ	measure of skewness
χ^2	chi square, value used in test for goodness of fit

THE GREEK ALPHABET

A	α	alpha	H	η	eta	N	ν	nu	T	τ	tau
B	β	beta	Θ	θ	theta	Ξ	ξ	xi	Υ	υ	upsilon
Γ	γ	gamma	I	ι	iota	O	o	omicron	Φ	ϕ	phi
Δ	δ	delta	K	κ	kappa	Π	π	pi	X	χ	chi
E	ϵ	epsilon	Λ	λ	lambda	P	ρ	rho	Ψ	ψ	psi
Z	ζ	zeta	M	μ	mu	Σ	σ	sigma	Ω	ω	omega

Technical Appendix I

Proof That Sum of Deviations About Arithmetic Mean Equals Zero

By definition (page 13)

$$x = X - \overline{X}$$

and

$$\Sigma x = \Sigma(X - \overline{X})$$

then

$$\Sigma x = \Sigma X - \Sigma \overline{X}$$

Since $\overline{X}$ is a constant

$$\Sigma \overline{X} = N\overline{X}$$

and

$$\Sigma x = \Sigma X - N\overline{X}$$

By definition (page 12)

$$\overline{X} = \frac{\Sigma X}{N}$$

then

$$N\overline{X} = \Sigma X$$

and therefore

$$\Sigma x = \Sigma X - N\overline{X} = \Sigma X - \Sigma X = 0$$

Technical Appendix II

Derivation of Short Method of Computing Arithmetic Mean

UNGROUPED DATA
Let each value (X) equal

$$X = \overline{Z} + d$$

or the arbitrary starting point plus the deviation of the value from that point.

The total of all values will then be

$$\Sigma X = \Sigma \overline{Z} + \Sigma d$$

but since $\overline{Z}$ is a constant, this may be rewritten

$$\Sigma X = N\overline{Z} + \Sigma d$$

Dividing the total by N to obtain the arithmetic mean the result is

$$\frac{\Sigma X}{N} = \overline{Z} + \frac{\Sigma d}{N}$$

or

$$\overline{X} = \overline{Z} + \frac{\Sigma d}{N}$$

GROUPED DATA
The midpoint of each group may be measured as a deviation from the guessed mean.

$$M.P. = \overline{Z} + d$$

To obtain the total value of all cases in the class interval the midpoint of each group is multiplied by the number of cases in the group

$$f \times MP = f\overline{Z} + fd$$

Totaling up for all class intervals

$$\Sigma(f \times MP) = \Sigma(f\overline{Z}) + \Sigma(fd)$$

or since $\overline{Z}$ is a constant

$$\Sigma(f \times MP) = \overline{Z}\Sigma(f) + \Sigma(fd)$$

and since $\Sigma(f) = N$

$$\Sigma(f \times MP) = N\overline{Z} + \Sigma(fd)$$

Dividing by N to obtain the arithmetic mean

$$\frac{\Sigma(f \times MP)}{N} = \overline{Z} + \frac{\Sigma(fd)}{N}$$

or

$$\overline{X} = \overline{Z} + \frac{\Sigma(fd)}{N}$$

Technical Appendix III

Derivation of Short Formula for Standard Deviation

GROUPED DATA

If d is the deviation of a given point from an arbitrary origin $(\overline{Z})$

$$d = X - \overline{Z}$$

and if the difference between the mean and this origin is termed c

$$c = \overline{X} - \overline{Z}$$

then

$$d - c = (X - \overline{Z}) - (\overline{X} - \overline{Z})$$
$$= X - \overline{Z} - \overline{X} + \overline{Z}$$
$$= X - \overline{X} \text{ or } x$$

where x is the deviation of a value from the arithmetic mean. But

$$\overline{X} = \overline{Z} + \frac{\Sigma(fd)}{N}$$

$$\therefore \overline{X} - \overline{Z} = \frac{\Sigma(fd)}{N} = c$$

Since

$$d - c = x$$
$$d = x + c$$
$$d^2 = x^2 + 2cx + c^2$$

and

$$f(d^2) = f(x^2) + 2cfx + f(c^2)$$

and

$$\Sigma f(d^2) = \Sigma f(x^2) + 2c\Sigma(fx) + c^2 \Sigma f$$
$$= \Sigma f(x^2) + 2c\Sigma(fx) + Nc^2$$

For

$$\Sigma f = N$$

But the sum of the deviations about the arithmetic mean is zero

$$\Sigma(fx) = 0$$

and the formula reduces to

$$\Sigma f(d^2) = \Sigma f(x^2) + Nc^2$$

or

$$\Sigma f(x^2) = \Sigma f(d^2) - Nc^2$$

and

$$\frac{\Sigma f(x^2)}{N} = \frac{\Sigma f(d^2)}{N} - c^2$$

but

$$\sigma = \sqrt{\frac{\Sigma f(x^2)}{N}} \quad \text{(see page 39)}$$

$$\therefore \sigma = \sqrt{\frac{\Sigma f(d^2)}{N} - c^2}$$

but as above

$$c = \frac{\Sigma(fd)}{N}$$

and

$$\sigma = \sqrt{\frac{\Sigma f(d^2)}{N} - \left(\frac{\Sigma fd}{N}\right)^2}$$

UNGROUPED DATA

A simple formula may be arrived at for ungrouped data by selecting zero as the arbitrary origin, so that $\overline{Z} = 0$

$$d = X - \overline{Z} = X$$

and

$$c = \overline{X} - \overline{Z} = \overline{X}$$

since

$$d^2 = x^2 + 2cx + c^2 \quad \text{(see page 256)}$$

and

$$d = X$$

$$\therefore X^2 = x^2 + 2cx + c^2$$

and

$$\Sigma(X^2) = \Sigma(x^2) + 2c\Sigma(x) + Nc^2$$

but

$$\Sigma(x) = 0 \text{ and } c = \overline{X}$$

$$\therefore \Sigma(X^2) = \Sigma(x^2) + N\overline{X}^2$$

and

$$\Sigma(x^2) = \Sigma(X^2) - N\overline{X}^2$$

$$\frac{\Sigma(x^2)}{N} = \frac{\Sigma(X^2)}{N} - (\overline{X}^2) = \frac{\Sigma(X^2)}{N} - \left(\frac{\Sigma X}{N}\right)^2$$

Since

$$\sigma = \sqrt{\frac{\Sigma(x^2)}{N}}$$

$$\therefore \sigma = \sqrt{\frac{\Sigma(X^2)}{N} - \left(\frac{\Sigma X}{N}\right)^2}$$

Technical Appendix IV
Derivation of Normal Equations for Least Squares Straight Line

The formula for any straight line is

$$Y_c = a + bX$$

where Y_c represents the computed or theoretical value for Y obtained by substituting the appropriate value in the formula.

The problem is to determine a line which will fulfill the conditions of the principle of least squares; i.e., the sums of the squares of the deviations of the actual from the theoretical values will be a minimum.

The letter d may be used to represent the difference between the actual and theoretical values. The purpose is then to obtain a line so that $\Sigma(d^2)$ is a minimum. $d = Y - Y_c$ so $\Sigma(Y - Y_c)^2$ must be a minimum. We obtain the partial derivatives with respect to a and b and equate to zero to obtain a minimum.

$$\frac{d(Y - Y_c)^2}{da} = 2\,Na - 2\Sigma(Y) + 2b(X)$$

and

$$\frac{d(Y - Y_c)^2}{db} = 2b\Sigma(X^2) - 2\Sigma(XY) + 2a\Sigma(X)$$

Equating to zero

(I) $2\,Na - 2\Sigma(Y) + 2b\Sigma(X) = 0$

(II) $2\,b\Sigma(X^2) - 2\Sigma(XY) + 2a\Sigma(X) = 0$

or

(I) $\Sigma(Y) = Na + b\Sigma(X)$

(II) $\Sigma(XY) = a\Sigma(X) + b\Sigma(X^2)$

Technical Appendix V

Derivation of Product Moment Formula for Coefficient of Correlation

The original formula for r is

$$r = \sqrt{1 - \frac{S_y^2}{\sigma_y^2}} \qquad (1)$$

where

$$S_y = \sqrt{\frac{\Sigma(d^2)}{N}}$$

$$S_y^2 = \frac{\Sigma(d^2)}{N}$$

Assuming a straight line regression

$$Y_c = a + bX \qquad (2)$$

With Y_c used for the theoretical value obtained from the equation and

$$d = Y - Y_c$$
$$\therefore d = Y - (a + bX)$$

multiplying by d

$$d^2 = dY - ad - bdX \tag{3}$$

Since there is one d for each value a summation is made for all points

$$\Sigma(d^2) = \Sigma(dY) - a\Sigma(d) - b\Sigma(dX) \tag{4}$$

Since the regression line is fitted by the least squares method

$$\Sigma(d) = 0$$
$$\Sigma(dX) = 0$$
$$\therefore \Sigma(d^2) = \Sigma(dY)$$

Multiplying

$$d = Y - a - bX$$

by Y and summing up

$$\Sigma(dY) = \Sigma(Y^2) - a\Sigma(Y) - b\Sigma(XY) \tag{5}$$

but since

$$\Sigma(d^2) = \Sigma(dY)$$
$$\Sigma(d^2) = \Sigma(Y^2) - a\Sigma(Y) - b\Sigma(XY) \tag{6}$$

and

$$\frac{\Sigma(d^2)}{N} = \frac{\Sigma(Y^2) - a\Sigma(Y) - b\Sigma(XY)}{N} \tag{7}$$

but

$$S_y^2 = \frac{\Sigma d^2}{N}$$

$$\therefore S_y^2 = \frac{\Sigma(Y^2) - a\Sigma(Y) - b\Sigma(XY)}{N} \tag{8}$$

and

$$\sigma_y = \sqrt{\frac{\Sigma(Y^2)}{N} - c_y^2} \tag{9}$$

while

$$\sigma_y^2 = \frac{\Sigma(Y^2)}{N} - c_y^2$$

Substituting

$$r^2 = 1 - \frac{S_y^2}{\sigma_y^2}$$

$$r^2 = 1 - \frac{\dfrac{\Sigma(Y^2) - a\Sigma(Y) - b\Sigma(XY)}{N}}{\dfrac{\Sigma(Y^2)}{N} - c_y^2} \tag{10}$$

Multiplying numerator and denominator of fraction by N

$$r^2 = 1 - \frac{\Sigma(Y^2) - a\Sigma(Y) - b\Sigma(XY)}{\Sigma(Y^2) - Nc_y^2}$$

This formula may be reduced to[1]

$$r^2 = \frac{a\Sigma(Y) + b\Sigma(XY) - Nc_y^2}{\Sigma(Y^2) - Nc_y^2} \tag{11}$$

The two normal equations for the line of regression are

(I) $\Sigma(Y) = Na + b\Sigma(X)$

(II) $\Sigma(XY) = a\Sigma(X) + b\Sigma(X^2)$

If the point of averages ($\overline{X}$ and $\overline{Y}$) is used as an origin all values will be reduced to deviations from their respective means (x and y) where

$$x = X - \overline{X}$$

$$y = Y - \overline{Y}$$

so that the equations will read

(I) $\Sigma(y) = Na + b\Sigma(x)$ $\qquad\qquad$ (12)

(II) $\Sigma(xy) = a\Sigma(x) + b\Sigma(x^2)$ $\qquad$ (13)

---

[1] This formula is known as the "least squares" formula for the coefficient of correlation.

but since the sum of the deviations about the arithmetic mean equals zero

$$\Sigma(x) = 0$$
$$\Sigma(y) = 0$$

and the normal equation will reduce to

$$\text{(I)}\ \ Na = 0$$
$$\therefore a = 0$$
$$\text{(II)}\ \ \Sigma(xy) = b\Sigma(x^2)$$
$$\therefore b = \frac{\Sigma(xy)}{\Sigma(x^2)}$$

reducing equation (11) into terms of deviations from the point of averages[1]

$$r^2 = \frac{a\Sigma(y) + b\Sigma(xy) - Nc_y^2}{\Sigma(y^2) - Nc_y^2} \tag{14}$$

but

$$\Sigma(y) = 0$$
$$a\Sigma(y) = 0$$
$$\text{and } c_y = 0$$

The equation thus reduces to

$$r^2 = \frac{b\Sigma(xy)}{\Sigma(y^2)} \tag{15}$$

but from (13)

$$b = \frac{\Sigma(xy)}{\Sigma(x^2)}$$

$$\therefore r^2 = \frac{\Sigma(xy)}{\Sigma(x^2)} \cdot \frac{\Sigma(xy)}{\Sigma(y^2)}$$

$$r^2 = \frac{[\Sigma(xy)]^2}{\Sigma(x^2) \cdot \Sigma(y^2)}$$

---

[1] Since "a" (the Y intercept) equals zero the line will pass through the origin or the point of averages.

Dividing numerator and denominator by N^2

$$r^2 = \frac{\left(\dfrac{\Sigma(xy)}{N}\right)^2}{\dfrac{\Sigma(x^2)}{N} \cdot \dfrac{\Sigma(y^2)}{N}} \tag{16}$$

but

$$\sigma_x^2 = \frac{\Sigma(x^2)}{N} \text{ (see Chapter IV)}$$

$$\sigma_y^2 = \frac{\Sigma(y^2)}{N} \text{ (see Chapter IV)}$$

$$\therefore r = \frac{p}{\sigma_x \; \sigma_y} \tag{17}$$

where

$$p = \frac{\Sigma(xy)}{N}$$

Using the values X and Y as deviations from an arbitrary origin (in the case of ungrouped data, zero, so that the original values may be used) p may be computed from

$$p = \frac{\Sigma(x'y')}{N} - c_x c_y \tag{18}$$

where x' and y' are deviations from arbitrary selected points for

$$x' = x + c_x$$

where c_x is the difference between the true mean and an arbitrary origin $\left(\dfrac{\Sigma(fd)}{N}\right.$ for grouped data and $\dfrac{\Sigma(X)}{N}$ for ungrouped where zero is selected as an origin$\left.\right)$

$$y' = y + c_y$$

$$x'y' = xy + c_x y + c_y x + c_x c_y$$

Summing up for all points

$$\Sigma(x'y') = \Sigma(xy) + c_x \Sigma(y) + c_y \Sigma(x) + N c_x c_y \tag{19}$$

but since the sum of deviations about the means total up to zero

$$\Sigma(y) = 0$$

$$\Sigma(x) = 0$$

and equation (19) reduces to

$$\Sigma(x'y') = \Sigma(xy) + Nc_x c_y$$

Dividing by N

$$\frac{\Sigma(x'y')}{N} = \frac{\Sigma(xy)}{N} + c_x c_y$$

$$\therefore \frac{\Sigma(xy)}{N} = \frac{\Sigma(x'y')}{N} - c_x c_y = p$$

If the arbitrary origin used is zero

$$x' = X$$
$$y' = Y$$

and

$$p = \frac{\Sigma(xy)}{N} = \frac{\Sigma(XY)}{N} - c_x c_y$$

Technical Appendix VI

Derivation of Formula for Line of Regression

Since the regression line is assumed to be straight its formula will be of the type

$$Y = a + bX$$

with the two normal equations (see Chapter VI)

(I) $\Sigma(Y) = Na + b\Sigma(X)$ (1)

(II) $\Sigma(XY) = a\Sigma(X) + b\Sigma(X^2)$

If the origin of the line is assumed to be at the point of averages the normal equations will read

(I) $\Sigma(y) = Na + b\Sigma(x)$

(II) $\Sigma(xy) = \Sigma(x) = b\Sigma(x^2)$

but

$$\Sigma(y) = 0$$
$$\Sigma(x) = 0$$

$$\therefore \text{(I)} \quad Na = 0 \text{ and } a = 0$$

$$\text{(II)} \quad \Sigma(xy) = b\Sigma(x^2) \text{ and } b = \frac{\Sigma(xy)}{\Sigma(x^2)}$$

Equation (1) will reduce to

$$y = bx$$

where

$$b = \frac{\Sigma(xy)}{\Sigma(x^2)}$$

Dividing numerator and denominator by N

$$y = \frac{\Sigma(xy)}{N\frac{\Sigma(x^2)}{N}}x$$

but

$$\frac{\Sigma(x^2)}{N} = \sigma_x^2$$

$$\therefore y = \frac{\Sigma(xy)}{N\sigma_x^2}x$$

but

$$\frac{\Sigma(xy)}{N\sigma_x^2} = \frac{\Sigma(xy)}{N\sigma_x\sigma_y}\frac{\sigma_y}{\sigma_x}$$

and

$$r = \frac{\Sigma(xy)}{N\sigma_x\sigma_y} \quad \text{(product moment formula)}$$

$$\therefore y = r\frac{\sigma_y}{\sigma_x}x$$

Technical Appendix VII

Multiple Correlation Regression

The normal equations for three independent variables, linear correlation with the type formula

$$X_1 = a + b_{12\cdot34} \; X_2 + b_{13\cdot24} \; X_3 + b_{14\cdot23} \; X_4$$

are

(I) $\Sigma(X_1) = Na + b_{13\cdot24}\ \Sigma(X_2) + b_{13\cdot24}\ \Sigma(X_3) + b_{14\cdot23}\ \Sigma(X_4)$

(II) $\Sigma(X_1X_2) = a\Sigma(X_2) + b_{12\cdot34}\ \Sigma(X_2^2) + b_{13\cdot24}\ \Sigma(X_2X_3)$

$$+ b_{14\cdot23}\ \Sigma(X_2X_4)$$

(III) $\Sigma(X_1X_3) = a\Sigma(X_3) + b_{12\cdot34}\ \Sigma(X_2X_3) + b_{13\cdot24}\ \Sigma(X_3^2)$

$$+ b_{14\cdot23}\ \Sigma(X_3X_4)$$

(IV) $\Sigma(X_1X_4) = a\Sigma(X_4) + b_{12\cdot34}\ \Sigma(X_2X_4) + b_{13\cdot24}\ \Sigma(X_3X_4)$

$$+ b_{14\cdot23}\ \Sigma(X_4^2)$$

These equations may be simplified by assuming the origin to be at the point of averages and dividing both sides of the equations by N

(I) $\dfrac{\Sigma(x_1)}{N} = a + b_{12\cdot34}\ \dfrac{\Sigma(x_2)}{N} + b_{13\cdot24}\ \dfrac{\Sigma(x_3)}{N} + b_{14\cdot23}\ \dfrac{\Sigma(x_4)}{N}$

(II) $\dfrac{\Sigma(x_1x_2)}{N} = \dfrac{a\Sigma(x_2)}{N} + b_{12\cdot34}\ \dfrac{\Sigma(x_2^2)}{N} +$

$$b_{13\cdot24}\ \dfrac{\Sigma(x_2x_3)}{N} + b_{14\cdot23}\ \dfrac{\Sigma(x_2x_4)}{N}$$

(III) $\dfrac{\Sigma(x_1x_3)}{N} = \dfrac{a\Sigma(x_3)}{N} + b_{12\cdot34}\ \dfrac{\Sigma(x_2x_3)}{N} +$

$$b_{13\cdot24}\ \dfrac{\Sigma(x_3^2)}{N} + b_{14\cdot23}\ \dfrac{\Sigma(x_3x_4)}{N}$$

(IV) $\dfrac{\Sigma(x_1x_4)}{N} = \dfrac{a\Sigma(x_4)}{N} + b_{12\cdot34}\ \dfrac{\Sigma(x_2x_4)}{N} +$

$$b_{13\cdot24}\ \dfrac{\Sigma(x_3x_4)}{N} + b_{14\cdot32}\ \dfrac{\Sigma(x_4^2)}{N}$$

where x_1, x_2, x_3, x_4, represent deviations from the respective means, $\overline{X}_1, \overline{X}_2, \overline{X}_3,$ and $\overline{X}_4$. Since the sum of the deviations about the arithmetic mean is zero

$$\frac{\Sigma(x_1)}{N} = 0,\ \frac{\Sigma(x_2)}{N} = 0,\ \frac{\Sigma(x_3)}{N} = 0,\ \frac{\Sigma(x_4)}{N} = 0$$

and[1]

$$\sigma_2 = \sqrt{\frac{\Sigma(x_2^2)}{N}}, \sigma_3 = \sqrt{\frac{\Sigma(x_3^2)}{N}}, \sigma_4 = \sqrt{\frac{\Sigma(x_4^2)}{N}}$$

or

$$\sigma_2^2 = \frac{\Sigma(x_2^2)}{N}, \sigma_3^2 = \frac{\Sigma(x_3^2)}{N}, \sigma_4^2 = \frac{\Sigma(x_4^2)}{N}$$

while

$$\frac{\Sigma(x_1 x_2)}{N} = p_{12} \text{ (the product moment)}$$

$$\frac{\Sigma(x_1 x_3)}{N} = p_{13} \text{ etc.}$$

where the value of the product moment may be computed from

$$p_{12} = \frac{\Sigma(X_1 X_2)}{N} - \frac{\Sigma(X_1)}{N} \frac{\Sigma(X_2)}{N}$$

The normal equations will now read

$$p_{12} = b_{12 \cdot 34} \, \sigma_2^2 + b_{13 \cdot 24} \, p_{23} + b_{14 \cdot 23} \, p_{24}$$

$$p_{13} = b_{12 \cdot 34} \, p_{23} + b_{13 \cdot 24} \, \sigma_3^2 + b_{14 \cdot 23} \, p_{34}$$

$$p_{14} = b_{12 \cdot 34} \, p_{24} + b_{13 \cdot 24} \, p_{34} + b_{14 \cdot 23} \, \sigma_4^2$$

Technical Appendix VIII

Standard Error of Estimate— Multiple Correlation

The formula for the standard error for multiple correlation may be derived in the same fashion as the least squares formula for simple correlation

$$S_{1 \cdot 23}^2 = \frac{\Sigma(X_1^2) - b_{12 \cdot 34} \, \Sigma(X_1 X_2) - b_{13 \cdot 24} \, \Sigma(X_1 X_3) - b_{14 \cdot 23} \, \Sigma(X_1 X_4)}{N}$$

---

[1] See page 263.

By reducing this to deviations from the respective means this will read

$$S_{1\cdot23}^2 = \frac{\Sigma(x_1^2)}{N} - b_{12\cdot34}\frac{\Sigma(x_1 x_2)}{N} - b_{13\cdot24}\frac{\Sigma(x_1 x_3)}{N} - b_{14\cdot23}\frac{\Sigma(x_1 x_4)}{N}$$

or

$$S_{1\cdot23}^2 = \sigma_1^2 - b_{12\cdot34}\, p_{12} - b_{13\cdot24}\, p_{13} - b_{14\cdot23}\, p_{14}$$

Technical Appendix IX

Derivation[1] of Standard Error of the Arithmetic Mean

N random samples of n items each are drawn and the individual values expressed as deviations from the true arithmetic mean of the universe. This may be written as follows:

Item Number	Sample #1	Sample #2	Sample #3	Sample N
1	x'	x'	x'	x'
2	x''	x''	x''	x''
3	x'''	x'''	x'''	x'''
....	....	....	....	....
....	....	....	....	....
n	x_n	x_n	x_n	x_n
	Σx_1	Σx_2	Σx_3	Σx_4

If items #1 and #2 of each sample are added

$$x_{1+2} = x' + x''$$

But the standard deviation is equal to

$$\sigma = \sqrt{\frac{\Sigma(x^2)}{N}}$$

or in this instance

$$\sigma = \sqrt{\frac{\Sigma(x^2_{1+2})}{N}}$$

---

[1] This derivation follows Ezekial's in *Methods of Correlation Analysis*.

or

$$\sigma_{1+2}{}^2 = \frac{\Sigma(x^2{}_{1+2})}{N}$$

but

$$x_{1+2}{}^2 = (x' + x'')^2 =$$
$$(x')^2 + 2(x'x'') + (x'')^2$$

and

$$\Sigma(x_{1+2}{}^2) = \Sigma(x'^2) + 2\Sigma(x'x'') + \Sigma(x''^2)$$

Since the successive items of the sample are drawn at random they are uncorrelated ($r = o$) and therefore $\Sigma(x'x'') = 0$.

$$\therefore \Sigma(x^2{}_{1+2}) = \Sigma(x'^2) + \Sigma(x''^2)$$

or after dividing by N

$$\sigma^2{}_{1+2} = \sigma_x{}'^2 + \sigma_x{}''^2$$

But as the number (N) of the samples is increased $\sigma_x{}'$ will tend to approach the standard deviation of the universe from which the samples were drawn as will in a similar manner $\sigma_x{}''$

or $\qquad \sigma_x{}' = \sigma_x$

and $\qquad \sigma_x{}'' = \sigma_x$ $\qquad\qquad$ when N is very large

$$\therefore \sigma_{1+2}{}^2 = 2\sigma_x{}^2$$

For the sum of the first three items

$$\sigma^2{}_{1+2+3} = \sigma_x{}'^2 + \sigma_x{}''^2 = \sigma_x{}'''^2 = 3\sigma_x{}^2 \qquad \text{when } N \text{ is large}$$

And for the sum of n items

$$\sigma_{1+2+3}{}^2 \ldots n = \sigma_x{}'^2 + \sigma_x{}''^2 + \ldots + \sigma_{xn}{}^2 = n\sigma_x{}^2$$

when n is large

Dividing all items and totals by n gives

$$\sigma^2{}_{\underset{n}{x'}} = \frac{\sigma_x{}'^2}{n^2}$$

and since $\sigma_x{}'$ tends to equal o_x

$$\sigma_{\underset{n}{x'}}{}^2 = \frac{\sigma_x{}^2}{n^2} \text{—} \qquad\qquad \text{when } N \text{ is large}$$

and for x' and x''

$$\frac{\sigma_{x'}}{n} + \frac{x''^2}{n} = \frac{\sigma_x^2}{n^2} + \frac{\sigma_x^2}{n^2} = 2\frac{\sigma_x^2}{n^2}$$

or for the sum of n items

$$\frac{\sigma_{x'}}{n} + \frac{x'}{n} + .. + \frac{x n^2}{n} = \frac{\sigma_x^2}{n^2} + \frac{\sigma_x^2}{n^2} + ... + \frac{\sigma_x^2}{n^2} = \frac{n\sigma_x^2}{n^2} = \frac{\sigma_x^2}{n}$$

but since $\Sigma\left(\dfrac{x}{n}\right)$ for each sample $= \overline{X}$

$$\sigma_{\bar{x}}^2 = \frac{\sigma_x^2}{n}$$

$$\sigma_{\bar{x}} = \frac{\sigma_x}{\sqrt{n}}$$

where

σ_x is the standard deviation of the *universe* and not of the sample. However lacking this value the standard deviation of the sample (σ) is used as an estimate of this value.

Technical Appendix X

Proof That $\Sigma(x^2)$ Is a Minimum

If $\overline{Z}$ is a value other than the arithmetic mean ($\overline{X}$) for a specified set of data, it is shown that $\Sigma(X - \overline{X})^2 < \Sigma(X - \overline{Z})^2$ as follows.

$$d = X - \overline{Z}$$
$$x = X - \overline{X}$$

and

$$\Sigma d^2 = \Sigma(X - \overline{Z})^2$$
$$= \Sigma(X^2 - 2\overline{Z}X + \overline{Z}^2)$$
$$= \Sigma X^2 - 2\overline{Z}\Sigma X + \Sigma \overline{Z}^2$$

and since

$$\Sigma X = N\frac{\Sigma X}{N} = N\overline{X}$$
$$\Sigma \overline{Z}^2 = N\overline{Z}^2 \text{ as } \overline{Z} \text{ is a constant}$$

Then
$$\Sigma d^2 = \Sigma X^2 - 2\overline{Z}N\overline{X} + N\overline{Z}^2$$
$$= \Sigma X^2 - 2N\overline{Z}\overline{X} + N\overline{Z}^2$$

which may be written also as
$$\Sigma d^2 = \Sigma X^2 - 2N\overline{Z}\overline{X} + N\overline{Z}^2 + N\overline{X}^2 - N\overline{X}^2$$

or
$$\Sigma d^2 = \Sigma X^2 - N\overline{X}^2 + N\overline{Z}^2 - 2N\overline{Z}\overline{X} + N\overline{X}^2$$
$$= \Sigma X^2 - N\overline{X}^2 + N(\overline{Z}^2 - 2N\overline{Z}\overline{X} + \overline{X}^2)$$
$$= \Sigma X^2 - N\overline{X}^2 + N(\overline{Z} - \overline{X})^2$$

Thus if $\overline{X} = \overline{Z}$ then
$$\Sigma d^2 = \Sigma x^2 = \Sigma X^2 - N\overline{X}^2$$

But if
$$\overline{X} \neq \overline{Z}$$

then
$$\Sigma d^2 = \Sigma x^2 + \Sigma(\overline{Z} - \overline{X})^2$$

Let
$$\overline{Z} - \overline{X} = D$$

then
$$\Sigma d^2 = \Sigma x^2 + \Sigma(D^2)$$

and
$$\Sigma d^2 > \Sigma x^2$$

Therefore, the sum of the squares of the deviations about the arithmetic mean, $\Sigma(X - \overline{X})^2$ will be less than the sum of the squares of the deviations about any other value $\Sigma(X - \overline{Z})^2$.

Technical Appendix XI

Derivation of Formula for Combining Standard Deviations of Subgroups

By definition
$$\sigma = \sqrt{\frac{\Sigma(x^2)}{N}} = \sqrt{\frac{\Sigma(X - \overline{X})^2}{N}}$$

and

$$\sigma^2 = \frac{\Sigma(X - \overline{X})^2}{N}$$

When the observations are split in subgroups

$$X - \overline{X} = (X - \overline{X}_i) + (\overline{X}_i - \overline{X})$$

where

$\overline{X}$ = over-all arithmetic mean

$\overline{X}_i$ = mean of each subgroup

then

$$\Sigma(X - \overline{X})^2 = \Sigma\Sigma\,[(X - \overline{X}_i) + (\overline{X}_i - \overline{X})]^2$$
$$= \Sigma\Sigma\,[(X - \overline{X}_i)^2 + 2(\overline{X}_i - \overline{X})(X - \overline{X}_i) + (\overline{X}_i - \overline{X})^2]$$

where the double summation signifies totals within groups and then for all groups.

Therefore,[1]

$$\Sigma(X - \overline{X})^2 = \Sigma\,[\Sigma(X - \overline{X}_i)^2 + 2(\overline{X}_i - \overline{X})\,\Sigma(\overline{X} - \overline{X}_i) + \Sigma(\overline{X}_i - \overline{X})^2]$$

Since the sum of the deviations about the arithmetic mean equals zero (see Technical Appendix I)

$$\Sigma(X - \overline{X}_i) = 0$$

and the formula reduces to

$$\Sigma(X - \overline{X})^2 = \Sigma\,[\Sigma(X - \overline{X}_i)^2 + \Sigma(\overline{X}_i - \overline{X})^2]$$

Further for each group

$$\sigma_i^2 = \frac{\Sigma(X - \overline{X}_i)^2}{N_i}$$

and

$$N_i\sigma_i = \Sigma(X - \overline{X}_i)^2$$

Thus

$$\Sigma(X - \overline{X})^2 = \Sigma\,[N_i\sigma_i^2 + \Sigma(X_i - \overline{X})^2]$$

Since for each group $(\overline{X}_i - \overline{X})^2$ is computed for each observation in the group

$$\Sigma(\overline{X}_i - \overline{X})^2 = N_i(\overline{X}_i - \overline{X})^2$$

[1]Note that for each group $(\overline{X}_i - \overline{X})$ is a constant and may thus be placed outside the summation sign.

and

$$\Sigma(X - \overline{X})^2 = \Sigma\,[N_i\sigma_i{}^2 + N_i(\overline{X}_i - \overline{X})^2\,]$$

$$= \Sigma(N_i\sigma_i{}^2) + \Sigma\,[N_i(\overline{X}_i - \overline{X})^2\,]$$

Further

$$N\sigma^2 = \Sigma(X - \overline{X})^2$$

Therefore

$$N\sigma^2 = \Sigma(N_i\sigma_i{}^2) + \Sigma\,[N_i(\overline{X}_i - \overline{X})^2\,]$$

Technical Appendix XII
Proof of Formula for Standard Error of a Proportion

Consider a population of N observations, some of which have a characteristic A while the balance do not have the characteristic $(\overline{A})$.

Assign a value of $X = 1$ to each observation with characteristic A and a zero to each observation without characteristic A.

Then

$$\Sigma X = \text{the number of observations } A$$

since all observations $\overline{A}$ have a value of zero and all observations A have a value of one, and $1^2 = 1$,

and

$$\overline{X} = \frac{\Sigma X}{N}$$

Further, p, the percentage of the observations with characteristic A, is

$$p = \frac{\Sigma X}{N} = \frac{\Sigma X^2}{N}$$

while q, the percent without this characteristic is

$$q = 1 - p$$

The standard deviation may be computed from

$$\sigma = \sqrt{\frac{\Sigma(X^2)}{N} - \left(\frac{\Sigma X}{N}\right)^2}$$

$$\sigma = \sqrt{\frac{\Sigma X^2}{N} - \overline{X}^2} = \sqrt{\frac{\Sigma X^2 - N\overline{X}^2}{N}}$$

Thus,

$$\sigma = \sqrt{\frac{Np - Np^2}{N}}$$

Since

$$\sigma = \sqrt{\frac{N(p - p^2)}{N}} = \sqrt{p - p^2}$$

$$\sigma = \sqrt{p(1 - p)}$$

$$\sigma = \sqrt{pq}$$

Since

$$\sigma_{\overline{x}} = \frac{\sigma}{\sqrt{n}}$$

then, where n is the sample size

$$\sigma_{\overline{x}} = \frac{\sqrt{pq}}{\sqrt{n}} = \sqrt{\frac{pq}{n}}$$

and since in this case

$$\overline{x} = p = \%$$

$$\sigma_\% = \sqrt{\frac{pq}{n}}$$

Technical Appendix XIII

Formulae for Sums of Squares for Analysis of Variance

By definition

$$\sigma^2 = \frac{\Sigma(x^2)}{n} = \frac{\Sigma(X - \overline{X})^2}{n}$$

then

$$n\sigma^2 = \Sigma(X - \overline{X})^2$$

or sum of squares of deviations from arithmetic mean.

But (see Technical Appendix III)

$$\sigma^2 = \frac{\Sigma(X^2)}{n} - \left(\frac{\Sigma X}{n}\right)^2$$

and

$$n\sigma^2 = \Sigma(X^2) - \frac{(\Sigma X)^2}{n}$$

Let $\Sigma X = T$

$$n\sigma^2 = SS_{\text{Total}} = \Sigma(X^2) - \frac{T^2}{n}$$

which is the sum of squares for the total (over-all).
For the "within" variability:

$$SS_{\text{Within}} = \Sigma\left[n_i\sigma_i^2\right]$$

$$SS_{\text{Within}} = \Sigma\left[\Sigma(X^2) - \frac{T_i^2}{n_i}\right]$$

$$SS_{\text{Within}} = \Sigma\Sigma_i(X^2) - \Sigma\left(\frac{T_i^2}{n_i}\right)$$

$$= \Sigma(X^2) - \Sigma\left(\frac{T_i^2}{n_i}\right)$$

For the "among" variability:

It has been shown (see Technical Appendix XI) that

$$N\sigma^2 = \Sigma(n_i\sigma_i^2) + \Sigma\left[n_i(\overline{X}_i - \overline{X})^2\right]$$

but

$$n\sigma^2 = SS_{\text{Total}}$$

$$\Sigma(n_i\sigma_i^2) = SS_{\text{Within}}$$

$$\Sigma\left[n_i(\overline{X}_i - \overline{X})^2\right] = SS_{\text{Among}}$$

Therefore,

$$SS_{\text{Total}} = SS_{\text{Among}} + SS_{\text{Within}}$$

and

$$SS_{\text{Among}} = SS_{\text{Total}} - SS_{\text{Within}}$$

thus,

$$SS_{\text{Among}} = \left[\Sigma(X^2) - \frac{T^2}{n}\right] - \left[\Sigma(X^2) - \Sigma\frac{T_i^2}{n_i}\right]$$

$$= \Sigma(X^2) - \frac{T^2}{n} - \Sigma(X^2) + \Sigma\left(\frac{T_i^2}{n_i}\right)$$

$$= \Sigma\left(\frac{T_i^2}{n_i}\right) - \frac{T^2}{n}$$

Tables of Squares,
Square Roots, Cubes,
Cube Roots, and Logarithms

EXAMPLES OF HOW TO USE THE TABLES

1. To find the **square** of 490. On page 280, look down the extreme left-hand column till you find 490. Read the number opposite it in the first column to the right (headed by 0). The square of 490 is 240100.

2. To find the **square** of 493. On page 280, find 490 in the left-hand column. Read the number opposite it in the fourth column to the right (headed by 3). The square of 493 is 243049.

3. To find the **square root** of 1100. On page 282, find 1100 in the extreme left-hand column. Read the number opposite it in the first column to the right (headed by 0 and $\sqrt{N}$). This number is 3.1662. But at the top of this part of the column you will note the figure 30 +. Add 30 to the number you have found. The square root of 1100 is 33.1662.

4. To find the **square root** of 11000. On page 282, find 1100 in the left-hand column. Read the number in the second part of the first column to the right (headed by 0 and $\sqrt{10\,N}$). This number is 4.8809. But at the top of this part of the column you will note the figure 100 +. Add 100 to the number you have found. The square root of 11000 is 104.8809.

5. To find the **square root** of 1600. On page 284, find 1600 in the left-hand column, and read *0.0000 opposite it in the first column to the right. The asterisk means that the figure to be added has changed. Instead of adding the 30 which appears at the top of this column, add the 40 which appears at the bottom of the column (for all roots given after the asterisk has been used, up to the next asterisk which appears on page 288). The square root of 1600 is 40.0000.

6. To find the **cube** of 248. On page 324, find 248 in the left-hand column and read the number opposite it in the column headed n^3. The cube of 248 is 15252992.

7. To find the **cube root** of 2480. On page 324, find 248 in the left-hand column and read the number opposite it in the column headed $\sqrt[3]{10\,n}$. The cube root of 2480 is 13.53580.

8. If a decimal point is included in the number, separate it into triads on both sides of the decimal point and use the following table to select the appropriate column:

First Triad	First Figure of Cube Root
000-007	1
008-026	2
027-063	3
064-124	4
125-215	5
216-342	6
343-511	7
512-728	8
729-999	9

For example:

N	Triads	First Triad	Use	Cube Root
825.	825.	825	$\sqrt[3]{N}$	9.378887
82.5	082.500	082	$\sqrt[3]{100N}$	4.353294
8.25	008.250	008	$\sqrt[3]{10N}$	2.020620
.825	.825	825	$\sqrt[3]{N}$	.9378887

Decimal Point. The position of the decimal point is determined in the usual way. The following examples will indicate the method.

SQUARES

$(.03)^2 = .0009$
$(.3)^2 = .09$
$(3.)^2 = 9.$
$(30.)^2 = 900.$

SQUARE ROOTS

$\sqrt{.0009} = .03$
$\sqrt{.009} = .0948683$
$\sqrt{.09} = .3$
$\sqrt{.9} = .948683$
$\sqrt{9.} = 3.$
$\sqrt{90.} = 9.48683$
$\sqrt{900.} = 30.$
$\sqrt{9000.} = 94.8683$

CUBES

$(.02)^2 = .000008$
$(.2)^3 = .008$
$(2.)^3 = 8.$
$(20.)^3 = 8000.$

CUBE ROOTS

$\sqrt[3]{.000008} = .02$
$\sqrt[3]{.008} = .2$
$\sqrt[3]{8.} = 2.$
$\sqrt[3]{8000.} = 20.$

9. To find the **logarithm** of 413. On page 332 find 41 in the left-hand column. Read the number opposite it in the fourth column to the right (headed by 3). This (.61595) is the fractional part of the logarithm. It is preceded by a number one less than the number of digits before the decimal point, in this case 2. So the logarithm is 2.61595.

Table of Squares

	N² 0	N² 1	N² 2	N² 3	N² 4	N² 5	N² 6	N² 7	N² 8	N² 9
100	10000	10201	10404	10609	10816	11025	11236	11449	11664	11881
110	12100	12321	12544	12769	12996	13225	13456	13689	13924	14161
120	14400	14641	14884	15129	15376	15625	15876	16129	16384	16641
130	16900	17161	17424	17689	17956	18225	18496	18769	19044	19321
140	19600	19881	20164	20449	20736	21025	21316	21609	21904	22201
150	22500	22801	23104	23409	23716	24025	24336	24649	24964	25281
160	25600	25921	26244	26569	26896	27225	27556	27889	28224	28561
170	28900	29241	29584	29929	30276	30625	30976	31329	31684	32041
180	32400	32761	33124	33489	33856	34225	34596	34969	35344	35721
190	36100	36481	36864	37249	37636	38025	38416	38809	39204	39601
200	40000	40401	40804	41209	41616	42025	42436	42849	43264	43681
210	44100	44521	44944	45369	45796	46225	46656	47089	47524	47961
220	48400	48841	49284	49729	50176	50625	51076	51529	51984	52441
230	52900	53361	53824	54289	54756	55225	55696	56169	56644	57121
240	57600	58081	58564	59049	59536	60025	60516	61009	61504	62001
250	62500	63001	63504	64009	64516	65025	65536	66049	66564	67081
260	67600	68121	68644	69169	69696	70225	70756	71289	71824	72361
270	72900	73441	73984	74529	75076	75625	76176	76729	77284	77841
280	78400	78961	79524	80089	80656	81225	81796	82369	82944	83521
290	84100	84681	85264	85849	86436	87025	87616	88209	88804	89401
300	90000	90601	91204	91809	92416	93025	93636	94249	94864	95481
310	96100	96721	97344	97969	98596	99225	99856	100489	101124	101761
320	102400	103041	103684	104329	104976	105625	106276	106929	107584	108241
330	108900	109561	110224	110889	111556	112225	112896	113569	114244	114921
340	115600	116281	116964	117649	118336	119025	119716	120409	121104	121801
350	122500	123201	123904	124609	125316	126025	126736	127449	128164	128881
360	129600	130321	131044	131769	132496	133225	133956	134689	135424	136161
370	136900	137641	138384	139129	139876	140625	141376	142129	142884	143641
380	144400	145161	145924	146689	147456	148225	148996	149769	150544	151321
390	152100	152881	153664	154449	155236	156025	156816	157609	158404	159201
400	160000	160801	161604	162409	163216	164025	164836	165649	166464	167281
410	168100	168921	169744	170569	171396	172225	173056	173889	174724	175561
420	176400	177241	178084	178929	179776	180625	181476	182329	183184	184041
430	184900	185761	186624	187489	188356	189225	190096	190969	191844	192721
440	193600	194481	195364	196249	197136	198025	198916	199809	200704	201601
450	202500	203401	204304	205209	206116	207025	207936	208849	209764	210681
460	211600	212521	213444	214369	215296	216225	217156	218089	219024	219961
470	220900	221841	222784	223729	224676	225625	226576	227529	228484	229441
480	230400	231361	232324	233289	234256	235225	236196	237169	238144	239121
490	240100	241081	242064	243049	244036	245025	246016	247009	248004	249001
500	250000	251001	252004	253009	254016	255025	256036	257049	258064	259081
510	260100	261121	262144	263169	264196	265225	266256	267289	268324	269361
520	270400	271441	272484	273529	274576	275625	276676	277729	278784	279841
530	280900	281961	283024	284089	285156	286225	287296	288369	289444	290521
540	291600	292681	293764	294849	295936	297025	298116	299209	300304	301401

Table of Squares

	N^2 0	N^2 1	N^2 2	N^2 3	N^2 4	N^2 5	N^2 6	N^2 7	N^2 8	N^2 9
550	302500	303601	304704	305809	306916	308025	309136	310249	311364	312481
560	313600	314721	315844	316969	318096	319225	320356	321489	322624	323761
570	324900	326041	327184	328329	329476	330625	331776	332929	334084	335241
580	336400	337561	338724	339889	341056	342225	343396	344569	345744	346921
590	348100	349281	350464	351649	352836	354025	355216	356409	357604	358801
600	360000	361201	362404	363609	364816	366025	367236	368449	369664	370881
610	372100	373321	374544	375769	376996	378225	379456	380689	381924	383161
620	384400	385641	386884	388129	389376	390625	391876	393129	394384	395641
630	396900	398161	399424	400689	401956	403225	404496	405769	407044	408321
640	409600	410881	412164	413449	414736	416025	417316	418609	419904	421201
650	422500	423801	425104	426409	427716	429025	430336	431649	432964	434281
660	435600	436921	438244	439569	440896	442225	443556	444889	446224	447561
670	448900	450241	451584	452929	454276	455625	456976	458329	459684	461041
680	462400	463761	465124	466489	467856	469225	470596	471969	473344	474721
690	476100	477481	478864	480249	481636	483025	484416	485809	487204	488601
700	490000	491401	492804	494209	495616	497025	498436	499849	501264	502681
710	504100	505521	506944	508369	509796	511225	512656	514089	515524	516961
720	518400	519841	521284	522729	524176	525625	527076	528529	529984	531441
730	532900	534361	535824	537289	538756	540225	541696	543169	544644	546121
740	547600	549081	550564	552049	553536	555025	556516	558009	559504	561001
750	562500	564001	565504	567009	568516	570025	571536	573049	574564	576081
760	577600	579121	580644	582169	583696	585225	586756	588289	589824	591361
770	592900	594441	595984	597529	599076	600625	602176	603729	605284	606841
780	608400	609961	611524	613089	614656	616225	617796	619369	620944	622521
790	624100	625681	627264	628849	630436	632025	633616	635209	636804	638401
800	640000	641601	643204	644809	646416	648025	649636	651249	652864	654481
810	656100	657721	659344	660969	662596	664225	665856	667489	669124	670761
820	672400	674041	675684	677329	678976	680625	682276	683929	685584	687241
830	688900	690561	692224	693889	695556	697225	698896	700569	702244	703921
840	705600	707281	708964	710649	712336	714025	715716	717409	719104	720801
850	722500	724201	725904	727609	729316	731025	732736	734449	736164	737881
860	739600	741321	743044	744769	746496	748225	749956	751689	753424	755161
870	756900	758641	760384	762129	763876	765625	767376	769129	770884	772641
880	774400	776161	777924	779689	781456	783225	784996	786769	788544	790321
890	792100	793881	795664	797449	799236	801025	802816	804609	806404	808201
900	810000	811801	813604	815409	817216	819025	820836	822649	824464	826281
910	828100	829921	831744	833569	835396	837225	839056	840889	842724	844561
920	846400	848241	850084	851929	853776	855625	857476	859329	861184	863041
930	864900	866761	868624	870489	872356	874225	876096	877969	879844	881721
940	883600	885481	887364	889249	891136	893025	894916	896809	898704	900601
950	902500	904401	906304	908209	910116	912025	913936	915849	917764	919681
960	921600	923521	925444	927369	929296	931225	933156	935089	937024	938961
970	940900	942841	944784	946729	948676	950625	952576	954529	956484	958441
980	960400	962361	964324	966289	968256	970225	972196	974169	976144	978121
990	980100	982081	984064	986049	988036	990025	992016	994009	996004	998001

Tables of Square Roots

N	√N 0	√10N	√N 1	√10N	√N 2	√10N	√N 3	√10N	√N 4	√10N
	30+	100+	30+	100+	30+	100+	30+	100+	30+	100+
1000	1.6228	0.0000	1.6386	0.0500	1.6544	0.1000	1.6702	0.1499	1.6860	0.1998
1010	1.7805	0.4988	1.7962	0.5485	1.8119	0.5982	1.8277	0.6479	1.8434	0.6976
1020	1.9374	0.9950	1.9531	1.0445	1.9687	1.0940	1.9844	1.1435	2.0000	1.1929
1030	2.0936	1.4889	2.1092	1.5382	2.1248	1.5874	2.1403	1.6366	2.1559	1.6858
1040	2.2490	1.9804	2.2645	2.0294	2.2800	2.0784	2.2955	2.1274	2.3110	2.1763
1050	2.4037	2.4695	2.4191	2.5183	2.4345	2.5671	2.4500	2.6158	2.4654	2.6645
1060	2.5576	2.9563	2.5730	3.0049	2.5883	3.0534	2.6037	3.1019	2.6190	3.1504
1070	2.7109	3.4408	2.7261	3.4891	2.7414	3.5374	2.7567	3.5857	2.7719	3.6340
1080	2.8634	3.9230	2.8786	3.9712	2.8938	4.0192	2.9090	4.0673	2.9242	4.1153
1090	3.0151	4.4031	3.0303	4.4510	3.0454	4.4988	3.0606	4.5466	3.0757	4.5945
1100	3.1662	4.8809	3.1813	4.9285	3.1964	4.9762	3.2114	5.0238	3.2265	5.0714
1110	3.3167	5.3565	3.3317	5.4040	3.3467	5.4514	3.3617	5.4988	3.3766	5.5462
1120	3.4664	5.8301	3.4813	5.8773	3.4963	5.9245	3.5112	5.9717	3.5261	6.0189
1130	3.6155	6.3015	3.6303	6.3485	3.6452	6.3955	3.6601	6.4425	3.6749	6.4894
1140	3.7639	6.7708	3.7787	6.8176	3.7935	6.8644	3.8083	6.9112	3.8231	6.9579
1150	3.9117	7.2381	3.9264	7.2847	3.9411	7.3313	3.9559	7.3778	3.9706	7.4244
1160	4.0588	7.7033	4.0735	7.7497	4.0881	7.7961	4.1028	7.8425	4.1174	7.8888
1170	4.2053	8.1665	4.2199	8.2128	4.2345	8.2589	4.2491	8.3051	4.2637	8.3513
1180	4.3511	8.6278	4.3657	8.6738	4.3802	8.7198	4.3948	8.7658	4.4093	8.8118
1190	4.4964	9.0871	4.5109	9.1329	4.5254	9.1788	4.5398	9.2245	4.5543	9.2703
1200	4.6410	9.5445	4.6554	9.5901	4.6699	9.6358	4.6843	9.6814	4.6987	9.7269
1210	4.7851	10.0000	4.7994	10.0454	4.8138	10.0909	4.8282	10.1363	4.8425	10.1817
1220	4.9285	10.4536	4.9428	10.4989	4.9571	10.5441	4.9714	10.5893	4.9857	10.6345
1230	5.0714	10.9054	5.0856	10.9504	5.0999	10.9955	5.1141	11.0405	5.1283	11.0856
1240	5.2136	11.3553	5.2278	11.4002	5.2420	11.4451	5.2562	11.4899	5.2704	11.5347
1250	5.3553	11.8034	5.3695	11.8481	5.3836	11.8928	5.3977	11.9375	5.4119	11.9821
1260	5.4965	12.2497	5.5106	12.2943	5.5246	12.3388	5.5387	12.3833	5.5528	12.4278
1270	5.6371	12.6943	5.6511	12.7386	5.6651	12.7830	5.6791	12.8273	5.6931	12.8716
1280	5.7771	13.1371	5.7911	13.1813	5.8050	13.2254	5.8190	13.2696	5.8329	13.3137
1290	5.9166	13.5782	5.9305	13.6222	5.9444	13.6662	5.9583	13.7102	5.9722	13.7541
1300	6.0555	14.0175	6.0694	14.0614	6.0832	14.1052	6.0971	14.1490	6.1109	14.1928
1310	6.1939	14.4552	6.2077	14.4989	6.2215	14.5426	6.2353	14.5862	6.2491	14.6298
1320	6.3318	14.8913	6.3456	14.9348	6.3593	14.9783	6.3731	15.0217	6.3868	15.0652
1330	6.4692	15.3256	6.4829	15.3690	6.4966	15.4123	6.5103	15.4556	6.5240	15.4989
1340	6.6060	15.7584	6.6197	15.8016	6.6333	15.8447	6.6470	15.8879	6.6606	15.9310
1350	6.7423	16.1895	6.7560	16.2325	6.7696	16.2755	6.7831	16.3185	6.7967	16.3615
1360	6.8782	16.6190	6.8917	16.6619	6.9053	16.7048	6.9188	16.7476	6.9324	16.7904
1370	7.0135	17.0470	7.0270	17.0897	7.0405	17.1324	7.0540	17.1751	7.0675	17.2177
1380	7.1484	17.4734	7.1618	17.5160	7.1753	17.5585	7.1887	17.6010	7.2022	17.6435
1390	7.2827	17.8983	7.2961	17.9407	7.3095	17.9831	7.3229	18.0254	7.3363	18.0678
1400	7.4166	18.3216	7.4299	18.3638	7.4433	18.4061	7.4566	18.4483	7.4700	18.4905
1410	7.5500	18.7434	7.5633	18.7855	7.5766	18.8276	7.5899	18.8697	7.6032	18.9117
1420	7.6829	19.1638	7.6962	19.2057	7.7094	19.2476	7.7227	19.2896	7.7359	19.3315
1430	7.8153	19.5826	7.8286	19.6244	7.8418	19.6662	7.8550	19.7080	7.8682	19.7497
1440	7.9473	20.0000	7.9605	20.0417	7.9737	20.0833	7.9868	20.1249	8.0000	20.1666
	30+	100+	30+	100+	30+	100+	30+	100+	30+	100+

√N	√10N	√N	√10N	√N	√10N	√N	√10N	√N	√10N	N
	5		6		7		8		9	
30+	100+	30+	100+	30+	100+	30+	100+	30+	100+	
1.7017	0.2497	1.7175	0.2996	1.7333	0.3494	1.7490	0.3992	1.7648	0.4490	1000
1.8591	0.7472	1.8748	0.7968	1.8904	0.8464	1.9061	0.8960	1.9218	0.9455	1010
2.0156	1.2423	2.0312	1.2917	2.0468	1.3410	2.0624	1.3903	2.0780	1.4396	1020
2.1714	1.7350	2.1870	1.7841	2.2025	1.8332	2.2180	1.8823	2.2335	1.9313	1030
2.3265	2.2252	2.3419	2.2741	2.3574	2.3230	2.3728	2.3719	2.3883	2.4207	1040
2.4808	2.7132	2.4962	2.7619	2.5115	2.8105	2.5269	2.8591	2.5423	2.9077	1050
2.6343	3.1988	2.6497	3.2473	2.6650	3.2957	2.6803	3.3441	2.6956	3.3925	1060
2.7872	3.6822	2.8024	3.7304	2.8177	3.7786	2.8329	3.8268	2.8481	3.8749	1070
2.9393	4.1633	2.9545	4.2113	2.9697	4.2593	2.9848	4.3072	3.0000	4.3552	1080
3.0908	4.6422	3.1059	4.6900	3.1210	4.7378	3.1361	4.7855	3.1512	4.8332	1090
3.2415	5.1190	3.2566	5.1665	3.2716	5.2141	3.2866	5.2616	3.3017	5.3091	1100
3.3916	5.5936	3.4066	5.6409	3.4216	5.6882	3.4365	5.7355	3.4515	5.7828	1110
3.5410	6.0660	3.5559	6.1131	3.5708	6.1603	3.5857	6.2073	3.6006	6.2544	1120
3.6898	6.5364	3.7046	6.5833	3.7194	6.6302	3.7343	6.6771	3.7491	6.7239	1130
3.8378	7.0047	3.8526	7.0514	3.8674	7.0981	3.8821	7.1448	3.8969	7.1914	1140
3.9853	7.4709	4.0000	7.5174	4.0147	7.5639	4.0294	7.6104	4.0441	7.6569	1150
4.1321	7.9352	4.1467	7.9815	4.1614	8.0278	4.1760	8.0740	4.1906	8.1203	1160
4.2783	8.3974	4.2929	8.4435	4.3074	8.4896	4.3220	8.5357	4.3366	8.5818	1170
4.4238	8.8577	4.4384	8.9036	4.4529	8.9495	4.4674	8.9954	4.4819	9.0413	1180
4.5688	9.3161	4.5832	9.3618	4.5977	9.4075	4.6121	9.4532	4.6266	9.4989	1190
4.7131	9.7725	4.7275	9.8180	4.7419	9.8636	4.7563	9.9091	4.7707	9.9545	1200
4.8559	10.2270	4.8712	10.2724	4.8855	10.3177	4.8999	10.3630	4.9142	10.4083	1210
5.0000	10.6797	5.0143	10.7249	5.0286	10.7700	5.0428	10.8152	5.0571	10.8603	1220
5.1426	11.1306	5.1568	11.1755	5.1710	11.2205	5.1852	11.2654	5.1994	11.3104	1230
5.2846	11.5796	5.2987	11.6244	5.3129	11.6692	5.3270	11.7139	5.3412	11.7587	1240
5.4260	12.0268	5.4401	12.0714	5.4542	12.1160	5.4683	12.1606	5.4824	12.2052	1250
5.5668	12.4722	5.5809	12.5167	5.5949	12.5611	5.6090	12.6055	5.6230	12.6499	1260
5.7071	12.9159	5.7211	12.9602	5.7351	13.0044	5.7491	13.0487	5.7631	13.0929	1270
5.8469	13.3578	5.8608	13.4019	5.8748	13.4460	5.8887	13.4901	5.9026	13.5341	1280
5.9861	13.7981	6.0000	13.8420	6.0139	13.8859	6.0278	13.9298	6.0416	13.9737	1290
6.1248	14.2366	6.1386	14.2804	6.1525	14.3241	6.1663	14.3678	6.1801	14.4115	1300
6.2629	14.6734	6.2767	14.7170	6.2905	14.7606	6.3043	14.8042	6.3180	14.8477	1310
6.4005	15.1086	6.4143	15.1521	6.4280	15.1955	6.4417	15.2389	6.4555	15.2823	1320
6.5377	15.5422	6.5513	15.5855	6.5650	15.6287	6.5787	15.6720	6.5923	15.7152	1330
6.6742	15.9741	6.6879	16.0172	6.7015	16.0603	6.7151	16.1034	6.7287	16.1465	1340
6.8103	16.4045	6.8239	16.4474	6.8375	16.4903	6.8511	16.5333	6.8646	16.5762	1350
6.9459	16.8332	6.9594	16.8760	6.9730	16.9188	6.9865	16.9615	7.0000	17.0043	1360
7.0810	17.2604	7.0945	17.3030	7.1080	17.3456	7.1214	17.3882	7.1349	17.4308	1370
7.2156	17.6860	7.2290	17.7285	7.2424	17.7710	7.2559	17.8134	7.2693	17.8558	1380
7.3497	18.1101	7.3631	18.1524	7.3765	18.1948	7.3898	18.2371	7.4032	18.2793	1390
7.4833	18.5327	7.4967	18.5749	7.5100	18.6170	7.5233	18.6592	7.5366	18.7013	1400
7.6165	18.9538	7.6298	18.9958	7.6431	19.0378	7.6563	19.0798	7.6696	19.1218	1410
7.7492	19.3734	7.7624	19.4152	7.7757	19.4571	7.7889	19.4990	7.8021	19.5408	1420
7.8814	19.7915	7.8946	19.8332	7.9078	19.8749	7.9210	19.9166	7.9342	19.9583	1430
8.0132	20.2082	8.0263	20.2497	8.0395	20.2913	8.0526	20.3329	8.0657	20.3744	1440
30+	100+	30+	100+	30+	100+	30+	100+	30+	100+	

Tables of Square Roots

N	0 √N	0 √10N	1 √N	1 √10N	2 √N	2 √10N	3 √N	3 √10N	4 √N	4 √10N
	30.+	100+	30+	100+	30+	100+	30+	100+	30+	100+
1450	8.0789 20.4159		8.0920 20.4575		8.1051 20.4990		8.1182 20.5405		8.1314 20.5819	
1460	8.2099 20.8305		8.2230 20.8718		8.2361 20.9132		8.2492 20.9545		8.2623 20.9959	
1470	8.3406 21.2436		8.3536 21.2848		8.3667 21.3260		8.3797 21.3672		8.3927 21.4084	
1480	8.4708 21.6553		8.4838 21.6963		8.4968 21.7374		8.5097 21.7785		8.5227 21.8195	
1490	8.6005 22.0656		8.6135 22.1065		8.6264 22.1475		8.6394 22.1884		8.6523 22.2293	
1500	8.7298 22.4745		8.7427 22.5153		8.7556 22.5561		8.7685 22.5969		8.7814 22.6377	
1510	8.8587 22.8821		8.8716 22.9227		8.8844 22.9634		8.8973 23.0041		8.9102 23.0447	
1520	8.9872 23.2883		9.0000 23.3288		9.0128 23.3694		9.0256 23.4099		9.0384 23.4504	
1530	9.1152 23.6932		9.1280 23.7336		9.1408 23.7740		9.1535 23.8144		9.1663 23.8548	
1540	9.2428 24.0967		9.2556 24.1370		9.2683 24.1773		9.2810 24.2176		9.2938 24.2578	
1550	9.3700 24.4990		9.3827 24.5392		9.3954 24.5793		9.4081 24.6194		9.4208 24.6595	
1560	9.4968 24.9000		9.5095 24.9400		9.5221 24.9800		9.5348 25.0200		9.5474 25.0600	
1570	9.6232 25.2996		9.6358 25.3395		9.6485 25.3794		9.6611 25.4193		9.6737 25.4592	
1580	9.7492 25.6981		9.7618 25.7378		9.7744 25.7776		9.7869 25.8173		9.7995 25.8571	
1590	9.8748 26.0952		9.8873 26.1348		9.8999 26.1745		9.9124 26.2141		9.9249 26.2537	
1600	*0.0000 26.4911		*0.0125 26.5306		*0.0250 26.5701		*0.0375 26.6096		*0.0500 26.6491	
1610	0.1248 26.8858		0.1373 26.9252		0.1497 26.9646		0.1622 27.0039		0.1746 27.0433	
1620	0.2492 27.2792		0.2616 27.3185		0.2741 27.3578		0.2865 27.3970		0.2989 27.4363	
1630	0.3733 27.6715		0.3856 27.7106		0.3980 27.7498		0.4104 27.7889		0.4228 27.8280	
1640	0.4969 28.0625		0.5093 28.1015		0.5216 28.1405		0.5339 28.1796		0.5463 28.2186	
1650	0.6202 28.4523		0.6325 28.4912		0.6448 28.5302		0.6571 28.5690		0.6694 28.6079	
1660	0.7431 28.8410		0.7554 28.8798		0.7676 28.9186		0.7799 28.9574		0.7922 28.9961	
1670	0.8656 29.2285		0.8779 29.2672		0.8901 29.3058		0.9023 29.3445		0.9145 29.3832	
1680	0.9878 29.6148		1.0000 29.6534		1.0122 29.6919		1.0244 29.7305		1.0366 29.7690	
1690	1.1096 30.0000		1.1218 30.0385		1.1339 30.0769		1.1461 30.1153		1.1582 30.1538	
1700	1.2311 30.3840		1.2432 30.4224		1.2553 30.4607		1.2674 30.4990		1.2795 30.5374	
1710	1.3521 30.7670		1.3642 30.8052		1.3763 30.8434		1.3884 30.8816		1.4005 30.9198	
1720	1.4729 31.1488		1.4849 31.1869		1.4970 31.2250		1.5090 31.2631		1.5211 31.3012	
1730	1.5933 31.5295		1.6053 31.5675		1.6173 31.6055		1.6293 31.6435		1.6413 31.6814	
1740	1.7133 31.9091		1.7253 31.9470		1.7373 31.9848		1.7493 32.0227		1.7612 32.0606	
1750	1.8330 32.2876		1.8450 32.3254		1.8569 32.3631		1.8688 32.4009		1.8808 32.4387	
1760	1.9524 32.6650		1.9643 32.7027		1.9762 32.7403		1.9881 32.7780		2.0000 32.8157	
1770	2.0714 33.0413		2.0833 33.0789		2.0951 33.1165		2.1070 33.1540		2.1189 33.1916	
1780	2.1900 33.4166		2.2019 33.4541		2.2137 33.4916		2.2256 33.5290		2.2374 33.5665	
1790	2.3084 33.7909		2.3202 33.8282		2.3320 33.8656		2.3438 33.9030		2.3556 33.9403	
1800	2.4264 34.1641		2.4382 34.2013		2.4500 34.2386		2.4617 34.2758		2.4735 34.3131	
1810	2.5441 34.5362		2.5558 34.5734		2.5676 34.6105		2.5793 34.6477		2.5911 34.6848	
1820	2.6615 34.9074		2.6732 34.9444		2.6849 34.9815		2.6966 35.0185		2.7083 35.0555	
1830	2.7785 35.2775		2.7902 35.3144		2.8019 35.3514		2.8135 35.3883		2.8252 35.4253	
1840	2.8952 35.6466		2.9069 35.6835		2.9185 35.7203		2.9302 35.7571		2.9418 35.7940	
1850	3.0116 36.0147		3.0233 36.0515		3.0349 36.0882		3.0465 36.1249		3.0581 36.1617	
1860	3.1277 36.3818		3.1393 36.4185		3.1509 36.4551		3.1625 36.4918		3.1741 36.5284	
1870	3.2435 36.7479		3.2551 36.7845		3.2666 36.8211		3.2782 36.8576		3.2897 36.8941	
1880	3.3590 37.1131		3.3705 37.1496		3.3820 37.1860		3.3935 37.2224		3.4051 37.2589	
1890	3.4741 37.4773		3.4856 37.5136		3.4971 37.5500		3.5086 37.5863		3.5201 37.6227	
	40+ 100+		40+ 100+		40+ 100+		40+ 100+		40+ 100+	

Tables of Square Roots

√N (5)	√10N	√N (6)	√10N	√N (7)	√10N	√N (8)	√10N	√N (9)	√10N	N
30+	100+	30+	100+	30+	100+	30+	100+	30+	100+	
8.1445	20.6234	8.1576	20.6648	8.1707	20.7063	8.1838	20.7477	8.1969	20.7891	1450
8.2753	21.0372	8.2884	21.0785	8.3014	21.1198	8.3145	21.1611	8.3275	21.2023	1460
8.4057	21.4496	8.4187	21.4907	8.4318	21.5319	8.4418	21.5730	8.4578	21.6141	1470
8.5357	21.8606	8.5487	21.9016	8.5616	21.9426	8.5746	21.9836	8.5876	22.0246	1480
8.6652	22.2702	8.6782	22.3111	8.6911	22.3520	8.7040	22.3928	8.7169	22.4337	1490
8.7943	22.6784	8.8072	22.7192	8.8201	22.7599	8.8330	22.8007	8.8458	22.8414	1500
8.9230	23.0853	8.9358	23.1260	8.9487	23.1666	8.9615	23.2071	8.9744	23.2477	1510
9.0512	23.4909	9.0641	23.5314	9.0768	23.5718	9.0896	23.6123	9.1024	23.6527	1520
9.1791	23.8951	9.1918	23.9355	9.2046	23.9758	9.2173	24.0161	9.2301	24.0564	1530
9.3065	24.2980	9.3192	24.3382	9.3319	24.3785	9.3446	24.4186	9.3573	24.4588	1540
9.4335	24.6996	9.4462	24.7397	9.4588	24.7798	9.4715	24.8199	9.4842	24.8599	1550
9.5601	25.1000	9.5727	25.1399	9.5854	25.1799	9.5980	25.2198	9.6106	25.2597	1560
9.6863	25.4990	9.6989	25.5388	9.7115	25.5787	9.7240	25.6185	9.7366	25.6583	1570
9.8121	25.8968	9.8246	25.9365	9.8372	25.9762	9.8497	26.0159	9.8623	26.0555	1580
9.9375	26.2933	9.9500	26.3329	9.9625	26.3725	9.9750	26.4120	9.9875	26.4516	1590
0.0625	26.6886	*0.0749	26.7281	*0.0874	26.7675	*0.0999	26.8069	*0.1123	26.8464	1600
0.1871	27.0827	0.1995	27.1220	0.2119	27.1613	0.2244	27.2006	0.2368	27.2399	1610
0.3113	27.4755	0.3237	27.5147	0.3361	27.5539	0.3485	27.5931	0.3609	27.6323	1620
0.4351	27.8671	0.4475	27.9062	0.4599	27.9453	0.4722	27.9844	0.4846	28.0234	1630
0.5586	28.2576	0.5709	28.2965	0.5832	28.3355	0.5956	28.3745	0.6079	28.4134	1640
0.6817	28.6468	0.6940	28.6857	0.7063	28.7245	0.7185	28.7633	0.7308	28.8022	1650
0.8044	29.0349	0.8167	29.0736	0.8289	29.1124	0.8412	29.1511	0.8534	29.1898	1660
0.9268	29.4218	0.9390	29.4604	0.9512	29.4990	0.9634	29.5376	0.9756	29.5762	1670
1.0488	29.8076	1.0609	29.8461	1.0731	29.8846	1.0853	29.9231	1.0974	29.9615	1680
1.1704	30.1922	1.1825	30.2306	1.1947	30.2690	1.2068	30.3073	1.2189	30.3457	1690
1.2916	30.5756	1.3038	30.6139	1.3159	30.6522	1.3280	30.6905	1.3401	30.7287	1700
1.4126	30.9580	1.4246	30.9962	1.4367	31.0343	1.4488	31.0725	1.4608	31.1106	1710
1.5331	31.3393	1.5452	31.3773	1.5572	31.4154	1.5692	31.4534	1.5812	31.4914	1720
1.6533	31.7194	1.6653	31.7574	1.6773	31.7953	1.6893	31.8332	1.7013	31.8711	1730
1.7732	32.0984	1.7852	32.1363	1.7971	32.1741	1.8091	32.2120	1.8210	32.2498	1740
1.8927	32.4764	1.9047	32.5142	1.9166	32.5519	1.9285	32.5896	1.9404	32.6273	1750
2.0119	32.8533	2.0238	32.8909	2.0357	32.9286	2.0476	32.9662	2.0595	33.0038	1760
2.1307	33.2291	2.1426	33.2667	2.1545	33.3042	2.1663	33.3417	2.1782	33.3792	1770
2.2493	33.6039	2.2611	33.6413	2.2729	33.6787	2.2847	33.7161	2.2956	33.7535	1780
2.3674	33.9776	2.3792	34.0149	2.3910	34.0522	2.4028	34.0895	2.4146	34.1268	1790
2.4853	34.3503	2.4971	34.3875	2.5088	34.4247	2.5206	34.4619	2.5323	34.4991	1800
2.6028	34.7219	2.6146	34.7590	2.6263	34.7961	2.6380	34.8332	2.6497	34.8703	1810
2.7200	35.0926	2.7317	35.1296	2.7434	35.1666	2.7551	35.2036	2.7668	35.2405	1820
2.8369	35.4622	2.8486	35.4991	2.8602	35.5360	2.8719	35.5729	2.8836	35.6097	1830
2.9535	35.8308	2.9651	35.8676	2.9767	35.9044	2.9884	35.9412	3.0000	35.9779	1840
3.0697	36.1984	3.0813	36.2351	3.0929	36.2718	3.1045	36.3085	3.1161	36.3452	1850
3.1856	36.5650	3.1972	36.6016	3.2088	36.6382	3.2204	36.6748	3.2319	36.7114	1860
3.3013	36.9306	3.3128	36.9671	3.3244	37.0037	3.3359	37.0401	3.3474	37.0766	1870
3.4166	37.2953	3.4281	37.3317	3.4396	37.3681	3.4511	37.4045	3.4626	37.4409	1880
3.5316	37.6590	3.5431	37.6953	3.5546	37.7316	3.5660	37.7679	3.5775	37.8042	1890
40+	100+	40+	100+	40+	100+	40+	100+	40+	100+	

Tables of Square Roots

N	0 √N	0 √10N	1 √N	1 √10N	2 √N	2 √10N	3 √N	3 √10N	4 √N	4 √10N
	40+	100+	40+	100+	40+	100+	40+	100+	40+	100+
1900	3.5890	37.8405	3.6005	37.8768	3.6119	37.9130	3.6234	37.9493	3.6348	37.9855
1910	3.7035	38.2028	3.7150	38.2389	3.7264	38.2751	3.7379	38.3112	3.7493	38.3474
1920	3.8178	38.5641	3.8292	38.6001	3.8406	38.6362	3.8520	38.6723	3.8634	38.7083
1930	3.9318	38.9244	3.9431	38.9604	3.9545	38.9964	3.9659	39.0324	3.9773	39.0683
1940	4.0454	39.2839	4.0568	39.3198	4.0681	39.3557	4.0795	39.3915	4.0908	39.4274
1950	4.1588	39.6424	4.1701	39.6782	4.1814	39.7140	4.1928	39.7498	4.2041	39.7856
1960	4.2719	40.0000	4.2832	40.0357	4.2945	40.0714	4.3058	40.1071	4.3170	40.1428
1970	4.3847	40.3567	4.3959	40.3923	4.4072	40.4279	4.4185	40.4635	4.4297	40.4991
1980	4.4972	40.7125	4.5084	40.7480	4.5197	40.7835	4.5309	40.8190	4.5421	40.8545
1990	4.6094	41.0674	4.6206	41.1028	4.6318	41.1382	4.6430	41.1737	4.6542	41.2091
2000	4.7214	41.4214	4.7325	41.4567	4.7437	41.4920	4.7549	41.5274	4.7661	41.5627
2010	4.8330	41.7745	4.8442	41.8097	4.8553	41.8450	4.8665	41.8802	4.8776	41.9155
2020	4.9444	42.1267	4.9555	42.1619	4.9667	42.1970	4.9778	42.2322	4.9889	42.2674
2030	5.0555	42.4781	5.0666	42.5132	5.0777	42.5482	5.0888	42.5833	5.0999	42.6184
2040	5.1664	42.8286	5.1774	42.8636	5.1885	42.8986	5.1996	42.9336	5.2106	42.9685
2050	5.2769	43.1782	5.2880	43.2131	5.2990	43.2480	5.3100	43.2829	5.3211	43.3178
2060	5.3872	43.5270	5.3982	43.5618	5.4093	43.5967	5.4203	43.6315	5.4313	43.6663
2070	5.4973	43.8749	5.5082	43.9097	5.5192	43.9444	5.5302	43.9792	5.5412	44.0139
2080	5.6070	44.2221	5.6180	44.2567	5.6289	44.2914	5.6399	44.3260	5.6508	44.3607
2090	5.7165	44.5683	5.7275	44.6029	5.7384	44.6375	5.7493	44.6720	5.7602	44.7066
2100	5.8258	44.9138	5.8367	44.9483	5.8476	44.9828	5.8585	45.0172	5.8694	45.0517
2110	5.9347	45.2584	5.9456	45.2928	5.9565	45.3272	5.9674	45.3616	5.9783	45.3960
2120	6.0435	45.6022	6.0543	45.6365	6.0652	45.6709	6.0760	45.7052	6.0869	45.7395
2130	6.1519	45.9452	6.1628	45.9795	6.1736	46.0137	6.1844	46.0479	6.1952	46.0822
2140	6.2601	46.2874	6.2709	46.3216	6.2817	46.3557	6.2925	46.3899	6.3033	46.4240
2150	6.3681	46.6288	6.3789	46.6629	6.3897	46.6970	6.4004	46.7310	6.4112	46.7651
2160	6.4758	46.9694	6.4866	47.0034	6.4973	47.0374	6.5081	47.0714	6.5188	47.1054
2170	6.5833	47.3092	6.5940	47.3431	6.6047	47.3771	6.6154	47.4110	6.6262	47.4449
2180	6.6905	47.6482	6.7012	47.6821	6.7119	47.7159	6.7226	47.7498	6.7333	47.7836
2190	6.7974	47.9865	6.8081	48.0203	6.8188	48.0540	6.8295	48.0878	6.8402	48.1216
2200	6.9042	48.3240	6.9148	48.3577	6.9255	48.3914	6.9361	48.4251	6.9468	48.4587
2210	7.0106	48.6607	7.0213	48.6943	7.0319	48.7279	7.0425	48.7616	7.0532	48.7952
2220	7.1169	48.9966	7.1275	49.0302	7.1381	49.0637	7.1487	49.0973	7.1593	49.1308
2230	7.2229	49.3318	7.2335	49.3653	7.2440	49.3988	7.2546	49.4323	7.2652	49.4657
2240	7.3286	49.6663	7.3392	49.6997	7.3498	49.7331	7.3603	49.7665	7.3709	49.7999
2250	7.4342	50.0000	7.4447	50.0333	7.4552	50.0667	7.4658	50.1000	7.4763	50.1333
2260	7.5395	50.3330	7.5500	50.3662	7.5605	50.3995	7.5710	50.4327	7.5815	50.4659
2270	7.6445	50.6652	7.6550	50.6984	7.6655	50.7315	7.6760	50.7647	7.6865	50.7979
2280	7.7493	50.9967	7.7598	51.0298	7.7703	51.0629	7.7807	51.0960	7.7912	51.1291
2290	7.8539	51.3275	7.8644	51.3605	7.8748	51.3935	7.8853	51.4266	7.8957	51.4596
2300	7.9583	51.6575	7.9687	51.6905	7.9792	51.7234	7.9896	51.7564	8.0000	51.7893
2310	8.0625	51.9868	8.0729	52.0197	8.0833	52.0526	8.0937	52.0855	8.1041	52.1184
2320	8.1664	52.3155	8.1768	52.3483	8.1971	52.3911	8.1975	52.4139	8.2079	52.4467
2330	8.2701	52.6434	8.2904	52.6761	8.2908	52.7089	8.3011	52.7416	8.3115	52.7743
2340	8.3735	52.9706	8.3839	53.0033	8.3942	53.0359	8.4045	53.0686	8.4149	53.1013
	40+	100+	40+	100+	40+	100+	40+	100+	40+	100+

Tables of Square Roots

$\sqrt{N}$	5 $\sqrt{10N}$	$\sqrt{N}$	6 $\sqrt{10N}$	$\sqrt{N}$	7 $\sqrt{10N}$	$\sqrt{N}$	8 $\sqrt{10N}$	$\sqrt{N}$	9 $\sqrt{10N}$	N
40+	100+	40+	100+	40+	100+	40+	100+	40+	100+	
3.6463	38.0217	3.6578	38.0580	3.6692	38.0942	3.6807	38.1304	3.6921	38.1666	1900
3.7607	38.3835	3.7721	38.4197	3.7836	38.4558	3.7950	38.4919	3.8064	38.5280	1910
3.8748	38.7444	3.8862	38.7804	3.8976	38.8164	3.9090	38.8524	3.9204	38.8884	1920
3.9886	39.1043	4.0000	39.1402	4.0114	39.1761	4.0227	39.2121	4.0341	39.2480	1930
4.1022	39.4633	4.1135	39.4991	4.1248	39.5349	4.1362	39.5708	4.1475	39.6066	1940
4.2154	39.8213	4.2267	39.8571	4.2380	39.8928	4.2493	39.9286	4.2606	39.9643	1950
4.3283	40.1785	4.3396	40.2141	4.3509	40.2498	4.3621	40.2854	4.3734	40.3211	1960
4.4410	40.5347	4.4522	40.5703	4.4635	40.6058	4.4747	40.6414	4.4860	40.6769	1970
4.5533	40.8900	4.5646	40.9255	4.5758	40.9610	4.5870	40.9965	4.5982	41.0319	1980
4.6654	41.2445	4.6766	41.2799	4.6878	41.3153	4.6990	41.3506	4.7102	41.3860	1990
4.7772	41.5980	4.7884	41.6333	4.7996	41.6686	4.8107	41.7039	4.8219	41.7392	2000
4.8888	41.9507	4.8999	41.9859	4.9110	42.0211	4.9222	42.0563	4.9333	42.0915	2010
5.0000	42.3025	5.0111	42.3376	5.0222	42.3728	5.0333	42.4079	5.0444	42.4430	2020
5.1110	42.6534	5.1221	42.6885	5.1331	42.7235	5.1442	42.7585	5.1553	42.7936	2030
5.2217	43.0035	5.2327	43.0385	5.2438	43.0734	5.2548	43.1084	5.2659	43.1433	2040
5.3321	43.3527	5.3431	43.3876	5.3542	43.4225	5.3652	43.4573	5.3762	43.4922	2050
5.4423	43.7011	5.4533	43.7359	5.4643	43.7707	5.4753	43.8054	5.4863	43.8402	2060
5.5522	44.0486	5.5631	44.0833	5.5741	44.1180	5.5851	44.1527	5.5961	44.1874	2070
5.6618	44.3953	5.6727	44.4299	5.6837	44.4645	5.6946	44.4991	5.7056	44.5337	2080
5.7712	44.7411	5.7821	44.7757	5.7930	44.8102	5.8039	44.8447	5.8148	44.8793	2090
5.8803	45.0862	5.8912	45.1206	5.9021	45.1551	5.9130	45.1895	5.9239	45.2240	2100
5.9891	45.4304	6.0000	45.4648	6.0109	45.4991	6.0217	45.5335	6.0326	45.5679	2110
6.0977	45.7738	6.1086	45.8081	6.1194	45.8424	6.1303	45.8767	6.1411	45.9109	2120
6.2061	46.1164	6.2169	46.1506	6.2277	46.1848	6.2385	46.2190	6.2493	46.2532	2130
6.3141	46.4582	6.3249	46.4923	6.3357	46.5264	6.3465	46.5606	6.3573	46.5947	2140
6.4220	46.7992	6.4327	46.8332	6.4435	46.8673	6.4543	46.9013	6.4650	46.9354	2150
6.5296	47.1394	6.5403	47.1734	6.5510	47.2073	6.5618	47.2413	6.5725	47.2753	2160
6.6369	47.4788	6.6476	47.5127	6.6583	47.5466	6.6690	47.5805	6.6798	47.6144	2170
6.7440	47.8175	6.7547	47.8513	6.7654	47.8851	6.7761	47.9189	6.7868	47.9527	2180
6.8508	48.1553	6.8615	48.1891	6.8722	48.2228	6.8828	48.2565	6.8935	48.2903	2190
6.9574	48.4924	6.9681	48.5261	6.9787	48.5598	6.9894	48.5934	7.0000	48.6271	2200
7.0638	48.8288	7.0744	48.8624	7.0850	48.8959	7.0956	48.9295	7.1063	48.9631	2210
7.1699	49.1643	7.1805	49.1979	7.1911	49.2314	7.2017	49.2649	7.2123	49.2984	2220
7.2758	49.4992	7.2864	49.5326	7.2969	49.5660	7.3075	49.5995	7.3181	49.6329	2230
7.3814	49.8332	7.3920	49.8666	7.4025	49.9000	7.4131	49.9333	7.4236	49.9667	2240
7.4868	50.1666	7.4974	50.1999	7.5079	50.2332	7.5184	50.2664	7.5289	50.2997	2250
7.5920	50.4992	7.6025	50.5324	7.6130	50.5656	7.6235	50.5988	7.6340	50.6320	2260
7.6970	50.8310	7.7074	50.8642	7.7179	50.8973	7.7284	50.9304	7.7389	50.9636	2270
7.8017	51.1622	7.8121	51.1952	7.8226	51.2283	7.8330	51.2614	7.8435	51.2944	2280
7.9062	51.4926	7.9166	51.5256	7.9270	51.5586	7.9375	51.5916	7.9479	51.6245	2290
8.0104	51.8223	8.0208	51.8552	8.0312	51.8881	8.0416	51.9210	8.0521	51.9539	2300
8.1144	52.1512	8.1248	52.1841	8.1352	52.2170	8.1456	52.2498	8.1560	52.2826	2310
8.2183	52.4795	8.2286	52.5123	8.2390	52.5451	8.2494	52.5778	8.2597	52.6106	2320
8.3218	52.8071	8.3322	52.8398	8.3425	52.8725	8.3529	52.9052	8.3632	52.9379	2330
8.4252	53.1339	8.4355	53.1666	8.4458	53.1992	8.4562	53.2319	8.4665	53.2645	2340
40+	100+	40+	100+	40+	100+	40+	100+	40+	100+	

Tables of Square Roots

N	$\sqrt{N}$ 0	$\sqrt{10N}$	$\sqrt{N}$ 1	$\sqrt{10N}$	$\sqrt{N}$ 2	$\sqrt{10N}$	$\sqrt{N}$ 3	$\sqrt{10N}$	$\sqrt{N}$ 4	$\sqrt{10N}$
	40+	100+	40+	100+	40+	100+	40+	100+	40+	100+
2350	8.4768	53.2971	8.4871	53.3297	8.4974	53.3623	8.5077	53.3949	8.5180	53.4275
2360	8.5798	53.6229	8.5901	53.6555	8.6004	53.6880	8.6107	53.7205	8.6210	53.7530
2370	8.6826	53.9480	8.6929	53.9805	8.7032	54.0130	8.7134	54.0454	8.7237	54.0779
2380	8.7852	54.2725	8.7955	54.3049	8.8057	54.3373	8.8160	54.3697	8.8262	54.4021
2390	8.8876	54.5962	8.8979	54.6286	8.9081	54.6609	8.9183	54.6932	8.9285	54.7256
2400	8.9898	54.9193	9.0000	54.9516	9.0102	54.9839	9.0204	55.0161	9.0306	55.0484
2410	9.0918	55.2417	9.1019	55.2740	9.1121	55.3061	9.1223	55.3383	9.1325	55.3705
2420	9.1935	55.5635	9.2037	55.5956	9.2138	55.6278	9.2240	55.6599	9.2341	55.6920
2430	9.2950	55.8846	9.3052	55.9166	9.3153	55.9487	9.3255	55.9808	9.3356	56.0128
2440	9.3964	56.2050	9.4065	56.2370	9.4166	56.2690	9.4267	56.3010	9.4368	56.3330
2450	9.4975	56.5248	9.5076	56.5567	9.5177	56.5886	9.5278	56.6206	9.5379	56.6525
2460	9.5984	56.8439	9.6085	56.8757	9.6185	56.9076	9.6286	56.9395	9.6387	56.9713
2470	9.6991	57.1623	9.7092	57.1941	9.7192	57.2260	9.7293	57.2578	9.7393	57.2895
2480	9.7996	57.4802	9.8096	57.5119	9.8197	57.5436	9.8297	57.5754	9.8397	57.6071
2490	9.8999	57.7973	9.9099	57.8290	9.9199	57.8607	9.9300	57.8924	9.9400	57.9240
2500	*0.0000	58.1139	*0.0100	58.1455	*0.0200	58.1771	*0.0300	58.2087	*0.0400	58.2403
2510	0.0999	58.4298	0.1099	58.4614	0.1199	58.4929	0.1298	58.5244	0.1398	58.5560
2520	0.1996	58.7451	0.2096	58.7766	0.2195	58.8081	0.2295	58.8395	0.2394	58.8710
2530	0.2991	59.0597	0.3090	59.0912	0.3190	59.1226	0.3289	59.1540	0.3389	59.1854
2540	0.3984	59.3738	0.4083	59.4051	0.4183	59.4365	0.4282	59.4679	0.4381	59.4992
2550	0.4975	59.6872	0.5074	59.7185	0.5173	59.7498	0.5272	59.7811	0.5371	59.8124
2560	0.5964	60.0000	0.6063	60.0312	0.6162	60.0625	0.6261	60.0937	0.6360	60.1250
2570	0.6952	60.3122	0.7050	60.3434	0.7149	60.3746	0.7247	60.4057	0.7346	60.4369
2580	0.7937	60.6238	0.8035	60.6549	0.8134	60.6860	0.8232	60.7171	0.8331	60.7483
2590	0.8920	60.9348	0.9019	60.9658	0.9117	60.9969	0.9215	61.0279	0.9313	61.0590
2600	0.9902	61.2452	1.0000	61.2762	1.0098	61.3072	1.0196	61.3382	1.0294	61.3691
2610	1.0882	61.5549	1.0979	61.5859	1.1077	61.6168	1.1175	61.6478	1.1273	61.6787
2620	1.1859	61.8641	1.1957	61.8950	1.2055	61.9259	1.2152	61.9568	1.2250	61.9877
2630	1.2835	62.1727	1.2933	62.2036	1.3030	62.2344	1.3128	62.2652	1.3225	62.2960
2640	1.3809	62.4808	1.3907	62.5115	1.4004	62.5423	1.4101	62.5731	1.4198	62.6038
2650	1.4782	62.7882	1.4879	62.8189	1.4976	62.8496	1.5073	62.8803	1.5170	62.9110
2660	1.5752	63.0951	1.5849	63.1257	1.5946	63.1564	1.6043	63.1870	1.6140	63.2176
2670	1.6720	63.4013	1.6817	63.4319	1.6914	63.4625	1.7011	63.4931	1.7107	63.5237
2680	1.7687	63.7071	1.7784	63.7376	1.7880	63.7681	1.7977	63.7987	1.8073	63.8292
2690	1.8652	64.0122	1.8748	64.0427	1.8845	64.0732	1.8941	64.1036	1.9038	64.1341
2700	1.9615	64.3168	1.9711	64.3472	1.9808	64.3776	1.9904	64.4080	2.0000	64.4384
2710	2.0577	64.6208	2.0673	64.6511	2.0769	64.6815	2.0865	64.7119	2.0961	64.7422
2720	2.1536	64.9242	2.1632	64.9545	2.1728	64.9848	2.1824	65.0152	2.1920	65.0454
2730	2.2494	65.2271	2.2590	65.2574	2.2685	65.2876	2.2781	65.3179	2.2877	65.3481
2740	2.3450	65.5295	2.3546	65.5597	2.3641	65.5899	2.3737	65.6200	2.3832	65.6502
2750	2.4404	65.8312	2.4500	65.8614	2.4595	65.8915	2.4690	65.9217	2.4786	65.9518
2760	2.5357	66.1325	2.5452	66.1626	2.5547	66.1927	2.5642	66.2227	2.5738	66.2528
2770	2.6308	66.4332	2.6403	66.4632	2.6498	66.4932	2.6593	66.5233	2.6688	66.5533
2780	2.7257	66.7333	2.7352	66.7633	2.7447	66.7933	2.7541	66.8233	2.7636	66.8532
2790	2.8205	67.0329	2.8299	67.0629	2.8394	67.0928	2.8488	67.1227	2.8583	67.1526
	50+	100+	50+	100+	50+	100+	50+	100+	50+	100+

$\sqrt{N}$ 5 $\sqrt{10N}$		$\sqrt{N}$ 6 $\sqrt{10N}$		$\sqrt{N}$ 7 $\sqrt{10N}$		$\sqrt{N}$ 8 $\sqrt{10N}$		$\sqrt{N}$ 9 $\sqrt{10N}$		N
40+	100+	40+	100+	40+	100+	40+	100+	40+	100+	
8.5283	53.4601	8.5386	53.4927	8.5489	53.5252	8.5592	53.5578	8.5695	53.5904	2350
8.6313	53.7856	8.6415	53.8181	8.6518	53.8506	8.6621	53.8831	8.6724	53.9156	2360
8.7340	54.1104	8.7442	54.1428	8.7545	54.1752	8.7647	54.2077	8.7750	54.2401	2370
8.8365	54.4345	8.8467	54.4668	8.8569	54.4992	8.8672	54.5316	8.8774	54.5639	2380
8.9387	54.7579	8.9490	54.7902	8.9592	54.8225	8.9694	54.8548	8.9796	54.8871	2390
9.0408	55.0806	9.0510	55.1129	9.0612	55.1451	9.0714	55.1773	9.0816	55.2095	2400
9.1427	55.4027	9.1528	55.4349	9.1630	55.4670	9.1732	55.4992	9.1833	55.5313	2410
9.2443	55.7241	9.2544	55.7562	9.2646	55.7883	9.2747	55.8204	9.2849	55.8525	2420
9.3457	56.0449	9.3559	56.0769	9.3660	56.1089	9.3761	56.1410	9.3862	56.1730	2430
9.4469	56.3650	9.4571	56.3969	9.4672	56.4289	9.4773	56.4609	9.4874	56.4928	2440
9.5480	56.6844	9.5580	56.7163	9.5681	56.7482	9.5782	56.7801	9.5883	56.8120	2450
9.6488	57.0032	9.6588	57.0350	9.6689	57.0669	9.6790	57.0987	9.6890	57.1305	2460
9.7494	57.3213	9.7594	57.3531	9.7695	57.3849	9.7795	57.4166	9.7896	57.4484	2470
9.8498	57.6388	9.8598	57.6705	9.8698	57.7023	9.8799	57.7340	9.8899	57.7656	2480
9.9500	57.9557	9.9600	57.9873	9.9700	58.0190	9.9800	58.0506	9.9900	58.0823	2490
*0.0500	58.2719	*0.0600	58.3035	*0.0700	58.3351	*0.0799	58.3667	*0.0899	58.3982	2500
0.1498	58.5875	0.1597	58.6190	0.1697	58.6506	0.1797	58.6821	0.1896	58.7136	2510
0.2494	58.9025	0.2593	58.9339	0.2693	58.9654	0.2792	58.9969	0.2892	59.0283	2520
0.3488	59.2168	0.3587	59.2482	0.3686	59.2796	0.3786	59.3110	0.3885	59.3424	2530
0.4480	59.5306	0.4579	59.5619	0.4678	59.5932	0.4777	59.6246	0.4876	59.6559	2540
0.5470	59.8437	0.5569	59.8750	0.5668	59.9062	0.5767	59.9375	0.5866	59.9687	2550
0.6458	60.1562	0.6557	60.1874	0.6656	60.2186	0.6754	60.2498	0.6853	60.2810	2560
0.7445	60.4681	0.7543	60.4992	0.7642	60.5304	0.7740	60.5615	0.7839	60.5927	2570
0.8429	60.7794	0.8527	60.8104	0.8626	60.8415	0.8724	60.8726	0.8822	60.9037	2580
0.9411	61.0900	0.9510	61.1211	0.9608	61.1521	0.9706	61.1831	0.9804	61.2141	2590
1.0392	61.4001	1.0490	61.4311	1.0588	61.4621	1.0686	61.4930	1.0784	61.5240	2600
1.1371	61.7096	1.1468	61.7405	1.1566	61.7714	1.1664	61.8023	1.1762	61.8332	2610
1.2348	62.0185	1.2445	62.0494	1.2543	62.0802	1.2640	62.1111	1.2738	62.1419	2620
1.3323	62.3268	1.3420	62.3576	1.3517	62.3884	1.3615	62.4192	1.3712	62.4500	2630
1.4296	62.6346	1.4393	62.6653	1.4490	62.6960	1.4587	62.7268	1.4684	62.7575	2640
1.5267	62.9417	1.5364	62.9724	1.5461	63.0031	1.5558	63.0337	1.5655	63.0644	2650
1.6236	63.2483	1.6333	63.2789	1.6430	63.3095	1.6527	63.3401	1.6624	63.3707	2660
1.7204	63.5543	1.7301	63.5848	1.7397	63.6154	1.7494	63.6460	1.7591	63.6765	2670
1.8170	63.8597	1.8266	63.8902	1.8363	63.9207	1.8459	63.9512	1.8556	63.9817	2680
1.9134	64.1646	1.9230	64.1950	1.9326	64.2255	1.9423	64.2559	1.9519	64.2863	2690
2.0096	64.4688	2.0192	64.4992	2.0288	64.5296	2.0384	64.5600	2.0481	64.5904	2700
2.1057	64.7726	2.1153	64.8029	2.1249	64.8332	2.1344	64.8636	2.1440	64.8939	2710
2.2015	65.0757	2.2111	65.1060	2.2207	65.1363	2.2303	65.1666	2.2398	65.1969	2720
2.2972	65.3784	2.3068	65.4086	2.3163	65.4388	2.3259	65.4690	2.3355	65.4992	2730
2.3927	65.6804	2.4023	65.7106	2.4118	65.7408	2.4214	65.7709	2.4309	65.8011	2740
2.4881	65.9819	2.4976	66.0120	2.5071	66.0422	2.5167	66.0723	2.5262	66.1024	2750
2.5833	66.2829	2.5928	66.3130	2.6023	66.3430	2.6118	66.3731	2.6213	66.4031	2760
2.6783	66.5833	2.6878	66.6133	2.6972	66.6433	2.7067	66.6733	2.7162	66.7033	2770
2.7731	66.8832	2.7826	66.9132	2.7920	66.9431	2.8015	66.9731	2.8110	67.0030	2780
2.8678	67.1825	2.8772	67.2124	2.8867	67.2423	2.8961	67.2722	2.9056	67.3021	2790
50+	100+	50+	100+	50+	100+	50+	100+	50+	100+	

Tables of Square Roots

N	√N 0 (50+)	√10N (100+)	√N 1 (50+)	√10N (100+)	√N 2 (50+)	√10N (100+)	√N 3 (50+)	√10N (100+)	√N 4 (50+)	√10N (100+)
2800	2.9150	67.3320	2.9245	67.3619	2.9339	67.3918	2.9434	67.4216	2.9528	67.4515
2810	3.0094	67.6305	3.0189	67.6604	3.0283	67.6902	3.0377	67.7200	3.0471	67.7498
2820	3.1037	67.9286	3.1131	67.9583	3.1225	67.9881	3.1319	68.0179	3.1413	68.0476
2830	3.1977	68.2260	3.2071	68.2558	3.2165	68.2855	3.2259	68.3152	3.2353	68.3449
2840	3.2917	68.5230	3.3010	68.5527	3.3104	68.5823	3.3198	68.6120	3.3292	68.6416
2850	3.3854	68.8194	3.3948	68.8490	3.4041	68.8787	3.4135	68.9083	3.4228	68.9379
2860	3.4790	69.1153	3.4883	69.1449	3.4977	69.1745	3.5070	69.2040	3.5164	69.2336
2870	3.5724	69.4107	3.5817	69.4403	3.5910	69.4698	3.6004	69.4993	3.6097	69.5288
2880	3.6656	69.7056	3.6749	69.7351	3.6843	69.7645	3.6936	69.7940	3.7029	69.8234
2890	3.7587	70.0000	3.7680	70.0294	3.7773	70.0588	3.7866	70.0882	3.7959	70.1176
2900	3.8516	70.2939	3.8609	70.3232	3.8702	70.3526	3.8795	70.3819	3.8888	70.4113
2910	3.9444	70.5872	3.9537	70.6165	3.9630	70.6458	3.9722	70.6751	3.9815	70.7044
2920	4.0370	70.8801	4.0463	70.9093	4.0555	70.9386	4.0648	70.9678	4.0740	70.9971
2930	4.1295	71.1724	4.1387	71.2016	4.1479	71.2308	4.1572	71.2600	4.1664	71.2892
2940	4.2218	71.4643	4.2310	71.4934	4.2402	71.5226	4.2494	71.5517	4.2586	71.5809
2950	4.3139	71.7556	4.3231	71.7847	4.3323	71.8139	4.3415	71.8430	4.3507	71.8720
2960	4.4059	72.0465	4.4151	72.0756	4.4243	72.1046	4.4334	72.1337	4.4426	72.1627
2970	4.4977	72.3369	4.5069	72.3659	4.5161	72.3949	4.5252	72.4239	4.5344	72.4529
2980	4.5894	72.6268	4.5985	72.6557	4.6077	72.6847	4.6168	72.7136	4.6260	72.7426
2990	4.6809	72.9162	4.6900	72.9451	4.6992	72.9740	4.7083	73.0029	4.7175	73.0318
3000	4.7723	73.2051	4.7814	73.2339	4.7905	73.2628	4.7996	73.2917	4.8088	73.3205
3010	4.8635	73.4935	4.8726	73.5223	4.8817	73.5511	4.8908	73.5800	4.8999	73.6088
3020	4.9545	73.7815	4.9636	73.8102	4.9727	73.8390	4.9818	73.8678	4.9909	73.8965
3030	5.0454	74.0690	5.0545	74.0977	5.0636	74.1264	5.0727	74.1551	5.0818	74.1838
3040	5.1362	74.3560	5.1453	74.3846	5.1543	74.4133	5.1634	74.4420	5.1725	74.4706
3050	5.2268	74.6425	5.2359	74.6711	5.2449	74.6997	5.2540	74.7284	5.2630	74.7570
3060	5.3173	74.9286	5.3263	74.9571	5.3353	74.9857	5.3444	75.0143	5.3534	75.0429
3070	5.4076	75.2142	5.4166	75.2427	5.4256	75.2712	5.4346	75.2997	5.4437	75.3283
3080	5.4977	75.4993	5.5068	75.5278	5.5158	75.5563	5.5248	75.5847	5.5338	75.6132
3090	5.5878	75.7840	5.5968	75.8124	5.6058	75.8408	5.6147	75.8693	5.6237	75.8977
3100	5.6776	76.0682	5.6866	76.0966	5.6956	76.1250	5.7046	76.1533	5.7136	76.1817
3110	5.7674	76.3519	5.7763	76.3803	5.7853	76.4086	5.7943	76.4370	5.8032	76.4653
3120	5.8570	76.6352	5.8659	76.6635	5.8749	76.6918	5.8838	76.7201	5.8928	76.7484
3130	5.9464	76.9181	5.9553	76.9463	5.9643	76.9746	5.9732	77.0028	5.9821	77.0311
3140	6.0357	77.2005	6.0446	77.2287	6.0535	77.2569	6.0625	77.2851	6.0714	77.3133
3150	6.1249	77.4824	6.1338	77.5106	6.1427	77.5387	6.1516	77.5669	6.1605	77.5950
3160	6.2139	77.7639	6.2228	77.7920	6.2317	77.8201	6.2406	77.8483	6.2494	77.8764
3170	6.3028	78.0449	6.3116	78.0730	6.3205	78.1011	6.3294	78.1292	6.3383	78.1572
3180	6.3915	78.3255	6.4004	78.3536	6.4092	78.3816	6.4181	78.4096	6.4269	78.4377
3190	6.4801	78.6057	6.4889	78.6337	6.4978	78.6617	6.5066	78.6897	6.5155	78.7177
3200	6.5685	78.8854	6.5774	78.9134	6.5862	78.9413	6.5951	78.9693	6.6039	78.9972
3210	6.6569	79.1647	6.6657	79.1926	6.6745	79.2205	6.6833	79.2484	6.6922	79.2763
3220	6.7450	79.4436	6.7539	79.4714	6.7627	79.4993	6.7715	79.5272	6.7803	79.5550
3230	6.8331	79.7220	6.8419	79.7498	6.8507	79.7776	6.8595	79.8055	6.8683	79.8333
3240	6.9210	80.0000	6.9298	80.0278	6.9386	80.0555	6.9473	80.0833	6.9561	80.1111
	50+	100+	50+	100+	50+	100+	50+	100+	50+	100+

Tables of Square Roots

$\sqrt{N}$ 5 $\sqrt{10N}$		$\sqrt{N}$ 6 $\sqrt{10N}$		$\sqrt{N}$ 7 $\sqrt{10N}$		$\sqrt{N}$ 8 $\sqrt{10N}$		$\sqrt{N}$ 9 $\sqrt{10N}$		N
50+	100+	50+	100+	50+	100+	50+	100+	50+	100+	
2.9623	67.4813	2.9717	67.5112	2.9811	67.5410	2.9906	67.5709	3.0000	67.6007	2800
3.0566	67.7796	3.0660	67.8094	3.0754	67.8392	3.0848	67.8690	3.0943	67.8988	2810
3.1507	68.0774	3.1601	68.1071	3.1695	68.1368	3.1789	68.1666	3.1883	68.1963	2820
3.2447	68.3746	3.2541	68.4043	3.2635	68.4340	3.2729	68.4636	3.2823	68.4933	2830
3.3385	68.6713	3.3479	68.7009	3.3573	68.7306	3.3667	68.7602	3.3760	68.7898	2840
3.4322	68.9675	3.4416	68.9970	3.4509	69.0266	3.4603	69.0562	3.4696	69.0858	2850
3.5257	69.2631	3.5350	69.2926	3.5444	69.3222	3.5537	69.3517	3.5630	69.3812	2860
3.6190	69.5583	3.6284	69.5877	3.6377	69.6172	3.6470	69.6467	3.6563	69.6762	2870
3.7122	69.8529	3.7215	69.8823	3.7308	69.9117	3.7401	69.9412	3.7494	69.9706	2880
3.8052	70.1470	3.8145	70.1764	3.8238	70.2058	3.8331	70.2351	3.8424	70.2645	2890
3.8981	70.4406	3.9073	70.4699	3.9166	70.4993	3.9259	70.5286	3.9351	70.5579	2900
3.9907	70.7337	4.0000	70.7630	4.0093	70.7923	4.0185	70.8215	4.0278	70.8508	2910
4.0833	71.0263	4.0925	71.0555	4.1018	71.0848	4.1110	71.1140	4.1202	71.1432	2920
4.1756	71.3184	4.1849	71.3476	4.1941	71.3768	4.2033	71.4060	4.2125	71.4351	2930
4.2679	71.6100	4.2771	71.6392	4.2863	71.6683	4.2955	71.6974	4.3047	71.7265	2940
4.3599	71.9011	4.3691	71.9302	4.3783	71.9593	4.3875	71.9884	4.3967	72.0174	2950
4.4518	72.1918	4.4610	72.2208	4.4702	72.2498	4.4794	72.2788	4.4885	72.3079	2960
4.5436	72.4819	4.5527	72.5109	4.5619	72.5399	4.5711	72.5688	4.5802	72.5978	2970
4.6352	72.7715	4.6443	72.8005	4.6535	72.8294	4.6626	72.8583	4.6717	72.8872	2980
4.7266	73.0607	4.7357	73.0896	4.7449	73.1185	4.7540	73.1473	4.7631	73.1762	2990
4.8179	73.3494	4.8270	73.3782	4.8361	73.4070	4.8452	73.4359	4.8544	73.4647	3000
4.9090	73.6376	4.9181	73.6663	4.9272	73.6951	4.9363	73.7239	4.9454	73.7527	3010
5.0000	73.9253	5.0091	73.9540	5.0182	73.9828	5.0273	74.0115	5.0364	74.0402	3020
5.0908	74.2125	5.0999	74.2412	5.1090	74.2699	5.1181	74.2986	5.1271	74.3273	3030
5.1815	74.4993	5.1906	74.5279	5.1996	74.5566	5.2087	74.5852	5.2178	74.6139	3040
5.2721	74.7856	5.2811	74.8142	5.2901	74.8428	5.2992	74.8714	5.3082	74.9000	3050
5.3624	75.0714	5.3715	75.1000	5.3805	75.1285	5.3895	75.1571	5.3986	75.1856	3060
5.4527	75.3568	5.4617	75.3853	5.4707	75.4138	5.4797	75.4423	5.4887	75.4708	3070
5.5428	75.6417	5.5518	75.6701	5.5608	75.6986	5.5698	75.7271	5.5788	75.7555	3080
5.6327	75.9261	5.6417	75.9545	5.6507	75.9830	5.6597	76.0114	5.6687	76.0398	3090
5.7225	76.2101	5.7315	76.2385	5.7405	76.2668	5.7494	76.2952	5.7584	76.3236	3100
5.8122	76.4936	5.8211	76.5220	5.8301	76.5503	5.8391	76.5786	5.8480	76.6069	3110
5.9017	76.7767	5.9106	76.8050	5.9196	76.8333	5.9285	76.8615	5.9375	76.8898	3120
5.9911	77.0593	6.0000	77.0875	6.0089	77.1158	6.0179	77.1440	6.0268	77.1722	3130
6.0803	77.3415	6.0892	77.3697	6.0981	77.3979	6.1070	77.4260	6.1160	77.4542	3140
6.1694	77.6232	6.1783	77.6513	6.1872	77.6795	6.1961	77.7076	6.2050	77.7358	3150
6.2583	77.9045	6.2672	77.9326	6.2761	77.9607	6.2850	77.9888	6.2939	78.0169	3160
6.3471	78.1853	6.3560	78.2134	6.3649	78.2415	6.3738	78.2695	6.3826	78.2975	3170
6.4358	78.4657	6.4447	78.4937	6.4535	78.5217	6.4624	78.5497	6.4712	78.5777	3180
6.5243	78.7456	6.5332	78.7736	6.5420	78.8016	6.5509	78.8295	6.5597	78.8575	3190
6.6127	79.0251	6.6216	79.0531	6.6304	79.0810	6.6392	79.1089	6.6480	79.1368	3200
6.7010	79.3042	6.7098	79.3321	6.7186	79.3600	6.7274	79.3878	6.7362	79.4157	3210
6.7891	79.5829	6.7979	79.6107	6.8067	79.6385	6.8155	79.6664	6.8243	79.6942	3220
6.8771	79.8611	6.8859	79.8889	6.8946	79.9166	6.9034	79.9444	6.9122	79.9722	3230
6.9649	80.1388	6.9737	80.1666	6.9825	80.1943	6.9912	80.2221	7.0000	80.2498	3240
50+	100+	50+	100+	50+	100+	50+	100+	50+	100+	

N	√N 0	√10N	√N 1	√10N	√N 2	√10N	√N 3	√10N	√N 4	√10N
	50+	100+	50+	100+	50+	100+	50+	100+	50+	100+
3250	7.0088	80.2776	7.0175	80.3053	7.0263	80.3330	7.0351	80.3608	7.0438	80.3885
3260	7.0964	80.5547	7.1052	80.5824	7.1139	80.6101	7.1227	80.6378	7.1314	80.6654
3270	7.1839	80.8314	7.1927	80.8591	7.2014	80.8867	7.2101	80.9143	7.2189	80.9420
3280	7.2713	81.1077	7.2800	81.1353	7.2887	81.1629	7.2975	81.1905	7.3062	81.2181
3290	7.3585	81.3836	7.3672	81.4111	7.3760	81.4387	7.3847	81.4663	7.3934	81.4938
3300	7.4456	81.6590	7.4543	81.6865	7.4630	81.7141	7.4717	81.7416	7.4804	81.7691
3310	7.5326	81.9341	7.5413	81.9615	7.5500	81.9890	7.5587	82.0165	7.5674	82.0440
3320	7.6194	82.2087	7.6281	82.2361	7.6368	82.2635	7.6455	82.2910	7.6541	82.3184
3330	7.7062	82.4829	7.7148	82.5103	7.7335	82.5377	7.7321	82.5651	7.7408	82.5924
3340	7.7927	82.7567	7.8014	82.7840	7.8100	82.8114	7.8187	82.8387	7.8273	82.8661
3350	7.8792	83.0301	7.8878	83.0574	7.8965	83.0847	7.9051	83.1120	7.9137	83.1393
3360	7.9655	83.3030	7.9741	83.3303	7.9828	83.3576	7.9914	83.3848	8.0000	83.4121
3370	8.0517	83.5756	8.0603	83.6028	8.0689	83.6301	8.0775	83.6573	8.0861	83.6845
3380	8.1378	83.8478	8.1464	83.8750	8.1550	83.9021	8.1636	83.9293	8.1722	83.9565
3390	8.2237	84.1195	8.2323	84.1467	8.2409	84.1738	8.2495	84.2010	8.2580	84.2281
3400	8.3095	84.3909	8.3181	84.4180	8.3267	84.4451	8.3352	84.4722	8.3438	84.4993
3410	8.3952	84.6619	8.4038	84.6889	8.4123	84.7160	8.4209	84.7431	8.4294	84.7701
3420	8.4808	84.9324	8.4893	84.9595	8.4979	84.9865	8.5064	85.0135	8.5150	85.0405
3430	8.5662	85.2026	8.5747	85.2296	8.5833	85.2566	8.5918	85.2836	8.6003	85.3106
3440	8.6515	85.4724	8.6600	85.4993	8.6686	85.5263	8.6771	85.5532	8.6856	85.5802
3450	8.7367	85.7418	8.7452	85.7687	8.7537	85.7956	8.7622	85.8225	8.7707	85.8494
3460	8.8218	86.0108	8.8303	86.0376	8.8388	86.0645	8.8473	86.0914	8.8558	86.1182
3470	8.9067	86.2794	8.9152	86.3062	8.9237	86.3330	8.9322	86.3599	8.9406	86.3867
3480	8.9915	86.5476	9.0000	86.5744	9.0085	86.6012	9.0169	86.6280	9.0254	86.6548
3490	9.0762	86.8154	9.0847	86.8422	9.0931	86.8689	9.1016	86.8957	9.1101	86.9224
3500	9.1608	87.0829	9.1693	87.1096	9.1777	87.1363	9.1861	87.1630	9.1946	87.1897
3510	9.2453	87.3499	9.2537	87.3766	9.2621	87.4033	9.2706	87.4300	9.2790	87.4567
3520	9.3296	87.6166	9.3380	87.6433	9.3464	87.6699	9.3549	87.6966	9.3633	87.7232
3530	9.4138	87.8829	9.4222	87.9096	9.4306	87.9362	9.4390	87.9628	9.4475	87.9894
3540	9.4979	88.1489	9.5063	88.1755	9.5147	88.2020	9.5231	88.2286	9.5315	88.2551
3550	9.5819	88.4144	9.5903	88.4410	9.5987	88.4675	9.6070	88.4940	9.6154	88.5206
3560	9.6657	88.6796	9.6741	88.7061	9.6825	88.7326	9.6909	88.7591	9.6992	88.7856
3570	9.7495	88.9444	9.7578	88.9709	9.7662	88.9974	9.7746	89.0238	9.7829	89.0503
3580	9.8331	89.2089	9.8415	89.2353	9.8498	89.2617	9.8582	89.2881	9.8665	89.3146
3590	9.9166	89.4730	9.9250	89.4993	9.9333	89.5257	9.9416	89.5521	9.9500	89.5785
3600	*0.0000	89.7367	*0.0083	89.7630	*0.0167	89.7894	*0.0250	89.8157	*0.0333	89.8420
3610	0.0833	90.0000	0.0916	90.0263	0.0999	90.0526	0.1082	90.0789	0.1166	90.1052
3620	0.1664	90.2630	0.1747	90.2893	0.1831	90.3155	0.1914	90.3418	0.1997	90.3681
3630	0.2495	90.5256	0.2578	90.5518	0.2661	90.5781	0.2744	90.6043	0.2827	90.6305
3640	0.3324	90.7878	0.3407	90.8140	0.3490	90.8402	0.3573	90.8664	0.3656	90.8926
3650	0.4152	91.0497	0.4235	91.0759	0.4318	91.1021	0.4401	91.1282	0.4483	91.1544
3660	0.4979	91.3113	0.5062	91.3374	0.5145	91.3635	0.5227	91.3897	0.5310	91.4158
3670	0.5805	91.5724	0.5888	91.5985	0.5970	91.6246	0.6053	91.6507	0.6135	91.6768
3680	0.6630	91.8333	0.6712	91.8593	0.6795	91.8854	0.6877	91.9114	0.6960	91.9375
3690	0.7454	92.0937	0.7536	92.1198	0.7618	92.1458	0.7701	92.1718	0.7783	92.1978
	60+	100+	60+	100+	60+	100+	60+	100+	60+	100+

Tables of Square Roots

√N 5	√10N	√N 6	√10N	√N 7	√10N	√N 8	√10N	√N 9	√10N	N
50+	100+	50+	100+	50+	100+	50+	100+	59+	100+	
7.0526	80.4162	7.0614	80.4439	7.0701	80.4716	7.0789	80.4993	7.0877	80.5270	3250
7.1402	80.6931	7.1489	80.7208	7.1577	80.7484	7.1664	80.7761	7.1752	80.8038	3260
7.2276	80.9696	7.2364	80.9972	7.2451	81.0249	7.2538	81.0525	7.2626	81.0801	3270
7.3149	81.2457	7.3236	81.2733	7.3324	81.3009	7.3411	81.3284	7.3498	81.3560	3280
7.4021	81.5213	7.4108	81.5489	7.4195	81.5764	7.4282	81.6040	7.4369	81.6315	3290
7.4891	81.7966	7.4978	81.8241	7.5065	81.8516	7.5152	81.8791	7.5239	81.9066	3300
7.5760	82.0714	7.5847	82.0989	7.5934	82.1263	7.6021	82.1538	7.6108	82.1812	3310
7.6628	82.3458	7.6715	82.3732	7.6802	82.4007	7.6888	82.4281	7.6975	82.4555	3320
7.7495	82.6198	7.7581	82.6472	7.7668	82.6746	7.7754	92.7019	7.7841	82.7293	3330
7.8360	82.8934	7.8446	82.9207	7.8533	82.9481	7.8619	82.9754	7.8705	83.0027	3340
7.9224	83.1666	7.9310	83.1939	7.9396	83.2212	7.9483	83.2485	7.9569	83.2757	3350
8.0086	83.4394	8.0172	83.4666	8.0259	83.4939	8.0345	83.5211	8.0431	83.5484	3360
8.0948	83.7117	8.1034	83.7389	8.1120	83.7662	8.1206	83.7934	8.1292	83.8206	3370
8.1808	83.9837	8.1893	84.0109	8.1979	84.0380	8.2065	84.0652	8.2151	84.0924	3380
8.2666	84.2553	8.2752	84.2824	8.2838	84.3095	8.2924	84.3366	8.3009	84.3638	3390
8.3524	84.5264	8.3609	84.5535	8.3695	84.5806	8.3781	84.6077	8.3866	84.6348	3400
8.4380	84.7972	8.4466	84.8242	8.4551	84.8513	8.4637	84.8783	8.4722	84.9054	3410
8.5235	85.0676	8.5320	85.0946	8.5406	85.1216	8.5491	85.1486	8.5577	85.1756	3420
8.6089	85.3375	8.6174	85.3645	8.6259	85.3915	8.6345	85.4184	8.6430	85.4454	3430
8.6941	85.6071	8.7026	85.6340	8.7112	85.6610	8.7197	85.6879	8.7282	85.7148	3440
8.7792	85.8763	8.7878	85.9032	8.7963	85.9301	8.8048	85.9570	8.8133	85.9839	3450
8.8643	86.1451	8.8727	86.1720	8.8812	86.1988	8.8897	86.2257	8.8982	86.2525	3460
8.9491	86.4135	8.9576	86.4403	8.9661	86.4672	8.9746	86.4940	8.9830	86.5208	3470
9.0339	86.6815	9.0424	86.7083	9.0508	86.7351	9.0593	86.7619	9.0678	86.7887	3480
9.1185	86.9492	9.1270	86.9759	9.1354	87.0027	9.1439	87.0294	9.1523	87.0561	3490
9.2030	87.2165	9.2115	87.2431	9.2199	87.2699	9.2284	87.2966	9.2368	87.3233	3500
9.2874	87.4833	9.2959	87.5100	9.3043	87.5367	9.3127	87.5633	9.3212	87.5900	3510
9.3717	87.7498	9.3801	87.7765	9.3886	87.8031	9.3970	87.8297	9.4054	87.8563	3520
9.4559	88.0159	9.4643	88.0425	9.4727	88.0691	9.4811	88.0957	9.4895	88.1223	3530
9.5399	88.2817	9.5483	88.3083	9.5567	88.3348	9.5651	88.3614	9.5735	88.3879	3540
9.6238	88.5471	9.6322	88.5736	9.6406	88.6001	9.6490	88.6266	9.6574	88.6531	3550
9.7076	88.8121	9.7160	88.8386	9.7244	88.8650	9.7327	88.8915	9.7411	88.9180	3560
9.7913	89.0767	9.7997	89.1031	9.8080	89.1296	9.8164	89.1560	9.8247	89.1825	3570
9.8749	89.3410	9.8832	89.3674	9.8916	89.3938	9.8999	89.4202	9.9083	89.4466	3580
9.9583	89.6049	9.9667	89.6312	9.9750	89.6576	9.9833	89.6839	9.9917	89.7103	3590
0.0417	89.8684	*0.0500	89.8947	*0.0583	89.9210	*0.0666	89.9474	*0.0750	89.9737	3600
0.1249	90.1315	0.1332	90.1578	0.1415	90.1841	0.1498	90.2104	0.1581	90.2367	3610
0.2080	90.3943	0.2163	90.4206	0.2246	90.4468	0.2329	90.4731	0.2412	90.4993	3620
0.2910	90.6568	0.2993	90.6830	0.3075	90.7092	0.3158	90.7354	0.3241	90.7616	3630
0.3738	90.9188	0.3821	90.9450	0.3904	90.9712	0.3987	90.9974	0.4070	91.0236	3640
0.4566	91.1805	0.4649	91.2067	0.4731	91.2328	0.4814	91.2590	0.4897	91.2851	3650
0.5392	91.4419	0.5475	91.4680	0.5558	91.4941	0.5640	91.5202	0.5723	91.5463	3660
0.6218	91.7029	0.6300	91.7290	0.6383	91.7551	0.6465	91.7811	0.6548	91.8072	3670
0.7042	91.9635	0.7124	91.9896	0.7207	92.0156	0.7289	92.0417	0.7371	92.0677	3680
0.7865	92.2238	0.7947	92.2498	0.8030	92.2758	0.8112	92.3018	0.8194	92.3278	3690
60+	100+	60+	100+	60+	100+	60+	100+	60+	100+	

Tables of Square Roots

N	0 √N	0 √10N	1 √N	1 √10N	2 √N	2 √10N	3 √N	3 √10N	4 √N	4 √10N
	60+	100+	60+	100+	60+	100+	60+	100+	60+	100+
3700	0.8276	92.3538	0.8358	92.3798	0.8441	92.4058	0.8523	92.4318	0.8605	92.4578
3710	0.9098	92.6136	0.9180	92.6396	0.9262	92.6655	0.9344	92.6915	0.9426	92.7174
3720	0.9918	92.8730	1.0000	92.8989	1.0082	92.9249	1.0164	92.9508	1.0246	92.9767
3730	1.0737	93.1321	1.0819	93.1580	1.0901	93.1839	1.0983	93.2097	1.1065	93.2356
3740	1.1555	93.3908	1.1637	93.4166	1.1719	93.4425	1.1801	93.4683	1.1882	93.4942
3750	1.2372	93.6492	1.2454	93.6750	1.2536	93.7008	1.2617	93.7266	1.2699	93.7524
3760	1.3188	93.9072	1.3270	93.9330	1.3351	93.9588	1.3433	93.9845	1.3514	94.0103
3770	1.4003	94.1649	1.4085	94.1906	1.4166	94.2164	1.4248	94.2421	1.4329	94.2679
3780	1.4817	94.4222	1.4898	94.4479	1.4980	94.4736	1.5061	94.4994	1.5142	94.5251
3790	1.5630	94.6792	1.5711	94.7049	1.5792	94.7306	1.5873	94.7563	1.5955	94.7819
3800	1.6441	94.9359	1.6523	94.9615	1.6604	94.9872	1.6685	95.0128	1.6766	95.0385
3810	1.7252	95.1922	1.7333	95.2178	1.7414	95.2434	1.7495	95.2690	1.7576	95.2946
3820	1.8061	95.4482	1.8142	95.4738	1.8223	95.4994	1.8304	95.5249	1.8385	95.5505
3830	1.8870	95.7039	1.8951	95.7294	1.9032	95.7549	1.9112	95.7805	1.9193	95.8060
3840	1.9677	95.9592	1.9758	95.9847	1.9839	96.0102	1.9919	96.0357	2.0000	96.0612
3850	2.0484	96.2142	2.0564	96.2396	2.0645	96.2651	2.0725	96.2906	2.0806	96.3161
3860	2.1289	96.4688	2.1369	96.4943	2.1450	96.5197	2.1530	96.5452	2.1611	96.5706
3870	2.2093	96.7232	2.2174	96.7486	2.2254	96.7740	2.2334	96.7994	2.2415	96.8248
3880	2.2896	96.9772	2.2977	97.0025	2.3057	97.0279	2.3137	97.0533	2.3217	97.0787
3890	2.3699	97.2308	2.3779	97.2562	2.3859	97.2815	2.3939	97.3069	2.4019	97.3322
3900	2.4500	97.4842	2.4580	97.5095	2.4660	97.5348	2.4740	97.5601	2.4820	97.5854
3910	2.5300	97.7372	2.5380	97.7625	2.5460	97.7878	2.5540	97.8130	2.5620	97.8383
3920	2.6099	97.9899	2.6179	98.0152	2.6259	98.0404	2.6339	98.0656	2.6418	98.0909
3930	2.6897	98.2423	2.6977	98.2675	2.7057	98.2927	2.7136	98.3179	2.7216	98.3431
3940	2.7694	98.4943	2.7774	98.5195	2.7853	98.5447	2.7933	98.5699	2.8013	98.5951
3950	2.8490	98.7461	2.8570	98.7712	2.8649	98.7964	2.8729	98.8215	2.8808	98.8467
3960	2.9285	98.9975	2.9365	99.0226	2.9444	99.0477	2.9524	99.0729	2.9603	99.0980
3970	3.0079	99.2486	3.0159	99.2737	3.0238	99.2988	3.0317	99.3239	3.0397	99.3489
3980	3.0872	99.4994	3.0952	99.5244	3.1031	99.5495	3.1110	99.5745	3.1189	99.5996
3990	3.1664	99.7498	3.1744	99.7749	3.1823	99.7999	3.1902	99.8249	3.1981	99.8499
4000	3.2456	*0.0000	3.2535	*0.0250	3.2614	*0.0500	3.2693	*0.0750	3.2772	*0.1000
4010	3.3246	0.2498	3.3325	0.2748	3.3404	0.2998	3.3482	0.3247	3.3561	0.3497
4020	3.4035	0.4994	3.4114	0.5243	3.4192	0.5492	3.4271	0.5742	3.4350	0.5991
4030	3.4823	0.7486	3.4902	0.7735	3.4980	0.7984	3.5059	0.8233	3.5138	0.8482
4040	3.5610	0.9975	3.5689	1.0224	3.5767	1.0473	3.5846	1.0721	3.5925	1.0970
4050	3.6396	1.2461	3.6475	1.2710	3.6553	1.2958	3.6632	1.3206	3.6710	1.3455
4060	3.7181	1.4944	3.7260	1.5192	3.7338	1.5440	3.7417	1.5688	3.7495	1.5937
4070	3.7966	1.7424	3.8044	1.7672	3.8122	1.7920	3.8201	1.8167	3.8279	1.8415
4080	3.8749	1.9901	3.8827	2.0149	3.8905	2.0396	3.8984	2.0643	3.9062	2.0891
4090	3.9531	2.2375	3.9609	2.2622	3.9687	2.2869	3.9766	2.3116	3.9844	2.3364
4100	4.0312	2.4846	4.0391	2.5093	4.0469	2.5339	4.0547	2.5586	4.0625	2.5833
4110	4.1093	2.7313	4.1171	2.7560	4.1249	2.7807	4.1327	2.8053	4.1405	2.8300
4120	4.1872	2.9778	4.1950	3.0025	4.2028	3.0271	4.2106	3.0517	4.2184	3.0763
4130	4.2651	3.2240	4.2729	3.2486	4.2806	3.2732	4.2884	3.2978	4.2962	3.3224
4140	4.3428	3.4699	4.3506	3.4945	4.3584	3.5190	4.3661	3.5436	4.3739	3.5682
	60+	200+	60+	200+	60+	200+	60+	200+	60+	200+

Tables of Square Roots

$\sqrt{N}$	5 $\sqrt{10N}$	$\sqrt{N}$	6 $\sqrt{10N}$	$\sqrt{N}$	7 $\sqrt{10N}$	$\sqrt{N}$	8 $\sqrt{10N}$	$\sqrt{N}$	9 $\sqrt{10N}$	N
60+	100+	60+	100+	60+	100+	60+	100+	60+	100+	
0.8687	92.4838	0.8769	92.5097	0.8851	92.5357	0.8933	92.5617	0.9016	92.5876	3700
0.9508	92.7434	0.9590	92.7693	0.9672	92.7952	0.9754	92.8212	0.9836	92.8471	3710
1.0328	93.0026	1.0410	93.0285	1.0492	93.0544	1.0574	93.0803	1.0655	93.1062	3720
1.1146	93.2615	1.1228	93.2874	1.1310	93.3132	1.1392	93.3391	1.1474	93.3649	3730
1.1964	93.5200	1.2046	93.5459	1.2127	93.5717	1.2209	93.5975	1.2291	93.6233	3740
1.2781	93.7782	1.2862	93.8040	1.2944	93.8298	1.3025	93.8556	1.3107	93.8814	3750
1.3596	94.0361	1.3677	94.0618	1.3759	94.0876	1.3840	94.1134	1.3922	94.1391	3760
1.4410	94.2936	1.4492	94.3193	1.4573	94.3451	1.4654	94.3708	1.4736	94.3965	3770
1.5224	94.5508	1.5305	94.5765	1.5386	94.6022	1.5467	94.6279	1.5549	94.6535	3780
1.6036	94.8076	1.6117	94.8333	1.6198	94.8589	1.6279	94.8846	1.6360	94.9102	3790
1.6847	95.0641	1.6928	95.0897	1.7009	95.1154	1.7090	95.1410	1.7171	95.1666	3800
1.7657	95.3203	1.7738	95.3458	1.7819	95.3714	1.7900	95.3970	1.7981	95.4226	3810
1.8466	95.5761	1.8547	95.6016	1.8628	95.6272	1.8708	95.6528	1.8789	95.6783	3820
1.9274	95.8316	1.9355	95.8571	1.9435	95.8826	1.9516	95.9081	1.9597	95.9337	3830
2.0081	96.0867	2.0161	96.1122	2.0242	96.1377	2.0323	96.1632	2.0403	96.1887	3840
2.0886	96.3415	2.0967	96.3670	2.1048	96.3925	2.1128	96.4179	2.1209	96.4434	3850
2.1691	96.5960	2.1772	96.6215	2.1852	96.6469	2.1932	96.6723	2.2013	96.6977	3860
2.2495	96.8502	2.2575	96.8756	2.2656	96.9010	2.2736	96.9264	2.2816	96.9518	3870
2.3298	97.1040	2.3378	97.1294	2.3458	97.1548	2.3538	97.1801	2.3618	97.2055	3880
2.4099	97.3575	2.4179	97.3829	2.4260	97.4082	2.4340	97.4335	2.4420	97.4589	3890
2.4900	97.6107	2.4980	97.6360	2.5060	97.6613	2.5140	97.6866	2.5220	97.7119	3900
2.5700	97.8636	2.5780	97.8889	2.5859	97.9141	2.5939	97.9394	2.6019	97.9646	3910
2.6498	98.1161	2.6578	98.1414	2.6658	98.1666	2.6738	98.1918	2.6817	98.2171	3920
2.7296	98.3683	2.7375	98.3935	2.7455	98.4187	2.7535	98.4439	2.7615	98.4691	3930
2.8092	98.6202	2.8172	98.6454	2.8252	98.6706	2.8331	98.6957	2.8411	98.7209	3940
2.8888	98.8718	2.8967	98.8970	2.9047	98.9221	2.9126	98.9472	2.9206	98.9724	3950
2.9682	99.1231	2.9762	99.1482	2.9841	99.1733	2.9921	99.1984	3.0000	99.2235	3960
3.0476	99.3740	3.0555	99.3991	3.0635	99.4242	3.0714	99.4492	3.0793	99.4743	3970
3.1269	99.6246	3.1348	99.6497	3.1427	99.6747	3.1506	99.6998	3.1585	99.7248	3980
3.2060	99.8750	3.2139	99.9000	3.2218	99.9250	3.2297	99.9500	3.2376	99.9750	3990
3.2851	*0.1250	3.2930	*0.1499	3.3009	*0.1749	3.3088	*0.1999	3.3167	*0.2249	4000
3.3640	0.3746	3.3719	0.3996	3.3798	0.4245	3.3877	0.4495	3.3956	0.4744	4010
3.4429	0.6240	3.4508	0.6489	3.4586	0.6739	3.4665	0.6988	3.4744	0.7237	4020
3.5217	0.8731	3.5295	0.8980	3.5374	0.9229	3.5453	0.9478	3.5531	0.9726	4030
3.6003	1.1219	3.6082	1.1467	3.6160	1.1716	3.6239	1.1964	3.6318	1.2213	4040
3.6789	1.3703	3.6867	1.3951	3.6946	1.4200	3.7024	1.4448	3.7103	1.4696	4050
3.7574	1.6185	3.7652	1.6432	3.7730	1.6680	3.7809	1.6928	3.7887	1.7176	4060
3.8357	1.8663	3.8436	1.8911	3.8514	1.9158	3.8592	1.9406	3.8670	1.9653	4070
3.9140	2.1138	3.9218	2.1386	3.9296	2.1633	3.9375	2.1880	3.9453	2.2128	4080
3.9922	2.3611	4.0000	2.3858	4.0078	2.4105	4.0156	2.4352	4.0234	2.4599	4090
4.0703	2.6080	4.0781	2.6327	4.0859	2.6573	4.0937	2.6820	4.1015	2.7067	4100
4.1483	2.8546	4.1561	2.8793	4.1639	2.9039	4.1716	2.9286	4.1794	2.9532	4110
4.2262	3.1010	4.2339	3.1256	4.2417	3.1502	4.2495	3.1748	4.2573	3.1994	4120
4.3040	3.3470	4.3117	3.3716	4.3195	3.3962	4.3273	3.4207	4.3351	3.4453	4130
4.3817	3.5927	4.3894	3.6173	4.3972	3.6418	4.4050	3.6664	4.4127	3.6909	4140
60+	200+	60+	200+	60+	200+	60+	200+	60+	200+	

Tables of Square Roots

N	D √N 60+	√10N 200+	1 √N 60+	√10N 200+	2 √N 60+	√10N 200+	3 √N 60+	√10N 200+	4 √N 60+	√10N 200+
4150	4.4205	3.7155	4.4283	3.7400	4.4360	3.7646	4.4438	3.7891	4.4515	3.8136
4160	4.4981	3.9608	4.5058	3.9853	4.5136	4.0098	4.5213	4.0343	4.5291	4.0588
4170	4.5755	4.2058	4.5833	4.2303	4.5910	4.2547	4.5988	4.2792	4.6065	4.3037
4180	4.6529	4.4505	4.6607	4.4749	4.6684	4.4994	4.6761	4.5238	4.6838	4.5483
4190	4.7302	4.6949	4.7379	4.7193	4.7457	4.7437	4.7534	4.7682	4.7611	4.7926
4200	4.8074	4.9390	4.8151	4.9634	4.8228	4.9878	4.8305	5.0122	4.8383	5.0366
4210	4.8845	5.1828	4.8922	5.2072	4.8999	5.2316	4.9076	5.2559	4.9153	5.2803
4220	4.9615	5.4264	4.9692	5.4507	4.9769	5.4751	4.9846	5.4994	4.9923	5.5237
4230	5.0385	5.6696	5.0461	5.6939	5.0538	5.7183	5.0615	5.7426	5.0692	5.7669
4240	5.1153	5.9126	5.1230	5.9369	5.1306	5.9612	5.1383	5.9854	5.1460	6.0097
4250	5.1920	6.1553	5.1997	6.1795	5.2074	6.2038	5.2150	6.2280	5.2227	6.2523
4260	5.2687	6.3977	5.2763	6.4219	5.2840	6.4461	5.2917	6.4703	5.2993	6.4946
4270	5.3452	6.6398	5.3529	6.6640	5.3605	6.6882	5.3682	6.7124	5.3758	6.7365
4280	5.4217	6.8816	5.4294	6.9058	5.4370	6.9299	5.4446	6.9541	5.4523	6.9783
4290	5.4981	7.1232	5.5057	7.1473	5.5134	7.1714	5.5210	7.1956	5.5286	7.2197
4300	5.5744	7.3644	5.5820	7.3885	5.5896	7.4126	5.5973	7.4367	5.6049	7.4608
4310	5.6506	7.6054	5.6582	7.6295	5.6658	7.6536	5.6734	7.6776	5.6810	7.7017
4320	5.7267	7.8461	5.7343	7.8702	5.7419	7.8942	5.7495	7.9183	5.7571	7.9423
4330	5.8027	8.0865	5.8103	8.1105	5.8179	8.1346	5.8255	8.1586	5.8331	8.1826
4340	5.8787	8.3267	5.8863	8.3507	5.8939	8.3747	5.9014	8.3987	5.9090	8.4226
4350	5.9545	8.5665	5.9621	8.5905	5.9697	8.6145	5.9773	8.6384	5.9848	8.6624
4360	6.0303	8.8061	6.0379	8.8301	6.0454	8.8540	6.0530	8.8780	6.0606	8.9019
4370	6.1060	9.0455	6.1135	9.0694	6.1211	9.0933	6.1287	9.1172	6.1362	9.1411
4380	6.1816	9.2845	6.1891	9.3084	6.1967	9.3323	6.2042	9.3562	6.2118	9.3800
4390	6.2571	9.5233	6.2646	9.5471	6.2722	9.5710	6.2797	9.5948	6.2873	9.6187
4400	6.3325	9.7618	6.3400	9.7856	6.3476	9.8094	6.3551	9.8333	6.3626	9.8571
4410	6.4078	10.0000	6.4154	10.0238	6.4229	10.0476	6.4304	10.0714	6.4379	10.0952
4420	6.4831	10.2380	6.4906	10.2617	6.4981	10.2855	6.5056	10.3093	6.5132	10.3331
4430	6.5582	10.4757	6.5658	10.4994	6.5733	10.5232	6.5808	10.5469	6.5883	10.5707
4440	6.6333	10.7131	6.6408	10.7368	6.6483	10.7605	6.6558	10.7843	6.6633	10.8080
4450	6.7083	10.9502	6.7158	10.9739	6.7233	10.9976	6.7308	11.0213	6.7383	11.0450
4460	6.7832	11.1871	6.7907	11.2108	6.7982	11.2345	6.8057	11.2581	6.8132	11.2818
4470	6.8581	11.4237	6.8655	11.4474	6.8730	11.4710	6.8805	11.4947	6.8880	11.5183
4480	6.9328	11.6601	6.9403	11.6837	6.9477	11.7073	6.9552	11.7310	6.9627	11.7546
4490	7.0075	11.8962	7.0149	11.9198	7.0224	11.9434	7.0298	11.9670	7.0373	11.9906
4500	7.0820	12.1320	7.0895	12.1556	7.0969	12.1792	7.1044	12.2027	7.1118	12.2263
4510	7.1565	12.3676	7.1640	12.3911	7.1714	12.4147	7.1789	12.4382	7.1863	12.4618
4520	7.2309	12.6029	7.2384	12.6264	7.2458	12.6499	7.2533	12.6735	7.2607	12.6970
4530	7.3053	12.8380	7.3127	12.8615	7.3201	12.8849	7.3276	12.9084	7.3350	12.9319
4540	7.3795	13.0728	7.3869	13.0962	7.3944	13.1197	7.4018	13.1431	7.4092	13.1666
4550	7.4537	13.3073	7.4611	13.3307	7.4685	13.3542	7.4759	13.3776	7.4833	13.4010
4560	7.5278	13.5416	7.5352	13.5650	7.5426	13.5884	7.5500	13.6118	7.5574	13.6352
4570	7.6018	13.7756	7.6092	13.7990	7.6166	13.8224	7.6240	13.8457	7.6314	13.8691
4580	7.6757	14.0093	7.6831	14.0327	7.6905	14.0561	7.6979	14.0794	7.7052	14.1028
4590	7.7495	14.2429	7.7569	14.2662	7.7643	14.2895	7.7717	14.3129	7.7791	14.3362
	60+	200+	60+	200+	60+	200+	60+	200+	60+	200+

Tables of Square Roots

√N 5	√10N	√N 6	√10N	√N 7	√10N	√N 8	√10N	√N 9	√10N	N
60+	200+	60+	200+	60+	200+	60+	200+	60+	200+	
4.4593	3.8382	4.4670	3.8627	4.4748	3.8872	4.4826	3.9117	4.4903	3.9363	4150
4.5368	4.0833	4.5446	4.1078	4.5523	4.1323	4.5601	4.1568	4.5678	4.1813	4160
4.6142	4.3282	4.6220	4.3526	4.6297	4.3771	4.6375	4.4016	4.6452	4.4260	4170
4.6916	4.5727	4.6993	4.5972	4.7070	4.6216	4.7148	4.6460	4.7225	4.6705	4180
4.7688	4.8170	4.7765	4.8414	4.7843	4.8658	4.7920	4.8902	4.7997	4.9146	4190
4.8460	5.0610	4.8537	5.0853	4.8614	5.1097	4.8691	5.1341	4.8768	5.1585	4200
4.9230	5.3047	4.9307	5.3290	4.9384	5.3534	4.9461	5.3777	4.9538	5.4020	4210
5.0000	5.5480	5.0077	5.5724	5.0154	5.5967	5.0231	5.6210	5.0308	5.6453	4220
5.0769	5.7912	5.0846	5.8155	5.0922	5.8397	5.0999	5.8640	5.1076	5.8883	4230
5.1537	6.0340	5.1613	6.0582	5.1690	6.0825	5.1767	6.1068	5.1844	6.1310	4240
5.2304	6.2765	5.2380	6.3008	5.2457	6.3250	5.2534	6.3492	5.2610	6.3734	4250
5.3070	6.5188	5.3146	6.5430	5.3223	6.5672	5.3299	6.5914	5.3376	6.6156	4260
5.3835	6.7607	5.3911	6.7849	5.3988	6.8091	5.4064	6.8333	5.4141	6.8574	4270
5.4599	7.0024	5.4675	7.0266	5.4752	7.0507	5.4828	7.0749	5.4905	7.0990	4280
5.5363	7.2438	5.5439	7.2679	5.5515	7.2921	5.5591	7.3162	5.5668	7.3403	4290
5.6125	7.4849	5.6201	7.5090	5.6277	7.5331	5.6354	7.5572	5.6430	7.5813	4300
5.6887	7.7258	5.6963	7.7499	5.7039	7.7739	5.7115	7.7980	5.7191	7.8220	4310
5.7647	7.9663	5.7723	7.9904	5.7799	8.0144	5.7875	8.0385	5.7951	8.0625	4320
5.8407	8.2066	5.8483	8.2306	5.8559	8.2547	5.8635	8.2787	5.8711	8.3027	4330
5.9166	8.4466	5.9241	8.4706	5.9318	8.4946	5.9394	8.5186	5.9469	8.5426	4340
5.9924	8.6864	6.0000	8.7103	6.0076	8.7343	6.0152	8.7582	6.0227	8.7822	4350
6.0681	8.9258	6.0757	8.9493	6.0833	8.9737	6.0908	8.9976	6.0984	9.0215	4360
6.1438	9.1650	6.1513	9.1889	6.1589	9.2128	6.1665	9.2367	6.1740	9.2606	4370
6.2193	9.4039	6.2269	9.4278	6.2344	9.4517	6.2420	9.4755	6.2495	9.4994	4380
6.2948	9.6426	6.3023	9.6664	6.3099	9.6902	6.3174	9.7141	6.3250	9.7379	4390
6.3702	9.8809	6.3777	9.9047	6.3852	9.9286	6.3928	9.9524	6.4003	9.9762	4400
6.4455	10.1190	6.4530	10.1428	6.4605	10.1666	6.4680	10.1904	6.4756	10.2142	4410
6.5207	10.3568	6.5282	10.3806	6.5357	10.4044	6.5432	10.4281	6.5507	10.4519	4420
6.5958	10.5944	6.6033	10.6181	6.6108	10.6419	6.6183	10.6656	6.6258	10.6893	4430
6.6708	10.8317	6.6783	10.8554	6.6858	10.8791	6.6933	10.9028	6.7008	10.9265	4440
6.7458	11.0687	6.7533	11.0924	6.7608	11.1161	6.7683	11.1398	6.7757	11.1634	4450
6.8207	11.3055	6.8281	11.3291	6.8356	11.3528	6.8431	11.3764	6.8506	11.4001	4460
6.8954	11.5420	6.9029	11.5656	6.9104	11.5892	6.9179	11.6129	6.9253	11.6365	4470
6.9701	11.7782	6.9776	11.8018	6.9851	11.8254	6.9925	11.8490	7.0000	11.8726	4480
7.0448	12.0142	7.0522	12.0377	7.0597	12.0613	7.0671	12.0849	7.0746	12.1085	4490
7.1193	12.2499	7.1267	12.2734	7.1342	12.2970	7.1416	12.3205	7.1491	12.3441	4500
7.1938	12.4853	7.2012	12.5088	7.2086	12.5324	7.2161	12.5559	7.2235	12.5794	4510
7.2681	12.7205	7.2756	12.7440	7.2830	12.7675	7.2904	12.7910	7.2978	12.8145	4520
7.3424	12.9554	7.3498	12.9789	7.3573	13.0023	7.3647	13.0258	7.3721	13.0493	4530
7.4166	13.1901	7.4240	13.2135	7.4314	13.2370	7.4389	13.2604	7.4463	13.2838	4540
7.4907	13.4245	7.4981	13.4479	7.5056	13.4713	7.5130	13.4947	7.5204	13.5181	4550
7.5648	13.6586	7.5722	13.6820	7.5796	13.7054	7.5870	13.7288	7.5944	13.7522	4560
7.6387	13.8925	7.6461	13.9159	7.6535	13.9392	7.6609	13.9626	7.6683	13.9860	4570
7.7126	14.1261	7.7200	14.1495	7.7274	14.1728	7.7348	14.1962	7.7422	14.2195	4580
7.7864	14.3595	7.7938	14.3828	7.8012	14.4062	7.8086	14.4295	7.8159	14.4528	4590
60+	200+	60+	200+	60+	200+	60+	200+	60+	200+	

Tables of Square Roots

N	$\sqrt{N}$ 0	$\sqrt{10N}$	$\sqrt{N}$ 1	$\sqrt{10N}$	$\sqrt{N}$ 2	$\sqrt{10N}$	$\sqrt{N}$ 3	$\sqrt{10N}$	$\sqrt{N}$ 4	$\sqrt{10N}$
	60+	200+	60+	200+	60+	200+	60+	200+	60+	200+
4600	7.8233	14.4761	7.8307	14.4994	7.8380	14.5227	7.8454	14.5460	7.8528	14.5693
4610	7.8970	14.7091	7.9043	14.7324	7.9117	14.7557	7.9191	14.7790	7.9264	14.8022
4620	7.9706	14.9419	7.9779	14.9651	7.9853	14.9884	7.9926	15.0116	8.0000	15.0349
4630	8.0441	15.1743	8.0515	15.1976	8.0588	15.2208	8.0661	15.2440	8.0735	15.2673
4640	8.1175	15.4066	8.1249	15.4298	8.1322	15.4530	8.1396	15.4762	8.1469	15.4994
4650	8.1909	15.6386	8.1982	15.6618	8.2056	15.6850	8.2129	15.7081	8.2202	15.7313
4660	8.2642	15.8703	8.2715	15.8935	8.2788	15.9167	8.2862	15.9398	8.2935	15.9630
4670	8.3374	16.1018	8.3447	16.1250	8.3520	16.1481	8.3593	16.1712	8.3667	16.1944
4680	8.4105	16.3331	8.4178	16.3562	8.4251	16.3793	8.4324	16.4024	8.4398	16.4255
4690	8.4836	16.5641	8.4909	16.5872	8.4982	16.6102	8.5055	16.6333	8.5128	16.6564
4700	8.5565	16.7948	8.5638	16.8179	8.5711	16.8410	8.5784	16.8640	8.5857	16.8871
4710	8.6294	17.0253	8.6367	17.0484	8.6440	17.0714	8.6513	17.0944	8.6586	17.1175
4720	8.7022	17.2556	8.7095	17.2786	8.7168	17.3016	8.7241	17.3446	8.7314	17.3476
4730	8.7750	17.4856	8.7823	17.5086	8.7895	17.5316	8.7968	17.5546	8.8041	17.5776
4740	8.8477	17.7154	8.8549	17.7384	8.8622	17.7613	8.8694	17.7843	8.8767	17.8073
4750	8.9202	17.9449	8.9275	17.9679	8.9348	17.9908	8.9420	18.0138	8.9493	18.0367
4760	8.9928	18.1742	9.0000	18.1972	9.0072	18.2201	9.0145	18.2430	9.0217	18.2659
4770	9.0652	18.4033	9.0724	18.4262	9.0797	18.4491	9.0869	18.4720	9.0941	18.4949
4780	9.1375	18.6321	9.1448	18.6550	9.1520	18.6778	9.1592	18.7007	9.1665	18.7236
4790	9.2098	18.8607	9.2171	18.8835	9.2243	18.9064	9.2315	18.9292	9.2387	18.9521
4800	9.2820	19.0890	9.2892	19.1118	9.2965	19.1347	9.3037	19.1575	9.3109	19.1803
4810	9.3542	19.3171	9.3614	19.3399	9.3686	19.3627	9.3758	19.3855	9.3830	19.4083
4820	9.4262	19.5450	9.4334	19.5678	9.4406	19.5905	9.4478	19.6133	9.4550	19.6361
4830	9.4982	19.7726	9.5054	19.7954	9.5125	19.8181	9.5198	19.8409	9.5270	19.8636
4840	9.5701	20.0000	9.5773	20.0227	9.5845	20.0455	9.5917	20.0682	9.5989	20.0909
4850	9.6419	20.2272	9.6491	20.2499	9.6563	20.2726	9.6635	20.2953	9.6707	20.3180
4860	9.7137	20.4541	9.7209	20.4768	9.7280	20.4994	9.7352	20.5221	9.7424	20.5448
4870	9.7854	20.6808	9.7926	20.7034	9.7997	20.7261	9.8069	20.7487	9.8140	20.7714
4880	9.8570	20.9072	9.8642	20.9299	9.8713	20.9525	9.8785	20.9751	9.8856	20.9977
4890	9.9285	21.1334	9.9357	21.1561	9.9428	21.1787	9.9500	21.2013	9.9571	21.2239
4900	*0.0000	21.3594	*0.0071	21.3820	*0.0143	21.4046	*0.0214	21.4272	*0.0286	21.4498
4910	0.0714	21.5852	0.0785	21.6078	0.0857	21.6303	0.0928	21.6529	0.0999	21.6754
4920	0.1427	21.8107	0.1498	21.8333	0.1570	21.8558	0.1641	21.8783	0.1712	21.9009
4930	0.2140	22.0360	0.2211	22.0586	0.2282	22.0811	0.2353	22.1036	0.2424	22.1261
4940	0.2851	22.2611	0.2922	22.2836	0.2994	22.3061	0.3065	22.3286	0.3136	22.3511
4950	0.3562	22.4860	0.3633	22.5084	0.3704	22.5309	0.3776	22.5534	0.3847	22.5758
4960	0.4273	22.7106	0.4344	22.7330	0.4415	22.7555	0.4486	22.7779	0.4557	22.8004
4970	0.4982	22.9350	0.5053	22.9574	0.5124	22.9798	0.5195	23.0022	0.5266	23.0247
4980	0.5691	23.1591	0.5762	23.1815	0.5833	23.2039	0.5904	23.2263	0.5975	23.2487
4990	0.6399	23.3831	0.6470	23.4055	0.6541	23.4278	0.6612	23.4502	0.6682	23.4726
5000	0.7107	23.6068	0.7177	23.6292	0.7248	23.6515	0.7319	23.6739	0.7390	23.6962
5010	0.7814	23.8303	0.7884	23.8526	0.7955	23.8750	0.8025	23.8973	0.8096	23.9196
5020	0.8520	24.0536	0.8590	24.0759	0.8661	24.0982	0.8731	24.1205	0.8802	24.1428
5030	0.9225	24.2766	0.9295	24.2989	0.9366	24.3212	0.9436	24.3435	0.9507	24.3658
5040	0.9930	24.4994	1.0000	24.5217	1.0070	24.5440	1.0141	24.5662	1.0211	24.5885
	70+	200+	70+	200+	70+	200+	70+	200+	70+	200+

Tables of Square Roots

$\sqrt{N}$ (5)	$\sqrt{10N}$	$\sqrt{N}$ (6)	$\sqrt{10N}$	$\sqrt{N}$ (7)	$\sqrt{10N}$	$\sqrt{N}$ (8)	$\sqrt{10N}$	$\sqrt{N}$ (9)	$\sqrt{10N}$	N
60+	200+	60+	200+	60+	200+	60+	200+	60+	200+	
7.8602	14.5926	7.8675	14.6159	7.8749	14.6392	7.8823	14.6625	7.8896	14.6858	4600
7.9338	14.8255	7.9412	14.8488	7.9485	14.8721	7.9559	14.8953	7.9632	14.9186	4610
8.0074	15.0581	8.0147	15.0814	8.0221	15.1046	8.0294	15.1279	8.0368	15.1511	4620
8.0808	15.2905	8.0882	15.3137	8.0955	15.3369	8.1029	15.3602	8.1102	15.3834	4630
8.1542	15.5226	8.1616	15.5458	8.1689	15.5690	8.1762	15.5922	8.1836	15.6154	4640
8.2275	15.7545	8.2349	15.7777	8.2422	15.8008	8.2495	15.8240	8.2569	15.8472	4650
8.3008	15.9861	8.3081	16.0093	8.3154	16.0324	8.3228	16.0555	8.3301	16.0787	4660
8.3740	16.2175	8.3813	16.2406	8.3886	16.2637	8.3959	16.2868	8.4032	16.3100	4670
8.4471	16.4486	8.4544	16.4717	8.4617	16.4948	8.4690	16.5179	8.4763	16.5410	4680
8.5201	16.6795	8.5274	16.7026	8.5347	16.7256	8.5420	16.7487	8.5493	16.7718	4690
8.5930	16.9101	8.6003	16.9332	8.6076	16.9562	8.6149	16.9793	8.6222	17.0023	4700
8.6659	17.1405	8.6731	17.1635	8.6804	17.1866	8.6877	17.2096	8.6950	17.2326	4710
8.7386	17.3707	8.7459	17.3937	8.7532	17.4167	8.7605	17.4396	8.7677	17.4626	4720
8.8113	17.6006	8.8186	17.6235	8.8259	17.6465	8.8331	17.6695	8.8404	17.6924	4730
8.8840	17.8302	8.8912	17.8532	8.8985	17.8761	8.9057	17.8991	8.9130	17.9220	4740
8.9565	18.0596	8.9638	18.0826	8.9710	18.1055	8.9783	18.1284	8.9855	18.1513	4750
9.0290	18.2888	9.0362	18.3117	9.0435	18.3346	9.0507	18.3575	9.0579	18.3804	4760
9.1014	18.5177	9.1086	18.5406	9.1158	18.5635	9.1231	18.5864	9.1303	18.6092	4770
9.1737	18.7464	9.1809	18.7693	9.1881	18.7921	9.1954	18.8150	9.2026	18.8378	4780
9.2459	18.9749	9.2532	18.9977	9.2604	19.0205	9.2676	19.0434	9.2748	19.0662	4790
9.3181	19.2031	9.3253	19.2259	9.3325	19.2487	9.3397	19.2715	9.3470	19.2943	4800
9.3902	19.4311	9.3974	19.4539	9.4046	19.4767	9.4118	19.4994	9.4190	19.5222	4810
9.4622	19.6588	9.4694	19.6816	9.4766	19.7043	9.4838	19.7271	9.4910	19.7499	4820
9.5342	19.8863	9.5414	19.9091	9.5485	19.9318	9.5557	19.9545	9.5629	19.9773	4830
9.6060	20.1136	9.6132	20.1363	9.6204	20.1590	9.6276	20.1817	9.6348	20.2045	4840
9.6778	20.3406	9.6850	20.3633	9.6922	20.3860	9.6994	20.4087	9.7065	20.4314	4850
9.7496	20.5675	9.7567	20.5901	9.7639	20.6128	9.7711	20.6354	9.7782	20.6581	4860
9.8212	20.7940	9.8284	20.8167	9.8355	20.8393	9.8427	20.8619	9.8498	20.8846	4870
9.8928	21.0204	9.8999	21.0430	9.9071	21.0656	9.9142	21.0882	9.9214	21.1108	4880
9.9643	21.2465	9.9714	21.2691	9.9786	21.2917	9.9857	21.3143	9.9929	21.3368	4890
0.0357	21.4723	*0.0428	21.4949	*0.0500	21.5175	*0.0571	21.5401	*0.0643	21.5626	4900
0.1071	21.6980	0.1142	21.7205	0.1213	21.7431	0.1285	21.7656	0.1356	21.7882	4910
0.1783	21.9234	0.1855	21.9459	0.1926	21.9685	0.1997	21.9910	0.2068	22.0135	4920
0.2496	22.1486	0.2567	22.1711	0.2638	22.1936	0.2709	22.2161	0.2780	22.2386	4930
0.3207	22.3736	0.3278	22.3960	0.3349	22.4185	0.3420	22.4410	0.3491	22.4635	4940
0.3918	22.5983	0.3989	22.6208	0.4060	22.6432	0.4131	22.6657	0.4202	22.6981	4950
0.4628	22.8228	0.4699	22.8452	0.4769	22.8677	0.4840	22.8901	0.4911	22.9125	4960
0.5337	23.0471	0.5408	23.0695	0.5479	23.0919	0.5549	23.1143	0.5620	23.1367	4970
0.6045	23.2711	0.6116	23.2935	0.6187	23.3159	0.6258	23.3383	0.6329	23.3607	4980
0.6753	23.4950	0.6824	23.5173	0.6895	23.5397	0.6965	23.5621	0.7036	23.5844	4990
0.7460	23.7186	0.7531	23.7409	0.7602	23.7633	0.7672	23.7856	0.7743	23.8080	5000
0.8167	23.9420	0.8237	23.9643	0.8308	23.9866	0.8378	24.0089	0.8449	24.0312	5010
0.8872	24.1651	0.8943	24.1874	0.9013	24.2097	0.9084	24.2320	0.9154	24.2543	5020
0.9577	24.3881	0.9648	24.4103	0.9718	24.4326	0.9789	24.4549	0.9859	24.4772	5030
1.0282	24.6108	1.0352	24.6330	1.0422	24.6553	1.0493	24.6775	1.0563	24.6998	5040
70+	200+	70+	200+	70+	200+	70+	200+	70+	200+	

Tables of Square Roots

N	0 √N	0 √10N	1 √N	1 √10N	2 √N	2 √10N	3 √N	3 √10N	4 √N	4 √10N
	70+	200+	70+	200+	70+	200+	70+	200+	70+	200+
5050	1.0634 24.7221		1.0704 24.7443		1.0774 24.7665		1.0845 24.7888		1.0915 24.8110	
5060	1.1337 24.9444		1.1407 24.9667		1.1477 24.9889		1.1548 25.0111		1.1618 25.0333	
5070	1.2039 25.1666		1.2110 25.1888		1.2180 25.2110		1.2250 25.2332		1.2320 25.2554	
5080	1.2741 25.3886		1.2811 25.4107		1.2881 25.4329		1.2952 25.4551		1.3022 25.4773	
5090	1.3442 25.6103		1.3512 25.6324		1.3583 25.6546		1.3653 25.6768		1.3723 25.6989	
5100	1.4143 25.8318		1.4213 25.8539		1.4283 25.8761		1.4353 25.8982		1.4423 25.9203	
5110	1.4843 26.0531		1.4913 26.0752		1.4983 26.0973		1.5052 26.1194		1.5122 26.1415	
5120	1.5542 26.2742		1.5612 26.2963		1.5681 26.3184		1.5751 26.3405		1.5821 26.3625	
5130	1.6240 26.4950		1.6310 26.5171		1.6380 26.5392		1.6450 26.5613		1.6519 26.5933	
5140	1.6938 26.7157		1.7008 26.7377		1.7077 26.7598		1.7147 26.7818		1.7217 26.8039	
5150	1.7635 26.9361		1.7705 26.9581		1.7774 26.9802		1.7844 27.0022		1.7914 27.0242	
5160	1.8331 27.1563		1.8401 27.1783		1.8471 27.2004		1.8540 27.2224		1.8610 27.2444	
5170	1.9027 27.3763		1.9097 27.3983		1.9166 27.4203		1.9236 27.4423		1.9305 27.4643	
5180	1.9722 27.5961		1.9792 27.6181		1.9861 27.6401		1.9931 27.6620		2.0000 27.6840	
5190	2.0417 27.8157		2.0486 27.8377		2.0555 27.8596		2.0625 27.8815		2.0694 27.9035	
5200	2.1110 28.0351		2.1180 28.0570		2.1249 28.0789		2.1318 28.1009		2.1388 28.1228	
5210	2.1803 28.2542		2.1873 28.2761		2.1942 28.2981		2.2011 28.3200		2.2080 28.3418	
5220	2.2496 28.4732		2.2565 28.4951		2.2634 28.5170		2.2703 28.5388		2.2772 28.5607	
5230	2.3187 28.6919		2.3257 28.7138		2.3226 28.7357		2.3395 28.7575		2.3464 28.7794	
5240	2.3878 28.9105		2.3948 28.9323		2.4017 28.9541		2.4086 28.9760		2.4155 28.9978	
5250	2.4569 29.1288		2.4638 29.1506		2.4707 29.1724		2.4776 29.1942		2.4845 29.2161	
5260	2.5259 29.3469		2.5328 29.3687		2.5396 29.3905		2.5465 29.4123		2.5534 29.4341	
5270	2.5948 29.5648		2.6017 29.5866		2.6085 29.6084		2.6154 29.6301		2.6223 29.6519	
5280	2.6636 29.7825		2.6705 29.8043		2.6774 29.8260		2.6842 29.8478		2.6911 29.8695	
5290	2.7324 30.0000		2.7393 30.0217		2.7461 30.0435		2.7530 30.0652		2.7599 30.0869	
5300	2.8011 30.2173		2.8080 30.2390		2.8148 30.2607		2.8217 30.2824		2.8286 30.3041	
5310	2.8697 30.4344		2.8766 30.4561		2.8835 30.4778		2.8903 30.4995		2.8972 30.5211	
5320	2.9383 30.6513		2.9452 30.6729		2.9520 30.6946		2.9589 30.7163		2.9657 30.7379	
5330	3.0068 30.8679		3.0137 30.8896		3.0205 30.9112		3.0274 30.9329		3.0342 30.9545	
5340	3.0753 31.0844		3.0821 31.1060		3.0890 31.1277		3.0958 31.1493		3.1027 31.1709	
5350	3.1437 31.3007		3.1505 31.3223		3.1574 31.3439		3.1642 31.3655		3.1710 31.3871	
5360	3.2120 31.5167		3.2189 31.5383		3.2257 31.5599		3.2325 31.5815		3.2393 31.6031	
5370	3.2803 31.7326		3.2871 31.7542		3.2939 31.7758		3.3008 31.7973		3.3076 31.8189	
5380	3.3485 31.9483		3.3553 31.9698		3.3621 31.9914		3.3689 32.0129		3.3757 32.0345	
5390	3.4166 32.1637		3.4234 32.1853		3.4302 32.2068		3.4370 32.2283		3.4439 32.2499	
5400	3.4847 32.3790		3.4915 32.4005		3.4983 32.4220		3.5051 32.4435		3.5119 32.4651	
5410	3.5527 32.5941		3.5595 32.6156		3.5663 32.6371		3.5731 32.6585		3.5799 32.6800	
5420	3.6206 32.8089		3.6274 32.8304		3.6342 32.8519		3.6410 32.8734		3.6478 32.8948	
5430	3.6885 33.0236		3.6953 33.0451		3.7021 33.0665		3.7089 33.0880		3.7157 33.1094	
5440	3.7564 33.2381		3.7631 33.2595		3.7699 33.2809		3.7767 33.3024		3.7835 33.3238	
5450	3.8241 33.4524		3.8309 33.4738		3.8377 33.4952		3.8444 33.5166		3.8512 33.5380	
5460	3.8918 33.6664		3.8986 33.6878		3.9053 33.7092		3.9121 33.7306		3.9189 33.7520	
5470	3.9594 33.8803		3.9662 33.9017		3.9730 33.9231		3.9797 33.9444		3.9865 33.9658	
5480	4.0270 34.0940		4.0338 34.1154		4.0405 34.1367		4.0473 34.1581		4.0540 34.1794	
5490	4.0945 34.3075		4.1013 34.3288		4.1080 34.3502		4.1148 34.3715		4.1215 34.3928	
	70+	200+	70+	200+	70+	200+	70+	200+	70+	200+

√N	5 √10N	√N	6 √10N	√N	7 √10N	√N	8 √10N	√N	9 √10N	N
70+	200+	70+	200+	70+	200+	70+	200+	70+	200+	
1.0985	24.8333	1.1056	24.8555	1.1126	24.8777	1.1196	24.9000	1.1266	24.9222	5050
1.1688	25.0555	1.1758	25.0778	1.1829	25.1000	1.1899	25.1222	1.1969	25.1444	5060
1.2390	25.2776	1.2461	25.2998	1.2531	25.3220	1.2601	25.3442	1.2671	25.3664	5070
1.3092	25.4994	1.3162	25.5216	1.3232	25.5438	1.3302	25.5660	1.3372	25.5881	5080
1.3793	25.7211	1.3863	25.7432	1.3933	25.7654	1.4003	25.7875	1.4073	25.8097	5090
1.4493	25.9425	1.4563	25.9646	1.4633	25.9867	1.4703	26.0088	1.4773	26.0310	5100
1.5192	26.1637	1.5262	26.1858	1.5332	26.2079	1.5402	26.2300	1.5472	26.2521	5110
1.5891	26.3846	1.5961	26.4067	1.6031	26.4288	1.6101	26.4509	1.6170	26.4730	5120
1.6589	26.6054	1.6659	26.6274	1.6729	26.6495	1.6798	26.6716	1.6868	26.6936	5130
1.7287	26.8259	1.7356	26.8480	1.7426	26.8700	1.7496	26.8920	1.7565	26.9141	5140
1.7983	27.0463	1.8053	27.0683	1.8123	27.0903	1.8192	27.1123	1.8262	27.1343	5150
1.8679	27.2664	1.8749	27.2884	1.8818	27.3104	1.8888	27.3324	1.8958	27.3543	5160
1.9375	27.4863	1.9444	27.5082	1.9514	27.5302	1.9583	27.5522	1.9653	27.5742	5170
2.0069	27.7060	2.0139	27.7279	2.0208	27.7499	2.0278	27.7718	2.0347	27.7938	5180
2.0763	27.9254	2.0833	27.9474	2.0902	27.9693	2.0972	27.9912	2.1041	28.0132	5190
2.1457	28.1447	2.1526	28.1666	2.1595	28.1885	2.1665	28.2104	2.1734	28.2323	5200
2.2150	28.3637	2.2219	28.3856	2.2288	28.4075	2.2357	28.4294	2.2426	28.4513	5210
2.2842	28.5826	2.2911	28.6045	2.2980	28.6263	2.3049	28.6482	2.3118	28.6701	5220
2.3533	28.8012	2.3602	28.8231	2.3671	28.8449	2.3740	28.8668	2.3809	28.8886	5230
2.4224	29.0197	2.4293	29.0415	2.4362	29.0633	2.4431	29.0851	2.4500	29.1070	5240
2.4914	29.2379	2.4983	29.2597	2.5052	29.2815	2.5121	29.3033	2.5190	29.3251	5250
2.5603	29.4559	2.5672	29.4777	2.5741	29.4995	2.5810	29.5212	2.5879	29.5430	5260
2.6292	29.6737	2.6361	29.6955	2.6430	29.7172	2.6498	29.7390	2.6567	29.7607	5270
2.6980	29.8913	2.7049	29.9130	2.7118	29.9348	2.7186	29.9565	2.7255	29.9783	5280
2.7668	30.1087	2.7736	30.1304	2.7805	30.1521	2.7874	30.1738	2.7942	30.1956	5290
2.8354	30.3259	2.8423	30.3476	2.8492	30.3693	2.8560	30.3910	2.8629	30.4127	5300
2.9040	30.5428	2.9109	30.5645	2.9178	30.5862	2.9246	30.6079	2.9315	30.6296	5310
2.9726	30.7596	2.9794	30.7813	2.9863	30.8029	2.9932	30.8246	3.0000	30.8463	5320
3.0411	30.9762	3.0479	30.9978	3.0548	31.0195	3.0616	31.0411	3.0685	31.0628	5330
3.1095	31.1926	3.1163	31.2142	3.1232	31.2358	3.1300	31.2574	3.1369	31.2791	5340
3.1779	31.4087	3.1847	31.4303	3.1915	31.4519	3.1984	31.4735	3.2052	31.4951	5350
3.2462	31.6247	3.2530	31.6463	3.2598	31.6679	3.2666	31.6894	3.2735	31.7110	5360
3.3144	31.8405	3.3212	31.8620	3.3280	31.8836	3.3348	31.9052	3.3417	31.9267	5370
3.3826	32.0560	3.3894	32.0776	3.3962	32.0991	3.4030	32.1207	3.4098	32.1422	5380
3.4507	32.2714	3.4575	32.2929	3.4643	32.3144	3.4711	32.3360	3.4779	32.3575	5390
3.5187	32.4866	3.5255	32.5081	3.5323	32.5296	3.5391	32.5511	3.5459	32.5726	5400
3.5867	32.7015	3.5935	32.7230	3.6003	32.7445	3.6071	32.7660	3.6139	32.7875	5410
3.6546	32.9163	3.6614	32.9378	3.6682	32.9592	3.6750	32.9807	3.6817	33.0021	5420
3.7225	33.1309	3.7292	33.1523	3.7360	33.1738	3.7428	33.1952	3.7496	33.2166	5430
3.7902	33.3452	3.7970	33.3667	3.8038	33.3881	3.8106	33.4095	3.8173	33.4309	5440
3.8580	33.5594	3.8647	33.5808	3.8715	33.6022	3.8783	33.6236	3.8850	33.6450	5450
3.9256	33.7734	3.9324	33.7948	3.9392	33.8162	3.9459	33.8376	3.9527	33.8599	5460
3.9932	33.9872	4.0000	34.0085	4.0068	34.0299	4.0135	34.0513	4.0203	34.0726	5470
4.0608	34.2008	4.0675	34.2221	4.0743	34.2435	4.0810	34.2648	4.0878	34.2862	5480
4.1283	34.4142	4.1350	34.4355	4.1418	34.4568	4.1485	34.4781	4.1552	34.4995	5490
70+	200+	70+	200+	70+	200+	70+	200+	70+	200+	

Tables of Square Roots

N	0 √N	0 √10N	1 √N	1 √10N	2 √N	2 √10N	3 √N	3 √10N	4 √N	4 √10N
	70+	200+	70+	200+	70+	200+	70+	200+	70+	200+
5500	4.1620	34.5208	4.1687	34.5421	4.1755	34.5634	4.1822	34.5847	4.1889	34.6061
5510	4.2294	34.7339	4.2361	34.7552	4.2428	34.7765	4.2496	34.7978	4.2563	34.8191
5520	4.2967	34.9468	4.3034	34.9681	4.3102	34.9894	4.3169	35.0106	4.3236	35.0319
5530	4.3640	35.1595	4.3707	35.1808	4.3774	35.2020	4.3841	35.2233	4.3909	35.2446
5540	4.4312	35.3720	4.4379	35.3933	4.4446	35.4145	4.4513	35.4358	4.4580	35.4570
5550	4.4983	35.5844	4.5050	35.6056	4.5117	35.6268	4.5185	35.6480	4.5252	35.6693
5560	4.5654	35.7965	4.5721	35.8177	4.5788	35.8389	4.5855	35.8601	4.5922	35.8813
5570	4.6324	36.0085	4.6391	36.0297	4.6458	36.0508	4.6525	36.0720	4.6592	36.0932
5580	4.6994	36.2202	4.7061	36.2414	4.7128	36.2626	4.7195	36.2837	4.7262	36.3049
5590	4.7663	36.4318	4.7730	36.4530	4.7797	36.4741	4.7864	36.4952	4.7930	36.5164
5600	4.8331	36.6432	4.8398	36.6643	4.8465	36.6854	4.8532	36.7066	4.8599	36.7277
5610	4.8999	36.8544	4.9066	36.8755	4.9133	36.8966	4.9200	36.9177	4.9266	36.9388
5620	4.9667	37.0654	4.9733	37.0865	4.9800	37.1076	4.9867	37.1287	4.9933	37.1497
5630	5.0333	37.2762	5.0400	37.2973	5.0467	37.3184	5.0533	37.3394	5.0600	37.3605
5640	5.0999	37.4868	5.1066	37.5079	5.1132	37.5289	5.1199	37.5500	5.1266	37.5710
5650	5.1665	37.6973	5.1731	37.7183	5.1798	37.7394	5.1864	37.7604	5.1931	37.7814
5660	5.2330	37.9075	5.2396	37.9286	5.2463	37.9496	5.2529	37.9706	5.2596	37.9916
5670	5.2994	38.1176	5.3060	38.1386	5.3127	38.1596	5.3193	38.1806	5.3260	38.2016
5680	5.3658	38.3275	5.3724	38.3485	5.3790	38.3695	5.3857	38.3904	5.3923	38.4114
5690	5.4321	38.5372	5.4387	38.5582	5.4453	38.5791	5.4520	38.6001	5.4586	38.6210
5700	5.4983	38.7467	5.5050	38.7677	5.5116	38.7886	5.5182	38.8095	5.5248	38.8305
5710	5.5645	38.9561	5.5712	38.9770	5.5778	38.9979	5.5844	39.0188	5.5910	39.0397
5720	5.6307	39.1652	5.6373	39.1861	5.6439	39.2070	5.6505	39.2279	5.6571	39.2488
5730	5.6968	39.3742	5.7034	39.3951	5.7100	39.4160	5.7166	39.4368	5.7232	39.4577
5740	5.7628	39.5830	5.7694	39.6038	5.7760	39.6247	5.7826	39.6456	5.7892	39.6664
5750	5.8288	39.7916	5.8353	39.8124	5.8419	39.8333	5.8485	39.8541	5.8551	39.8750
5760	5.8947	40.0000	5.9013	40.0208	5.9078	40.0417	5.9144	40.0625	5.9210	40.0833
5770	5.9605	40.2082	5.9671	40.2291	5.9737	40.2499	5.9803	40.2707	5.9868	40.2915
5780	6.0263	40.4163	6.0329	40.4371	6.0395	40.4579	6.0460	40.4787	6.0526	40.4995
5790	6.0921	40.6242	6.0986	40.6450	6.1052	40.6657	6.1118	40.6865	6.1183	40.7073
5800	6.1577	40.8319	6.1643	40.8527	6.1709	40.8734	6.1774	40.8942	6.1840	40.9149
5810	6.2234	41.0394	6.2299	41.0602	6.2365	41.0809	6.2430	41.1016	6.2496	41.1224
5820	6.2889	41.2468	6.2955	41.2675	6.3020	41.2882	6.3086	41.3089	6.3151	41.3297
5830	6.3544	41.4539	6.3610	41.4746	6.3675	41.4953	6.3741	41.5160	6.3806	41.5367
5840	6.4199	41.6609	6.4264	41.6816	6.4330	41.7023	6.4395	41.7230	6.4461	41.7437
5850	6.4853	41.8677	6.4918	41.8884	6.4984	41.9091	6.5049	41.9297	6.5114	41.9504
5860	6.5506	42.0744	6.5572	42.0950	6.5637	42.1157	6.5702	42.1363	6.5768	42.1570
5870	6.6159	42.2808	6.6225	42.3015	6.6290	42.3221	6.6355	42.3427	6.6420	42.3634
5880	6.6812	42.4871	6.6877	42.5077	6.6942	42.5283	6.7007	42.5490	6.7072	42.5696
5890	6.7463	42.6932	6.7529	42.7138	6.7594	42.7344	6.7659	42.7550	6.7724	42.7756
5900	6.8115	42.8992	6.8180	42.9197	6.8245	42.9403	6.8310	42.9609	6.8375	42.9815
5910	6.8765	43.1049	6.8830	43.1255	6.8895	43.1460	6.8960	43.1666	6.9025	43.1872
5920	6.9415	43.3105	6.9480	43.3311	6.9545	43.3516	6.9610	43.3721	6.9675	43.3927
5930	7.0065	43.5159	7.0130	43.5364	7.0195	43.5570	7.0260	43.5775	7.0325	43.5980
5940	7.0714	43.7212	7.0779	43.7417	7.0844	43.7622	7.0909	43.7827	7.0973	43.8032
	70+	200+	70+	200+	70+	200+	70+	200+	70+	200+

Tables of Square Roots

5		6		7		8		9		N
$\sqrt{N}$	$\sqrt{10N}$	$\sqrt{N}$	$\sqrt{10N}$	$\sqrt{N}$	$\sqrt{10N}$	$\sqrt{N}$	$\sqrt{10N}$	$\sqrt{N}$	$\sqrt{10N}$	
70+	200+	70+	200+	70+	200+	70+	200+	70+	200+	
4.1957	34.6274	4.2024	34.6487	4.2092	34.6700	4.2159	34.6913	4.2226	34.7126	5500
4.2630	34.8404	4.2698	34.8617	4.2765	34.8830	4.2832	34.9042	4.2900	34.9255	5510
4.3303	35.0532	4.3371	35.0745	4.3438	35.0957	4.3505	35.1170	4.3572	35.1383	5520
4.3976	35.2658	4.4043	35.2871	4.4110	35.3083	4.4177	35.3296	4.4245	35.3508	5530
4.4648	35.4782	4.4715	35.4995	4.4782	35.5207	4.4849	35.5419	4.4916	35.5632	5540
4.5319	35.6905	4.5386	35.7117	4.5453	35.7329	4.5520	35.7541	4.5587	35.7753	5550
4.5989	35.9025	4.6056	35.9237	4.6123	35.9449	4.6190	35.9661	4.6257	35.9873	5560
4.6659	36.1144	4.6726	36.1386	4.6793	36.1567	4.6860	36.1779	4.6927	36.1991	5570
4.7329	36.3260	4.7395	36.3472	4.7462	36.3684	4.7529	36.3895	4.7596	36.4107	5580
4.7997	36.5375	4.8064	36.5587	4.8131	36.5798	4.8198	36.6009	4.8265	36.6221	5590
4.8665	36.7488	4.8732	36.7699	4.8799	36.7910	4.8866	36.8122	4.8933	36.8333	5600
4.9333	36.9599	4.9400	36.9810	4.9466	37.0021	4.9533	37.0232	4.9600	37.0443	5610
5.0000	37.1708	5.0067	37.1919	5.0133	37.2130	5.0200	37.2341	5.0267	37.2551	5620
5.0666	37.3815	5.0733	37.4026	5.0800	37.4237	5.0866	37.4447	5.0933	37.4658	5630
5.1332	37.5921	5.1399	37.6131	5.1465	37.6342	5.1532	37.6552	5.1598	37.6763	5640
5.1997	37.8024	5.2064	37.8235	5.2130	37.8445	5.2197	37.8655	5.2263	37.8865	5650
5.2662	38.0126	5.2728	38.0336	5.2795	38.0546	5.2861	38.0756	5.2928	38.0966	5660
5.3326	38.2226	5.3392	38.2436	5.3459	38.2646	5.3525	38.2855	5.3591	38.3065	5670
5.3989	38.4324	5.4056	38.4534	5.4122	38.4743	5.4188	38.4953	5.4255	38.5162	5680
5.4652	38.6420	5.4718	38.6629	5.4785	38.6839	5.4851	38.7048	5.4917	38.7258	5690
5.5315	38.8514	5.5381	38.8724	5.5447	38.8933	5.5513	38.9142	5.5579	38.9351	5700
5.5976	39.0607	5.6042	39.0816	5.6108	39.1025	5.6175	39.1234	5.6241	39.1443	5710
5.6637	39.2697	5.6703	39.2906	5.6769	39.3115	5.6836	39.3324	5.6902	39.3533	5720
5.7298	39.4786	5.7364	39.4995	5.7430	39.5204	5.7496	39.5412	5.7562	39.5621	5730
5.7958	39.6873	5.8024	39.7082	5.8090	39.7290	5.8156	39.7499	5.8222	39.7707	5740
5.8617	39.8958	5.8683	39.9167	5.8749	39.9375	5.8815	39.9583	5.8881	39.9792	5750
5.9276	40.1041	5.9342	40.1250	5.9408	40.1458	5.9474	40.1666	5.9539	40.1874	5760
5.9934	40.3123	6.0000	40.3331	6.0066	40.3539	6.0132	40.3747	6.0197	40.3955	5770
6.0592	40.5203	6.0658	40.5411	6.0723	40.5618	6.0789	40.5826	6.0855	40.6034	5780
6.1249	40.7281	6.1315	40.7488	6.1380	40.7696	6.1446	40.7904	6.1512	40.8111	5790
6.1906	40.9357	6.1971	40.9564	6.2037	40.9772	6.2102	40.9979	6.2168	41.0187	5800
6.2561	41.1431	6.2627	41.1638	6.2693	41.1846	6.2758	41.2053	6.2824	41.2260	5810
6.3217	41.3504	6.3282	41.3711	6.3348	41.3918	6.3413	41.4125	6.3479	41.4332	5820
6.3872	41.5574	6.3937	41.5781	6.4003	41.5988	6.4068	41.6195	6.4134	41.6402	5830
6.4526	41.7643	6.4591	41.7850	6.4657	41.8057	6.4722	41.8264	6.4788	41.8471	5840
6.5180	41.9711	6.5245	41.9917	6.5310	42.0124	6.5376	42.0331	6.5441	42.0537	5850
6.5833	42.1776	6.5898	42.1983	6.5963	42.2189	6.6029	42.2396	6.6094	42.2602	5860
6.6485	42.3840	6.6551	42.4046	6.6616	42.4252	6.6681	42.4459	6.6746	42.4665	5870
6.7138	42.5902	6.7203	42.6108	6.7268	42.6314	6.7333	42.6520	6.7398	42.6726	5880
6.7789	42.7962	6.7854	42.8168	6.7919	42.8374	6.7984	42.8580	6.8049	42.8786	5890
6.8440	43.0021	6.8505	43.0226	6.8570	42.0432	6.8635	43.0638	6.8700	43.0843	5900
6.9090	43.2077	6.9155	43.2283	6.9220	43.2488	6.9285	43.2694	6.9350	43.2900	5910
6.9740	43.4132	6.9805	43.4338	6.9870	43.4543	6.9935	43.4748	7.0000	43.4954	5920
7.0390	43.6186	7.0454	43.6391	7.0519	43.6596	7.0584	43.6801	7.0649	43.7006	5930
7.1038	43.8237	7.1103	43.8442	7.1168	43.8647	7.1233	43.8852	7.1298	43.9057	5940
70+	200+	70+	200+	70+	200+	70+	200+	70+	200+	

Tables of Square Roots

N	$\sqrt{N}$ 70+	$\sqrt{10N}$ 200+	$\sqrt{N}$ 70+	$\sqrt{10N}$ 200+	$\sqrt{N}$ 70+	$\sqrt{10N}$ 200+	$\sqrt{N}$ 70+	$\sqrt{10N}$ 200+	$\sqrt{N}$ 70+	$\sqrt{10N}$ 200+
	0		**1**		**2**		**3**		**4**	
5950	7.1362	43.9262	7.1427	43.9467	7.1492	43.9672	7.1557	43.9877	7.1622	44.0082
5960	7.2010	44.1311	7.2075	44.1516	7.2140	44.1721	7.2205	44.1925	7.2269	44.2130
5970	7.2658	44.3358	7.2722	44.3563	7.2787	44.3768	7.2852	44.3972	7.2917	44.4177
5980	7.3305	44.5404	7.3369	44.5608	7.3434	44.5813	7.3499	44.6017	7.3563	44.5222
5990	7.3951	44.7448	7.4016	44.7652	7.4080	44.7856	7.4145	44.8060	7.4209	44.8265
6000	7.4597	44.9490	7.4661	44.9694	7.4726	44.9898	7.4790	45.0102	7.4855	45.0306
6010	7.5242	45.1530	7.5306	45.1734	7.5371	45.1938	7.5435	45.2142	7.5500	45.2346
6020	7.5887	45.3569	7.5951	45.3773	7.6015	45.3976	7.6080	45.4180	7.6144	45.4384
6030	7.6531	45.5606	7.6595	45.5809	7.6660	45.6013	7.6724	45.6217	7.6788	45.6420
6040	7.7174	45.7641	7.7239	45.7845	7.7303	45.8048	7.7367	45.8251	7.7432	45.8455
6050	7.7817	45.9675	7.7881	45.9878	7.7946	46.0081	7.8010	46.0285	7.8075	46.0498
6060	7.8460	46.1707	7.8524	46.1910	7.8588	46.2113	7.8653	46.2316	7.8717	46.2519
6070	7.9102	46.3737	7.9166	46.3940	7.9230	46.4143	7.9295	46.4346	7.9359	46.4549
6080	7.9744	46.5766	7.9808	46.5968	7.9872	46.6171	7.9936	46.6374	8.0000	46.6577
6090	8.0385	46.7793	8.0449	46.7995	8.0513	46.8198	8.0577	46.8400	8.0641	46.8603
6100	8.1025	46.9818	8.1089	47.0020	8.1153	47.0223	8.1217	47.0425	8.1281	47.0627
6110	8.1665	47.1841	8.1729	47.2044	8.1793	47.2246	8.1857	47.2448	8.1921	47.2650
6120	8.2304	47.3863	8.2368	47.4065	8.2432	47.4268	8.2496	47.4470	8.2560	47.4672
6130	8.2943	47.5884	8.3007	47.6086	8.3071	47.6288	8.3135	47.6489	8.3199	47.6691
6140	8.3582	47.7902	8.3645	47.8104	8.3709	47.8306	8.3773	47.8508	8.3837	47.8709
6150	8.4219	47.9919	8.4283	48.0121	8.4347	48.0323	8.4411	48.0524	8.4474	48.0726
6160	8.4857	48.1935	8.4920	48.2136	8.4984	48.2338	8.5048	48.2539	8.5111	48.2740
6170	8.5493	48.3948	8.5557	48.4150	8.5621	48.4351	8.5684	48.4552	8.5748	48.4754
6180	8.6130	48.5961	8.6193	48.6162	8.6257	48.6363	8.6321	48.6564	8.6384	48.6765
6190	8.6766	48.7971	8.6829	48.8172	8.6893	48.8373	8.6956	48.8574	8.7020	48.8775
6200	8.7401	48.9980	8.7464	49.0181	8.7528	49.0382	8.7591	49.0582	8.7655	49.0783
6210	8.8036	49.1987	8.8099	49.2188	8.8162	49.2388	8.8226	49.2589	8.8289	49.2790
6220	8.8670	49.3993	8.8733	49.4193	8.8797	49.4394	8.8860	49.4594	8.8923	49.4795
6230	8.9303	49.5997	8.9367	49.6197	8.9430	49.6397	8.9494	49.6598	8.9557	49.6798
6240	8.9937	49.7999	9.0000	49.8199	9.0063	49.8399	9.0127	49.8600	9.0190	49.8800
6250	9.0569	50.0000	9.0633	50.0200	9.0696	50.0400	9.0759	50.0600	9.0822	50.0800
6260	9.1202	50.1999	9.1265	50.2199	9.1328	50.2399	9.1391	50.2599	9.1454	50.2798
6270	9.1833	50.3997	9.1896	50.4196	9.1960	50.4396	9.2023	50.4596	9.2086	50.4795
6280	9.2465	50.5993	9.2528	50.6192	9.2591	50.6392	9.2654	50.6591	9.2717	50.6791
6290	9.3095	50.7987	9.3158	50.8187	9.3221	50.8386	9.3284	50.8585	9.3347	50.8785
6300	9.3725	50.9980	9.3788	51.0179	9.3851	51.0378	9.3914	51.0578	9.3977	51.0777
6310	9.4355	51.1971	9.4418	51.2170	9.4481	51.2369	9.4544	51.2568	9.4607	51.2767
6320	9.4984	51.3961	9.5047	51.4160	9.5110	51.4359	9.5173	51.4558	9.5236	51.4756
6330	9.5613	51.5949	9.5676	51.6148	9.5739	51.6347	9.5801	51.6545	9.5864	51.6744
6340	9.6241	51.7936	9.6304	51.8134	9.6367	51.8333	9.6430	51.8531	9.6492	51.8730
6350	9.6869	51.9920	9.6932	52.0119	9.6994	52.0317	9.7057	52.0516	9.7120	52.0714
6360	9.7496	52.1904	9.7559	52.2102	9.7621	52.2301	9.7684	52.2499	9.7747	52.2697
6370	9.8123	52.3886	9.8185	52.4084	9.8248	52.4282	9.8311	52.4480	9.8373	52.4678
6380	9.8749	52.5866	9.8812	52.6064	9.8874	52.6262	9.8937	52.6460	9.8999	52.6658
6390	9.9375	52.7845	9.9437	52.8043	9.9500	52.8240	9.9562	52.8438	9.9625	52.8636
	70+	200+	70+	200+	70+	200+	70+	200+	70+	200+

Tables of Square Roots

$\sqrt{N}$ (5)	$\sqrt{10N}$	$\sqrt{N}$ (6)	$\sqrt{10N}$	$\sqrt{N}$ (7)	$\sqrt{10N}$	$\sqrt{N}$ (8)	$\sqrt{10N}$	$\sqrt{N}$ (9)	$\sqrt{10N}$	N
70+	200+	70+	200+	70+	200+	70+	200+	70+	200+	
7.1686	44.0287	7.1751	44.0492	7.1816	44.0697	7.1881	44.0901	7.1946	44.1106	5950
7.2334	44.2335	7.2399	44.2540	7.2464	44.2744	7.2528	44.2949	7.2593	44.3154	5960
7.2981	44.4381	7.3046	44.4586	7.3111	44.4790	7.3175	44.4995	7.3240	44.5199	5970
7.3628	44.6426	7.3692	44.6630	7.3757	44.6835	7.3822	44.7039	7.3886	44.7243	5980
7.4274	44.8469	7.4338	44.8673	7.4403	44.8877	7.4468	44.9081	7.4532	44.9286	5990
7.4919	45.0510	7.4984	45.0714	7.5048	45.0918	7.5113	45.1122	7.5177	45.1326	6000
7.5564	45.2550	7.5629	45.2754	7.5693	45.2957	7.5758	45.3161	7.5822	45.3365	6010
7.6209	45.4588	7.6273	45.4791	7.6338	45.4995	7.6402	45.5199	7.6466	45.5402	6020
7.6853	45.6624	7.6917	45.6827	7.6981	45.7031	7.7046	45.7234	7.7110	45.7438	6030
7.7496	45.8658	7.7560	45.8862	7.7625	45.9065	7.7689	45.9268	7.7753	45.9471	6040
7.8139	46.0691	7.8203	46.0894	7.8267	46.1097	7.8332	46.1300	7.8396	46.1504	6050
7.8781	46.2722	7.8845	46.2925	7.8909	46.3128	7.8974	46.3331	7.9038	46.3534	6060
7.9423	46.4752	7.9487	46.4954	7.9551	46.5157	7.9615	46.5360	7.9679	46.5563	6070
8.0064	46.6779	8.0128	46.6982	8.0192	46.7185	8.0256	46.7387	8.0320	46.7590	6080
8.0705	46.8805	8.0769	46.9008	8.0833	46.9210	8.0897	46.9413	8.0961	46.9615	6090
8.1345	47.0830	8.1409	47.1032	8.1473	47.1235	8.1537	47.1437	8.1601	47.1639	6100
8.1985	47.2853	8.2049	47.3055	8.2113	47.3257	8.2176	47.3459	8.2240	47.3661	6110
8.2624	47.4874	8.2688	47.5076	8.2752	47.5278	8.2815	47.5480	8.2879	47.5682	6120
8.3262	47.6893	8.3326	47.7095	8.3390	47.7297	8.3454	47.7499	8.3518	47.7701	6130
8.3901	47.8911	8.3964	47.9113	8.4028	47.9314	8.4092	47.9516	8.4156	47.9718	6140
8.4538	48.0927	8.4602	48.1129	8.4666	48.1330	8.4729	48.1532	8.4793	48.1733	6150
8.5175	48.2942	8.5239	48.3143	8.5302	48.3345	8.5366	48.3546	8.5430	48.3747	6160
8.5812	48.4955	8.5875	48.5156	8.5939	48.5357	8.6003	48.5558	8.6066	48.5759	6170
8.6448	48.6966	8.6511	48.7167	8.6575	48.7368	8.6638	48.7569	8.6702	48.7770	6180
8.7083	48.8976	8.7147	48.9177	8.7210	48.9377	8.7274	48.9578	8.7337	48.9779	6190
8.7718	49.0984	8.7782	49.1184	8.7845	49.1385	8.7909	49.1586	8.7972	49.1787	6200
8.8353	49.2990	8.8416	49.3191	8.8480	49.3391	8.8543	49.3592	8.8606	49.3792	6210
8.8987	49.4995	8.9050	49.5195	8.9113	49.5396	8.9177	49.5596	8.9240	49.5796	6220
8.9620	49.6998	8.9683	49.7198	8.9747	49.7399	8.9810	49.7599	8.9873	49.7799	6230
9.0253	49.9000	9.0316	49.9200	9.0380	49.9400	9.0443	49.9600	9.0506	49.9800	6240
9.0886	50.1000	9.0949	50.1200	9.1012	50.1400	9.1075	50.1599	9.1138	50.1799	6250
9.1518	50.2998	9.1581	50.3198	9.1644	50.3398	9.1707	50.3597	9.1770	50.3797	6260
9.2149	50.4995	9.2212	50.5195	9.2275	50.5394	9.2338	50.5594	9.2401	50.5793	6270
9.2780	50.6990	9.2843	50.7190	9.2906	50.7389	9.2969	50.7588	9.3032	50.7788	6280
9.3410	50.8984	9.3473	50.9183	9.3536	50.9382	9.3599	50.9582	9.3662	50.9781	6290
9.4040	51.0976	9.4103	51.1175	9.4166	51.1374	9.4229	51.1573	9.4292	51.1772	6300
9.4670	51.2966	9.4733	51.3165	9.4796	51.3364	9.4858	51.3563	9.4921	51.3762	6310
9.5299	51.4955	9.5362	51.5154	9.5424	51.5353	9.5487	51.5552	9.5550	51.5750	6320
9.5927	51.6943	9.5990	51.7141	9.6053	51.7340	9.6116	51.7538	9.6178	51.7737	6330
9.6555	51.8928	9.6618	51.9127	9.6681	51.9325	9.6743	51.9524	9.6806	51.9722	6340
9.7183	52.0913	9.7245	52.1111	9.7308	52.1309	9.7371	52.1507	9.7433	52.1706	6350
9.7810	52.2895	9.7872	52.3093	9.7935	52.3292	9.7997	52.3490	9.8060	52.3688	6360
9.8436	52.4876	9.8499	52.5074	9.8561	52.5272	9.8624	52.5470	9.8686	52.5668	6370
9.9062	52.6856	9.9125	52.7054	9.9187	52.7251	9.9250	52.7449	9.9312	52.7647	6380
9.9687	52.8834	9.9750	52.9031	9.9812	52.9229	9.9875	52.9427	9.9938	52.9624	6390
70+	200+	70+	200+	70+	200+	70+	200+	70+	200+	

N	0 $\sqrt{N}$	0 $\sqrt{10N}$	1 $\sqrt{N}$	1 $\sqrt{10N}$	2 $\sqrt{N}$	2 $\sqrt{10N}$	3 $\sqrt{N}$	3 $\sqrt{10N}$	4 $\sqrt{N}$	4 $\sqrt{10N}$
	80+	200+	80+	200+	80+	200+	80+	200+	80+	200+
6400	0.0000	52.9822	0.0063	53.0020	0.0125	53.0217	0.0187	53.0415	0.0250	53.0613
6410	0.0625	53.1798	0.0687	53.1995	0.0750	53.2193	0.0812	53.2390	0.0875	53.2588
6420	0.1249	53.3772	0.1311	53.3969	0.1374	53.4167	0.1436	53.4364	0.1499	53.4561
6430	0.1873	53.5744	0.1935	53.5942	0.1998	53.6139	0.2060	53.6336	0.2122	53.6533
6440	0.2496	53.7716	0.2558	53.7913	0.2621	53.8110	0.2683	53.8307	0.2745	53.8504
6450	0.3119	53.9685	0.3181	53.9882	0.3243	54.0079	0.3306	54.0276	0.3368	54.0472
6460	0.3741	54.1653	0.3803	54.1850	0.3866	54.2046	0.3928	54.2243	0.3990	54.2440
6470	0.4363	54.3619	0.4425	54.3816	0.4487	54.4013	0.4550	54.4209	0.4612	54.4406
6480	0.4984	54.5584	0.5047	54.5781	0.5109	54.5977	0.5171	54.6174	0.5233	54.6370
6490	0.5605	54.7548	0.5667	54.7744	0.5729	54.7940	0.5792	54.8137	0.5854	54.8333
6500	0.6226	54.9510	0.6288	54.9706	0.6350	54.9902	0.6412	55.0098	0.6474	55.0294
6510	0.6846	55.1470	0.6908	55.1666	0.6970	55.1862	0.7032	55.2058	0.7094	55.2254
6520	0.7465	55.3429	0.7527	55.3625	0.7589	55.3821	0.7651	55.4016	0.7713	55.4212
6530	0.8084	55.5386	0.8146	55.5582	0.8208	55.5778	0.8270	55.5973	0.8332	55.6169
6540	0.8703	55.7342	0.8764	55.7538	0.8826	55.7733	0.8888	55.7929	0.8950	55.8124
6550	0.9321	55.9297	0.9382	55.9492	0.9444	55.9687	0.9506	55.9883	0.9568	56.0078
6560	0.9938	56.1250	1.0000	56.1445	1.0062	56.1640	1.0123	56.1835	1.0185	56.2030
6570	1.0555	56.3201	1.0617	56.3396	1.0679	56.3591	1.0740	56.3786	1.0802	56.3981
6580	1.1172	56.5151	1.1234	56.5346	1.1295	56.5541	1.1357	56.5736	1.1419	56.5931
6590	1.1788	56.7100	1.1850	56.7294	1.1911	56.7489	1.1973	56.7684	1.2034	56.7879
6600	1.2404	56.9047	1.2465	56.9241	1.2527	56.9436	1.2588	56.9630	1.2650	56.9825
6610	1.3019	57.0992	1.3081	57.1187	1.3142	57.1381	1.3204	57.1575	1.3265	57.1770
6620	1.3634	57.2936	1.3695	57.3130	1.3757	57.3325	1.3818	57.3519	1.3880	57.3713
6630	1.4248	57.4879	1.4310	57.5073	1.4371	57.5267	1.4432	57.5461	1.4494	57.5655
6640	1.4862	57.6820	1.4923	57.7014	1.4985	57.7208	1.5046	57.7402	1.5107	57.7596
6650	1.5475	57.8759	1.5537	57.8953	1.5598	57.9147	1.5659	57.9341	1.5721	57.9535
6660	1.6088	58.0698	1.6150	58.0891	1.6211	58.1085	1.6272	58.1279	1.6333	58.1472
6670	1.6701	58.2634	1.6762	58.2828	1.6823	58.3021	1.6884	58.3215	1.6946	58.3409
6680	1.7313	58.4570	1.7374	58.4763	1.7435	58.4956	1.7496	58.5150	1.7557	58.5343
6690	1.7924	58.6503	1.7985	58.6697	1.8046	58.6890	1.8108	58.7083	1.8169	58.7277
6700	1.8535	58.8436	1.8596	58.8629	1.8657	58.8822	1.8719	58.9015	1.8780	58.9208
6710	1.9146	59.0367	1.9207	59.0560	1.9268	59.0753	1.9329	59.0946	1.9390	59.1139
6720	1.9756	59.2296	1.9817	59.2489	1.9878	59.2682	1.9939	59.2875	2.0000	59.3068
6730	2.0366	59.4224	2.0427	59.4417	2.0488	59.4610	2.0549	59.4803	2.0610	59.4995
6740	2.0975	59.6151	2.1036	59.6344	2.1097	59.6536	2.1158	59.6729	2.1219	59.6921
6750	2.1584	59.8076	2.1645	59.8269	2.1706	59.8461	2.1766	59.8654	2.1827	59.8846
6760	2.2192	60.0000	2.2253	60.0192	2.2314	60.0385	2.2375	60.0577	2.2435	60.0769
6770	2.2800	60.1922	2.2861	60.2115	2.2922	60.2307	2.2982	60.2499	2.3043	60.2691
6780	2.3408	60.3843	2.3468	60.4035	2.3529	60.4227	2.3590	60.4419	2.3650	60.4611
6790	2.4015	60.5763	2.4075	60.5955	2.4136	60.6147	2.4197	60.6338	2.4257	60.6530
6800	2.4621	60.7681	2.4682	60.7873	2.4742	60.8064	2.4803	60.8256	2.4864	60.8448
6810	2.5227	60.9598	2.5288	60.9789	2.5348	60.9981	2.5409	61.0172	2.5470	61.0364
6820	2.5833	61.1513	2.5893	61.1704	2.5954	61.1896	2.6015	61.2087	2.6075	61.2279
6830	2.6438	61.3427	2.6499	61.3618	2.6559	61.3809	2.6620	61.4001	2.6680	61.4192
6840	2.7043	61.5339	2.7103	61.5531	2.7164	61.5722	2.7224	61.5913	2.7285	61.6104
	80+	200+	80+	200+	80+	200+	80+	200+	80+	200+

Tables of Square Roots

√N	5 √10N	√N	6 √10N	√N	7 √10N	√N	8 √10N	√N	9 √10N	N
80+	200+	80+	200+	80+	200+	80+	200+	80+	200+	
0.0312	53.0810	0.0375	53.1008	0.0437	53.1205	0.0500	53.1403	0.0562	53.1600	6400
0.0937	53.2785	0.0999	53.2982	0.1062	53.3180	0.1124	53.3377	0.1187	53.3575	6410
0.1561	53.4758	0.1623	53.4956	0.1686	53.5153	0.1748	53.5350	0.1810	53.5547	6420
0.2185	53.6730	0.2247	53.6927	0.2309	53.7124	0.2371	53.7321	0.2434	53.7518	6430
0.2808	53.8700	0.2870	53.8897	0.2932	53.9094	0.2994	53.9291	0.3057	53.9488	6440
0.3430	54.0669	0.3492	54.0866	0.3555	54.1063	0.3617	54.1260	0.3679	54.1456	6450
0.4052	54.2636	0.4114	54.2833	0.4177	54.3030	0.4239	54.3226	0.4301	54.3423	6460
0.4674	54.4602	0.4736	54.4799	0.4798	54.4995	0.4860	54.5192	0.4922	54.5388	6470
0.5295	54.6566	0.5357	54.6763	0.5419	54.6959	0.5481	54.7155	0.5543	54.7352	6480
0.5916	54.8529	0.5978	54.8725	0.6040	54.8921	0.6102	54.9117	0.6164	54.9314	6490
0.6536	55.0490	0.6598	55.0686	0.6660	55.0882	0.6722	55.1078	0.6784	55.1274	6500
0.7156	55.2450	0.7217	55.2646	0.7279	55.2842	0.7341	55.3037	0.7403	55.3233	6510
0.7775	55.4408	0.7837	55.4604	0.7899	55.4799	0.7960	55.4995	0.8022	55.5191	6520
0.8393	55.6365	0.8455	55.6560	0.8517	55.6756	0.8579	55.6951	0.8641	55.7147	6530
0.9102	55.8320	0.9074	55.8515	0.9135	55.8711	0.9197	55.8906	0.9259	55.9101	6540
0.9630	56.0273	0.9691	56.0469	0.9753	56.0664	0.9815	56.0859	0.9877	56.1054	6550
1.0247	56.2226	1.0309	56.2421	1.0370	56.2616	1.0432	56.2811	1.0494	56.3006	6560
1.0864	56.4176	1.0925	56.4371	1.0987	56.4566	1.1049	56.4761	1.1110	56.4956	6570
1.1480	56.6125	1.1542	56.6320	1.1603	56.6515	1.1665	56.6710	1.1727	56.6905	6580
1.2096	56.8073	1.2158	56.8268	1.2219	56.8463	1.2281	56.8657	1.2342	56.8852	6590
1.2712	57.0019	1.2773	57.0214	1.2835	57.0409	1.2896	57.0603	1.2958	57.0798	6600
1.3327	57.1964	1.3388	57.2159	1.3449	57.2353	1.3511	57.2547	1.3572	57.2742	6610
1.3941	57.3908	1.4002	57.4102	1.4064	57.4296	1.4125	57.4490	1.4187	57.4684	6620
1.4555	57.5849	1.4616	57.6043	1.4678	57.6238	1.4739	57.6432	1.4801	57.6626	6630
1.5169	57.7790	1.5230	57.7984	1.5291	57.8178	1.5353	57.8372	1.5414	57.8565	6640
1.5782	57.9729	1.5843	57.9922	1.5904	58.0116	1.5966	58.0310	1.6027	58.0504	6650
1.6395	58.1666	1.6456	58.1860	1.6517	58.2053	1.6578	58.2247	1.6639	58.2441	6660
1.7007	58.3602	1.7068	58.3796	1.7129	58.3989	1.7190	58.4183	1.7251	58.4376	6670
1.7618	58.5537	1.7680	58.5730	1.7741	58.5923	1.7802	58.6117	1.7863	58.6310	6680
1.8230	58.7470	1.8291	58.7663	1.8352	58.7856	1.8413	58.8049	1.8474	58.8243	6690
1.8841	58.9401	1.8902	58.9595	1.8963	58.9788	1.9024	58.9981	1.9085	59.0174	6700
1.9451	59.1332	1.9512	59.1525	1.9573	59.1718	1.9634	59.1910	1.9695	59.2103	6710
2.0061	59.3261	2.0122	59.3453	2.0183	59.3646	2.0244	59.3839	2.0305	59.4032	6720
2.0670	59.5188	2.0731	59.5381	2.0792	59.5573	2.0853	59.5766	2.0914	59.5958	6730
2.1279	59.7114	2.1340	59.7306	2.1401	59.7499	2.1462	59.7691	2.1523	59.7884	6740
2.1888	59.9038	2.1949	59.9231	2.2010	59.9423	2.2071	59.9615	2.2131	59.9808	6750
2.2496	60.0961	2.2557	60.1154	2.2618	60.1346	2.2679	60.1538	2.2739	60.1730	6760
2.3104	60.2883	2.3165	60.3075	2.3225	60.3267	2.3286	60.3459	2.3347	60.3651	6770
2.3711	60.4803	2.3772	60.4995	2.3833	60.5187	2.3893	60.5379	2.3954	60.5571	6780
2.4318	60.6722	2.4379	60.6914	2.4439	60.7106	2.4500	60.7297	2.4560	60.7489	6790
2.4924	60.8639	2.4985	60.8831	2.5045	60.9023	2.5106	60.9214	2.5167	60.9406	6800
2.5530	61.0556	2.5591	61.0747	2.5651	61.0939	2.5712	61.1130	2.5772	61.1322	6810
2.6136	61.2470	2.6196	61.2661	2.6257	61.2853	2.6317	61.3044	2.6378	61.3236	6820
2.6741	61.4383	2.6801	61.4575	2.6862	61.4766	2.6922	61.4957	2.6982	61.5148	6830
2.7345	61.6295	2.7406	61.6486	2.7466	61.6677	2.7526	61.6868	2.7587	61.7059	6840
80+	200+	80+	200+	80+	200+	80+	200+	80+	200+	

Tables of Square Roots

N	√N 0	√10N	√N 1	√10N	√N 2	√10N	√N 3	√10N	√N 4	√10N
	80+	200+	80+	200+	80+	200+	80+	200+	80+	200+
6850	2.7647	61.7250	2.7708	61.7442	2.7768	61.7633	2.7828	61.7824	2.7889	61.8015
6860	2.8251	61.9160	2.8312	61.9351	2.8372	61.9542	2.8432	61.9733	2.8493	61.9924
6870	2.8855	62.1068	2.8915	62.1259	2.8975	62.1450	2.9036	62.1641	2.9096	62.1831
6880	2.9458	62.2975	2.9518	62.3166	2.9578	62.3357	2.9638	62.3547	2.9699	62.3738
6890	3.0060	62.4881	3.0120	62.5071	3.0181	62.5262	3.0241	62.5452	3.0301	62.5643
6900	3.0662	62.6785	3.0723	62.6975	3.0793	62.7166	3.0843	62.7356	3.0903	62.7546
6910	3.1264	62.8688	3.1324	62.8878	3.1384	62.9068	3.1445	62.9258	3.1505	62.9449
6920	3.1865	63.0589	3.1925	63.0779	3.1986	63.0969	3.2046	63.1159	3.2106	63.1349
6930	3.2466	63.2489	3.2526	63.2679	3.2586	63.2869	3.2646	63.3059	3.2706	63.3249
6940	3.3067	63.4388	3.3127	63.4578	3.3187	63.4768	3.3247	63.4957	3.3307	63.5147
6950	3.3667	63.6285	3.3727	63.6475	3.3787	63.6665	3.3847	63.6854	3.3906	63.7044
6960	3.4266	63.8181	3.4326	63.8371	3.4386	63.8560	3.4446	63.8750	3.4506	63.8939
6970	3.4865	64.0076	3.4925	64.0265	3.4985	64.0455	3.5045	64.0644	3.5105	64.0833
6980	3.5464	64.1969	3.5524	64.2158	3.5584	64.2347	3.5643	64.2537	3.5703	64.2726
6990	3.6062	64.3861	3.6122	64.4050	3.6182	64.4239	3.6342	64.4428	3.6301	64.4617
7000	3.6660	64.5751	3.6720	64.5940	3.6780	64.6129	3.6839	64.6318	3.6899	64.6507
7010	3.7257	64.7640	3.7317	64.7829	3.7377	64.8018	3.7437	64.8207	3.7496	64.8396
7020	3.7854	64.9528	3.7914	64.9717	3.7974	64.9906	3.8033	65.0094	3.8093	65.0283
7030	3.8451	65.1415	3.8511	65.1603	3.8570	65.1792	3.8630	65.1980	3.8689	65.2169
7040	3.9047	65.3300	3.9107	65.3488	3.9166	65.3677	3.9226	65.3865	3.9285	65.4054
7050	3.9643	65.5184	3.9702	65.5372	3.9762	65.5560	3.9821	65.5748	3.9881	65.5937
7060	4.0238	65.7066	4.0298	65.7254	4.0357	65.7442	4.0417	65.7631	4.0476	65.7819
7070	4.0833	65.8947	4.0892	65.9135	4.0952	65.9323	4.1011	65.9511	4.1071	65.9699
7080	4.1427	66.0827	4.1487	66.1015	4.1546	66.1203	4.1606	66.1391	4.1665	66.1578
7090	4.2021	66.2705	4.2081	66.2893	4.2140	66.3081	4.2200	66.3269	4.2259	66.3456
7100	4.2615	66.4583	4.2674	66.4770	4.2734	66.4958	4.2793	66.5145	4.2852	66.5333
7110	4.3208	66.6458	4.3267	66.6646	4.3327	66.6833	4.3386	66.7021	4.3445	66.7208
7120	4.3801	66.8333	4.3860	66.8520	4.3919	66.8708	4.3979	66.8895	4.4038	66.9082
7130	4.4393	67.0206	4.4452	67.0393	4.4512	67.0580	4.4571	67.0768	4.4630	67.0955
7140	4.4985	67.2078	4.5044	67.2265	4.5104	67.2452	4.5163	67.2639	4.5222	67.2826
7150	4.5577	67.3948	4.5636	67.4135	4.5695	67.4322	4.5754	67.4509	4.5813	67.4696
7160	4.6168	67.5818	4.6227	67.6004	4.6286	67.6191	4.6345	67.6378	4.6404	67.6565
7170	4.6759	67.7686	4.6818	67.7872	4.6877	67.8059	4.6936	67.8246	4.6995	67.8432
7180	4.7349	67.9552	4.7408	67.9739	4.7467	67.9925	4.7526	68.0112	4.7585	68.0298
7190	4.7939	68.1418	4.7998	68.1604	4.8057	68.1790	4.8116	68.1977	4.8175	68.2163
7200	4.8528	68.3282	4.8587	68.3468	4.8646	68.3654	4.8705	68.3841	4.8764	68.4027
7210	4.9117	68.5144	4.9176	68.5331	4.9235	68.5517	4.9294	68.5703	4.9353	68.5889
7220	4.9706	68.7006	4.9765	68.7192	4.9824	68.7378	4.9882	68.7564	4.9941	68.7750
7230	5.0294	68.8866	5.0353	68.9052	5.0412	68.9238	5.0470	68.9424	5.0529	68.9610
7240	5.0882	69.0725	5.0941	69.0911	5.0999	69.1096	5.1058	69.1282	5.1117	69.1468
7250	5.1469	69.2582	5.1528	69.2768	5.1587	69.2954	5.1645	69.3139	5.1704	69.3325
7260	5.2056	69.4439	5.2115	69.4624	5.2174	69.4810	5.2232	69.4995	5.2291	69.5181
7270	5.2643	69.6294	5.2702	69.6479	5.2760	69.6665	5.2819	69.6850	5.2877	69.7035
7280	5.3229	69.8148	5.3288	69.8333	5.3346	69.8518	5.3405	69.8703	5.3464	69.8889
7290	5.3815	70.0000	5.3874	70.0185	5.3932	70.0370	5.3991	70.0556	5.4049	70.0741
	80+	200+	80+	200+	80+	200+	80+	200+	80+	200+

Tables of Square Roots

√N	5 √10N	√N	6 √10N	√N	7 √10N	√N	8 √10N	√N	9 √10N	N
80+	200+	80+	200+	80+	200+	80+	200+	80+	200+	
2.7949	61.8205	2.8010	61.8396	2.8070	61.8587	2.8130	61.8778	2.8191	61.8969	6850
2.8553	62.0115	2.8613	62.0305	2.8674	62.0496	2.8734	62.0687	2.8794	62.0878	6860
2.9156	62.2022	2.9217	62.2213	2.9277	62.2403	2.9337	62.2594	2.9397	62.2785	6870
2.9759	62.3928	2.9819	62.4119	2.9880	62.4309	2.9940	62.4500	3.0000	62.4690	6880
3.0361	62.5833	3.0422	62.6024	3.0482	62.6214	3.0542	62.6404	3.0502	62.6595	6890
3.0963	62.7737	3.1023	62.7927	3.1084	62.8117	3.1144	62.8307	3.1204	62.8498	6900
3.1565	62.9639	3.1625	62.9829	3.1685	63.0019	3.1745	63.0209	3.1805	63.0399	6910
3.2166	63.1539	3.2226	63.1729	3.2286	63.1919	3.2346	63.2109	3.2406	63.2299	6920
3.2766	63.3439	3.2827	63.3629	3.2887	63.3819	3.2947	63.4008	3.3007	63.4198	6930
3.3367	63.5337	3.3427	63.5527	3.3487	63.5716	3.3547	63.5906	3.3607	63.6096	6940
3.3966	63.7233	3.4026	63.7423	3.4086	63.7613	3.4146	63.7802	3.4206	63.7992	6950
3.4566	63.9129	3.4626	63.9318	3.4686	63.9508	3.4745	63.9697	3.4805	63.9886	6960
3.5165	64.1023	3.5225	64.1212	3.5284	64.1401	3.5344	64.1590	3.5404	64.1780	6970
3.5763	64.2915	3.5823	64.3104	3.5883	64.3293	3.5943	64.3483	3.6002	64.3672	6980
3.6361	64.4806	3.6421	64.4995	3.6481	64.5184	3.6541	64.5373	3.6600	64.5562	6990
3.6959	64.6696	3.7019	64.6885	3.7078	64.7074	3.7138	64.7263	3.7198	64.7452	7000
3.7556	64.8585	3.7616	64.8773	3.7675	64.8962	3.7735	64.9151	3.7795	64.9340	7010
3.8153	65.0472	3.8212	65.0660	3.8272	65.0849	3.8332	65.1038	3.8391	65.1226	7020
3.8749	65.2357	3.8309	65.2546	3.8868	65.2734	3.8928	65.2923	3.8987	65.3111	7030
3.9345	65.4242	3.9405	65.4430	3.9464	65.4619	3.9524	65.4807	3.9583	65.4995	7040
3.9940	65.6125	4.0000	65.6313	4.0060	65.6501	4.0119	65.6690	4.0179	65.6878	7050
4.0536	65.8007	4.0595	65.8195	4.0655	65.8383	4.0714	65.8571	4.0773	65.8759	7060
4.1130	65.9887	4.1190	66.0075	4.1249	66.0263	4.1309	66.0451	4.1368	66.0639	7070
4.1724	66.1766	4.1784	66.1954	4.1843	66.2142	4.1903	66.2330	4.1962	66.2518	7080
4.2318	66.3644	4.2378	66.3832	4.2437	66.4020	4.2496	66.4207	4.2556	66.4395	7090
4.2912	66.5521	4.2971	66.5708	4.3030	66.5896	4.3090	66.6083	4.3149	66.6271	7100
4.3505	66.7396	4.3564	66.7583	4.3623	66.7771	4.3682	66.7958	4.3742	66.8145	7110
4.4097	66.9270	4.4156	66.9457	4.4216	66.9644	4.4275	66.9831	4.4334	67.0019	7120
4.4689	67.1142	4.4748	67.1329	4.4808	67.1516	4.4867	67.1704	4.4926	67.1891	7130
4.5281	67.3013	4.5340	67.3200	4.5399	67.3387	4.5458	67.3574	4.5518	67.3761	7140
4.5872	67.4883	4.5931	67.5070	4.5991	67.5257	4.6050	67.5444	4.6109	67.5631	7150
4.6463	67.6752	4.6522	67.6939	4.6581	67.7125	4.6640	67.7312	4.6699	67.7499	7160
4.7054	67.8619	4.7113	67.8806	4.7172	67.8992	4.7231	67.9179	4.7290	67.9366	7170
4.7644	68.0485	4.7703	68.0672	4.7762	68.0858	4.7821	68.1045	4.7880	68.1231	7180
4.8233	68.2350	4.8292	68.2536	4.8351	68.2723	4.8410	68.2909	4.8469	68.3095	7190
4.8823	68.4213	4.8882	68.4399	4.8941	68.4586	4.8999	68.4772	4.9058	68.4958	7200
4.9412	63.6075	4.9470	68.6261	4.9529	68.6447	4.9588	68.6634	4.9647	68.6820	7210
5.0000	68.7936	5.0059	68.8122	5.0118	68.8308	5.0176	68.8494	5.0235	68.8680	7220
5.0588	68.9796	5.0647	68.9981	5.0706	69.0167	5.0764	69.0353	5.0823	69.0539	7230
5.1176	69.1654	5.1234	69.1840	5.1293	69.2025	5.1352	69.2211	5.1411	69.2397	7240
5.1763	69.3511	5.1822	69.3696	5.1880	69.3882	5.1939	69.4068	5.1998	69.4253	7250
5.2350	69.5366	5.2408	69.5552	5.2467	69.5737	5.2526	69.5923	5.2584	69.6108	7260
5.2936	69.7221	5.2995	69.7406	5.3053	69.7592	5.3112	69.7777	5.3171	69.7962	7270
5.3522	69.9074	5.3581	69.9259	5.3639	69.9444	5.3698	69.9630	5.3756	69.9815	7280
5.4108	70.0926	5.4166	70.1111	5.4225	70.1296	5.4283	70.1481	5.4342	70.1666	7290
80+	200+	80+	200+	80+	200+	80+	200+	80+	200+	

Tables of Square Roots

N	$\sqrt{N}$ 0	$\sqrt{10N}$	$\sqrt{N}$ 1	$\sqrt{10N}$	$\sqrt{N}$ 2	$\sqrt{10N}$	$\sqrt{N}$ 3	$\sqrt{10N}$	$\sqrt{N}$ 4	$\sqrt{10N}$
	80+	200+	80+	200+	80+	200+	90+	200+	80+	200+
7300	5.4400	70.1851	5.4459	70.2036	5.4517	70.2221	5.4576	70.2406	5.4634	70.2591
7310	5.4985	70.3701	5.5044	70.3886	5.5102	70.4071	5.5161	70.4256	5.5219	70.4441
7320	5.5570	70.5550	5.5628	70.5735	5.5687	70.5919	5.5745	70.6104	5.5804	70.6289
7330	5.6154	70.7397	5.6213	70.7582	5.6271	70.7767	5.6329	70.7951	5.6388	70.8136
7340	5.6738	70.9243	5.6796	70.9428	5.6855	70.9613	5.6913	70.9797	5.6971	70.9982
7350	5.7321	71.1088	5.7380	71.1273	5.7438	71.1457	5.7496	71.1642	5.7555	71.1826
7360	5.7904	71.2932	5.7963	71.3116	5.8021	71.3301	5.8079	71.3485	5.8138	71.3669
7370	5.8487	71.4774	5.8545	71.4959	5.8604	71.5143	5.8662	71.5327	5.8720	71.5511
7380	5.9069	71.6616	5.9127	71.6800	5.9186	71.6984	5.9244	71.7168	5.9302	71.7352
7390	5.9651	71.8455	5.9709	71.8639	5.9767	71.8823	5.9826	71.9007	5.9884	71.9191
7400	6.0233	72.0294	6.0291	72.0478	6.0349	72.0662	6.0407	72.0845	6.0465	72.1029
7410	6.0814	72.2132	6.0872	72.2315	6.0930	72.2499	6.0988	72.2683	6.1046	72.2866
7420	6.1394	72.3968	6.1452	72.4151	6.1510	72.4335	6.1568	72.4518	6.1626	72.4702
7430	6.1974	72.5803	6.2032	72.5986	6.2090	72.6169	6.2148	72.6353	6.2206	72.6536
7440	6.2554	72.7636	6.2612	72.7820	6.2670	72.8003	6.2728	72.8186	6.2786	72.8369
7450	6.3134	72.9469	6.3192	72.9652	6.3250	72.9835	6.3308	73.0018	6.3366	73.0201
7460	6.3713	73.1300	6.3771	73.1483	6.3829	73.1666	6.3887	73.1849	6.3944	73.2032
7470	6.4292	73.3130	6.4349	73.3313	6.4407	73.3496	6.4465	73.3679	6.4523	73.3862
7480	6.4870	73.4959	6.4928	73.5142	6.4986	73.5324	6.5043	73.5507	6.5101	73.5690
7490	6.5448	73.6786	6.5506	73.6969	6.5563	73.7152	6.5621	73.7334	6.5679	73.7517
7500	6.6025	73.8613	6.6083	73.8795	6.6141	73.8978	6.6199	73.9160	6.6256	73.9343
7510	6.6603	74.0438	6.6660	74.0620	6.6718	74.0803	6.6776	74.0985	6.6833	74.1168
7520	6.7179	74.2262	6.7237	74.2444	6.7295	74.2626	6.7352	74.2809	6.7410	74.2991
7530	6.7756	74.4085	6.7813	74.4267	6.7871	74.4449	6.7929	74.4631	6.7986	74.4813
7540	6.8332	74.5906	6.8389	74.6088	6.8447	74.6270	6.8504	74.6452	6.8562	74.6634
7550	6.8907	74.7726	6.8965	74.7908	6.9022	74.8090	6.9080	74.8272	6.9138	74.8454
7560	6.9483	74.9545	6.9540	74.9727	6.9598	74.9909	6.9655	75.0091	6.9713	75.0273
7570	7.0057	75.1363	7.0115	75.1545	7.0172	75.1727	7.0230	75.1908	7.0287	75.2090
7580	7.0632	75.3180	7.0689	75.3362	7.0747	75.3543	7.0804	75.3725	7.0862	75.3906
7590	7.1206	75.4995	7.1263	75.5177	7.1321	75.5358	7.1378	75.5540	7.1436	75.5721
7600	7.1780	75.6810	7.1837	75.6991	7.1894	75.7172	7.1952	75.7354	7.2009	75.7535
7610	7.2353	75.8623	7.2410	75.8804	7.2468	75.8985	7.2525	75.9167	7.2582	75.9348
7620	7.2926	76.0435	7.2983	76.0616	7.3041	76.0797	7.3098	76.0978	7.3155	76.1159
7630	7.3499	76.2245	7.3556	76.2426	7.3613	76.2607	7.3670	76.2788	7.3728	76.2969
7640	7.4071	76.4055	7.4128	76.4236	7.4185	76.4417	7.4243	76.4598	7.4300	76.4778
7650	7.4643	76.5863	7.4700	76.6044	7.4757	76.6225	7.4814	76.6406	7.4871	76.6586
7660	7.5214	76.7671	7.5271	76.7851	7.5329	76.8032	7.5386	76.8212	7.5443	76.8393
7670	7.5785	76.9476	7.5842	76.9657	7.5900	76.9838	7.5957	77.0018	7.6014	77.0199
7680	7.6356	77.1281	7.6413	77.1462	7.6470	77.1642	7.6527	77.1823	7.6584	77.2003
7690	7.6926	77.3085	7.6983	77.3265	7.7040	77.3446	7.7097	77.3626	7.7154	77.3806
7700	7.7496	77.4887	7.7553	77.5068	7.7610	77.5248	7.7667	77.5228	7.7724	77.5608
7710	7.8056	77.6689	7.8123	77.6869	7.8180	77.7049	7.8237	77.7229	7.8294	77.7409
7720	7.8635	77.8489	7.8692	77.8669	7.8749	77.8849	7.8806	77.9029	7.8863	77.9209
7730	7.9204	78.0288	7.9261	78.0468	7.9318	78.0647	7.9375	78.0827	7.9432	78.1007
7740	7.9773	78.2086	7.9830	78.2265	7.9886	78.2445	7.9943	78.2625	8.0000	78.2804
	80+	200+	80+	200+	80+	200+	80+	200+	90+	200+

Tables of Square Roots

	5		6		7		8		9	N
√N	√10N	√N	√10N	√N	√10N	√N	√10N	√N	√10N	
5.4693	70.2776	5.4751	70.2961	5.4810	70.3146	5.4868	70.3331	5.4927	70.3516	7300
5.5278	70.4626	5.5336	70.4811	5.5395	70.4995	5.5453	70.5180	5.5512	70.5365	7310
5.5862	70.6474	5.5921	70.6658	5.5979	70.6843	5.6037	70.7028	5.6096	70.7213	7320
5.6446	70.8321	5.6505	70.8505	5.6563	70.8690	5.6621	70.8874	5.6680	70.9059	7330
5.7030	71.0166	5.7088	71.0351	5.7146	71.0535	5.7205	71.0719	5.7263	71.0904	7340
5.7613	71.2010	5.7671	71.2195	5.7730	71.2379	5.7788	71.2563	5.7846	71.2748	7350
5.8196	71.3853	5.8254	71.4038	5.8312	71.4222	5.8371	71.4406	5.8429	71.4590	7360
5.8778	71.5695	5.8836	71.5879	5.8895	71.6063	5.8953	71.6247	5.9011	71.6431	7370
5.9360	71.7536	5.9418	71.7720	5.9477	71.7904	5.9535	71.8088	5.9593	71.8272	7380
5.9942	71.9375	6.0000	71.9559	6.0058	71.9743	6.0116	71.9926	6.0174	72.0110	7390
6.0523	72.1213	6.0581	72.1397	6.0639	72.1580	6.0697	72.1764	6.0755	72.1948	7400
6.1104	72.3050	6.1162	72.3233	6.1220	72.3417	6.1278	72.3601	6.1336	72.3784	7410
6.1684	72.4885	6.1742	72.5069	6.1800	72.5252	6.1858	72.5436	6.1916	72.5619	7420
6.2264	72.6720	6.2322	72.6903	6.2380	72.7086	6.2438	72.7270	6.2496	72.7453	7430
6.2844	72.8553	6.2902	72.8736	6.2960	72.8919	6.3018	72.9102	6.3076	72.9286	7440
6.3423	73.0385	6.3481	73.0568	6.3539	73.0751	6.3597	73.0934	6.3655	73.1117	7450
6.4002	73.2215	6.4060	73.2398	6.4118	73.2581	6.4176	73.2764	6.4234	73.2947	7460
6.4581	73.4045	6.4639	73.4228	6.4696	73.4410	6.4754	73.4593	6.4812	73.4776	7470
6.5159	73.5873	6.5217	73.6056	6.5275	73.6238	6.5332	73.6421	6.5390	73.6604	7480
6.5737	73.7700	6.5794	73.7882	6.5852	73.8065	6.5910	73.8248	6.5968	73.8430	7490
6.6314	73.9526	6.6372	73.9708	6.6429	73.9891	6.6487	74.0073	6.6545	74.0255	7500
6.6891	74.1350	6.6949	74.1532	6.7006	74.1715	6.7064	74.1897	6.7122	74.2080	7510
6.7468	74.3173	6.7525	74.3356	6.7583	74.3538	6.7640	74.3720	6.7698	74.3902	7520
6.8044	74.4995	6.8101	74.5178	6.8159	74.5360	6.8217	74.5542	6.8274	74.5724	7530
6.8620	74.6816	6.8677	74.6998	6.8735	74.7180	6.8792	74.7362	6.8850	74.7544	7540
6.9195	74.8636	6.9253	74.8818	6.9310	74.9000	6.9368	74.9182	6.9425	74.9364	7550
6.9770	75.0455	6.9828	75.0636	6.9885	75.0818	6.9943	75.1000	7.0000	75.1182	7560
7.0345	75.2272	7.0402	75.2453	7.0460	75.2635	7.0517	75.2817	7.0575	75.2998	7570
7.0919	75.4088	7.0976	75.4269	7.1034	75.4451	7.1091	75.4632	7.1149	75.4814	7580
7.1493	75.5903	7.1550	75.6084	7.1608	75.6266	7.1665	75.6447	7.1722	75.6628	7590
7.2067	75.7716	7.2124	75.7898	7.2181	75.8079	7.2239	75.8260	7.2296	75.8442	7600
7.2640	75.9529	7.2697	75.9710	7.2754	75.9891	7.2812	76.0072	7.2869	76.0254	7610
7.3212	76.1340	7.3270	76.1521	7.3327	76.1702	7.3384	76.1883	7.3441	76.2064	7620
7.3785	76.3150	7.3842	76.3331	7.3899	76.3512	7.3957	76.3693	7.4014	76.3874	7630
7.4357	76.4959	7.4414	76.5140	7.4471	76.5321	7.4528	76.5502	7.4586	76.5683	7640
7.4929	76.6767	7.4986	76.6948	7.5043	76.7128	7.5100	76.7309	7.5157	76.7490	7650
7.5500	76.8574	7.5557	76.8754	7.5614	76.8935	7.5671	76.9115	7.5728	76.9296	7660
7.6071	77.0379	7.6128	77.0560	7.6185	77.0740	7.6242	77.0920	7.6299	77.1101	7670
7.6641	77.2183	7.6698	77.2364	7.6755	77.2544	7.6812	77.2724	7.6869	77.2905	7680
7.7211	77.3986	7.7268	77.4167	7.7325	77.4347	7.7382	77.4527	7.7439	77.4707	7690
7.7781	77.5788	7.7838	77.5968	7.7895	77.6148	7.7952	77.6329	7.8009	77.6509	7700
7.8351	77.7589	7.8408	77.7769	7.8465	77.7949	7.8521	77.8129	7.8578	77.8309	7710
7.8920	77.9388	7.8977	77.9568	7.9034	77.9748	7.9090	77.9928	7.9147	78.0108	7720
7.9488	78.1187	7.9545	78.1367	7.9602	78.1546	7.9659	78.1726	7.9716	78.1906	7730
8.0057	78.2984	8.0114	78.3164	8.0170	78.3343	8.0227	78.3523	8.0284	78.3703	7740
80+	200+	80+	200+	80+	200+	80+	200+	80+	200+	

Tables of Square Roots

N	√N 0	√10N	√N 1	√10N	√N 2	√10N	√N 3	√10N	√N 4	√10N
	80+	200+	80+	200+	80+	200+	80+	200+	80+	200+
7750	8.0341	78.3882	8.0398	78.4062	8.0454	78.4241	8.0511	78.4421	8.0568	78.4601
7760	8.0909	78.5678	8.0965	78.5857	8.1022	78.6037	8.1079	78.6216	8.1136	78.6396
7770	8.1476	78.7472	8.1533	78.7651	8.1589	78.7831	8.1646	78.8010	8.1703	78.8189
7780	8.2043	78.9265	8.2100	78.9444	8.2156	78.9624	8.2213	78.9803	8.2270	78.9982
7790	8.2610	79.1057	8.2666	79.1236	8.2723	79.1415	8.2780	79.1595	8.2836	79.1774
7800	8.3176	79.2848	8.3233	79.3027	8.3289	79.3206	8.3346	79.3385	8.3403	79.3564
7810	8.3742	79.4638	8.3799	79.4817	8.3855	79.4996	8.3912	79.5174	8.3968	79.5353
7820	8.4308	79.6426	8.4364	79.6605	8.4421	79.6784	8.4477	79.6963	8.4534	79.7141
7830	8.4873	79.8214	8.4929	79.8392	8.4986	79.8571	8.5042	79.8750	8.5099	79.8928
7840	8.5438	80.0000	8.5494	80.0179	8.5551	80.0357	8.5607	80.0536	8.5664	80.0714
7850	8.6002	80.1785	8.6059	80.1964	8.6115	80.2142	8.6172	80.2320	8.6228	80.2499
7860	8.6566	80.3569	8.6623	80.3747	8.6679	80.3926	8.6736	80.4104	8.6792	80.4282
7870	8.7130	80.5352	8.7187	80.5530	8.7243	80.5708	8.7299	80.5887	8.7356	80.6065
7880	8.7694	80.7134	8.7750	80.7312	8.7806	80.7490	8.7863	80.7668	8.7919	80.7846
7890	8.8257	80.8914	8.8313	80.9092	8.8369	80.9270	8.8426	80.9448	8.8482	80.9626
7900	8.8819	81.0694	8.8876	81.0872	8.8932	81.1050	8.8988	81.1227	8.9044	81.1405
7910	8.9382	81.2472	8.9438	81.2650	8.9494	81.2828	8.9550	81.3006	8.9607	81.3183
7920	8.9944	81.4249	9.0000	81.4427	9.0056	81.4605	9.0112	81.4782	9.0169	81.4960
7930	9.0505	81.6026	9.0562	81.6203	9.0618	81.6381	9.0674	81.6558	9.0730	81.6736
7940	9.1067	81.7800	9.1123	81.7978	9.1180	81.8155	9.1235	81.8333	9.1291	81.8510
7950	9.1628	81.9574	9.1684	81.9752	9.1740	81.9929	9.1796	82.0106	9.1852	82.0284
7960	9.2188	82.1347	9.2244	82.1524	9.2300	82.1702	9.2356	82.1879	9.2412	82.2056
7970	9.2749	82.3119	9.2805	82.3296	9.2861	82.3473	9.2917	82.3650	9.2973	82.3827
7980	9.3308	82.4889	9.3364	82.5066	9.3420	82.5243	9.3476	82.5420	9.3532	82.5597
7990	9.3868	82.6659	9.3924	82.6836	9.3980	82.7013	9.4036	82.7189	9.4092	82.7366
8000	9.4427	82.8427	9.4483	82.8604	9.4539	82.8781	9.4595	82.8957	9.4651	82.9134
8010	9.4986	83.0194	9.5042	83.0371	9.5098	83.0548	9.5154	83.0724	9.5209	83.0901
8020	9.5545	83.1960	9.5600	83.2137	9.5656	83.2314	9.5712	83.2490	9.5768	83.2667
8030	9.6103	83.3725	9.6158	83.3902	9.6214	83.4078	9.6270	83.4255	9.6326	83.4431
8040	9.6660	83.5489	9.6716	83.5666	9.6772	83.5842	9.6828	83.6018	9.6883	83.6195
8050	9.7218	83.7252	9.7274	83.7428	9.7329	83.7605	9.7385	83.7781	9.7441	83.7957
8060	9.7775	83.9014	9.7831	83.9190	9.7886	83.9366	9.7942	83.9542	9.7998	83.9718
8070	9.8332	84.0775	9.8387	84.0951	9.8443	84.1127	9.8499	84.1303	9.8554	84.1478
8080	9.8888	84.2534	9.8944	84.2710	9.8999	84.2886	9.9055	84.3062	9.9111	84.3238
8090	9.9444	84.4293	9.9500	84.4468	9.9555	84.4644	9.9611	84.4820	9.9667	84.4996
8100	*0.0000	84.6050	*0.0056	84.6226	*0.0111	84.6401	*0.0167	84.6577	*0.0222	84.6753
8110	0.0555	84.7806	0.0611	84.7982	0.0666	84.8157	0.0722	84.8333	0.0777	84.8508
8120	0.1110	84.9561	0.1166	84.9737	0.1221	84.9912	0.1277	85.0088	0.1332	85.0263
8130	0.1665	85.1315	0.1721	85.1491	0.1776	85.1666	0.1831	85.1842	0.1887	85.2017
8140	0.2219	85.3069	0.2275	85.3244	0.2330	85.3419	0.2386	85.3594	0.2441	85.3769
8150	0.2774	85.4820	0.2829	85.4996	0.2884	85.5171	0.2940	85.5346	0.2995	85.5521
8160	0.3327	85.6571	0.3383	85.6746	0.3438	85.6921	0.3493	85.7096	0.3549	85.7271
8170	0.3881	85.8321	0.3936	85.8496	0.3991	85.8671	0.4046	85.8846	0.4102	85.9021
8180	0.4434	86.0070	0.4489	86.0245	0.4544	86.0420	0.4599	86.0594	0.4655	86.0769
8190	0.4986	86.1818	0.5041	86.1992	0.5097	86.2167	0.5152	86.2342	0.5207	86.2516
	90+	200+	90+	200+	90+	200+	90+	200+	90+	200+

Tables of Square Roots

	5		6		7		8		9	N
√N	√10N	√N	√10N	√N	√10N	√N	√10N	√N	√10N	
80+	200+	80+	200+	80+	200+	80+	200+	80+	200+	
8.0625	78.4780	8.0682	78.4960	8.0738	78.5139	8.0795	78.5319	8.0852	78.5498	7750
8.1192	78.6575	8.1249	78.6754	8.1306	78.6954	8.1363	78.7113	8.1419	78.7293	7760
8.1760	78.8369	8.1816	78.8548	8.1873	78.8727	8.1930	78.8907	8.1986	78.9086	7770
8.2326	79.0161	8.2383	79.0340	8.2440	79.0520	8.2496	79.0699	8.2553	79.0878	7780
8.2893	79.1953	8.2950	79.2132	8.3006	79.2311	8.3063	79.2490	8.3119	79.2669	7790
8.3459	79.3743	8.3516	79.3922	8.3572	79.4101	8.3629	79.4280	8.3685	79.4459	7800
8.4025	79.5532	8.4081	79.5711	8.4138	79.5890	8.4195	79.6069	8.4251	79.6247	7810
8.4590	79.7320	8.4647	79.7499	8.4703	79.7678	8.4760	79.7856	8.4816	79.8035	7820
8.5155	79.9107	8.5212	79.9286	8.5268	79.9464	8.5325	79.9643	8.5381	79.9821	7830
8.5720	80.0893	8.5777	80.1071	8.5833	80.1250	8.5889	80.1428	8.5946	80.1607	7840
8.6284	80.2677	8.6341	80.2856	8.6397	80.3034	8.6454	80.3212	8.6510	80.3391	7850
8.6848	80.4461	8.6905	80.4639	8.6961	80.4817	8.7017	80.4996	8.7074	80.5174	7860
8.7412	80.6243	8.7468	80.6421	8.7525	80.6599	8.7581	80.6778	8.7637	80.6956	7870
8.7975	80.8024	8.8032	80.8202	8.8088	80.8380	8.8144	80.8558	8.8200	80.8736	7880
8.8538	80.9804	8.8594	80.9982	8.8651	81.0160	8.8707	81.0338	8.8763	81.0516	7890
8.9101	81.1583	8.9157	81.1761	8.9213	81.1939	8.9269	81.2117	8.9326	81.2294	7900
8.9663	81.3361	8.9719	81.3539	8.9775	81.3716	8.9831	81.3894	8.9888	81.4072	7910
9.0225	81.5138	9.0281	81.5315	9.0337	81.5493	9.0393	81.5670	9.0449	81.5848	7920
9.0786	81.6913	9.0842	81.7091	9.0898	81.7268	9.0955	81.7446	9.1011	81.7623	7930
9.1347	81.8688	9.1403	81.8865	9.1459	81.9042	9.1516	81.9220	9.1572	81.9397	7940
9.1908	82.0461	9.1964	82.0638	9.2020	82.0815	9.2076	82.0993	9.2132	82.1170	7950
9.2468	82.2233	9.2525	82.2410	9.2581	82.2587	9.2637	82.2765	9.2693	82.2942	7960
9.3029	82.4004	9.3085	82.4181	9.3141	82.4358	9.3197	82.4535	9.3252	82.4712	7970
9.3588	82.5774	9.3644	82.5951	9.3700	82.6128	9.3756	82.6305	9.3812	82.6482	7980
9.4148	82.7543	9.4204	82.7720	9.4259	82.7897	9.4315	82.8074	9.4371	82.8250	7990
9.4707	82.9311	9.4763	82.9488	9.4818	82.9664	9.4874	82.9841	9.4930	83.0018	8000
9.5265	83.1078	9.5321	83.1254	9.5377	83.1431	9.5433	83.1607	9.5489	83.1784	8010
9.5824	83.2843	9.5879	83.3020	9.5935	83.3196	9.5991	83.3373	9.6047	83.3549	8020
9.6382	83.4608	9.6437	83.4784	9.6493	83.4960	9.6549	83.5137	9.6605	83.5313	8030
9.6939	83.6371	9.6995	83.6547	9.7051	83.6723	9.7106	83.6900	9.7162	83.7076	8040
9.7497	83.8133	9.7552	83.8309	9.7608	83.8486	9.7664	83.8662	9.7719	83.8838	8050
9.8053	83.9894	9.8109	84.0070	9.8165	84.0246	9.8220	84.0423	9.8276	84.0599	8060
9.8610	84.1654	9.8666	84.1830	9.8721	84.2006	9.8777	84.2182	9.8833	84.2358	8070
9.9166	84.3413	9.9222	84.3589	9.9277	84.3765	9.9333	84.3941	9.9389	84.4117	8080
9.9722	84.5171	9.9778	84.5347	9.9833	84.5523	9.9889	84.5699	9.9944	84.5874	8090
0.0278	84.6928	*0.0333	84.7104	*0.0389	84.7279	*0.0444	84.7455	*0.0500	84.7631	8100
0.0833	84.8684	0.0888	84.8859	0.0944	84.9035	0.0999	84.9210	0.1055	84.9386	8110
0.1388	85.0439	0.1443	85.0614	0.1499	85.0789	0.1654	85.0965	0.1610	85.1140	8120
0.1942	85.2192	0.1998	85.2367	0.2053	85.2543	0.2109	85.2718	0.2164	85.2893	8130
0.2497	85.3945	0.2552	85.4120	0.2607	85.4295	0.2663	85.4470	0.2718	85.4645	8140
0.3050	85.5696	0.3106	85.5871	0.3161	85.6046	0.3216	85.6221	0.3272	85.6396	8150
0.3604	85.7446	0.3659	85.7621	0.3715	85.7796	0.3770	85.7971	0.3825	85.8146	8160
0.4157	85.9196	0.4212	85.9371	0.4268	85.9545	0.4323	85.9720	0.4378	85.9895	8170
0.4710	86.0944	0.4765	86.1119	0.4820	86.1293	0.4876	86.1468	0.4931	86.1643	8180
0.5262	86.2691	0.5318	86.2866	0.5373	86.3040	0.5428	86.3215	0.5483	86.3390	8190
90+	200+	90+	200+	90+	200+	90+	200+	90+	200+	

Tables of Square Roots

N	0 √N	0 √10N	1 √N	1 √10N	2 √N	2 √10N	3 √N	3 √10N	4 √N	4 √10N
	90+	200+	90+	200+	90+	200+	90+	200+	90+	200+
8200	0.5539	86.3564	0.5594	86.3739	0.5649	86.3913	0.5704	86.4088	0.5759	86.4263
8210	0.6091	86.5310	0.6146	86.5484	0.6201	86.5659	0.6256	86.5833	0.6311	86.6008
8220	0.6642	86.7054	0.6697	86.7229	0.6752	86.7403	0.6808	86.7577	0.6863	86.7752
8230	0.7193	86.8798	0.7249	86.8972	0.7304	86.9146	0.7359	86.9320	0.7414	86.9495
8240	0.7744	87.0540	0.7800	87.0714	0.7855	87.0888	0.7910	87.1063	0.7965	87.1237
8250	0.8295	87.2281	0.8350	87.2455	0.8405	87.2629	0.8460	87.2804	0.8515	87.2978
8260	0.8845	87.4022	0.8900	97.4196	0.8955	87.4370	0.9010	87.4543	0.9065	87.4717
8270	0.9395	87.5761	0.9450	87.5935	0.9505	87.6108	0.9560	87.6282	0.9615	87.6456
8280	0.9945	87.7499	1.0000	87.7673	1.0055	87.7846	1.0110	87.8020	1.0165	87.8194
8290	1.0494	87.9236	1.0549	87.9410	1.0604	87.9583	1.0659	87.9757	1.0714	87.9931
8300	1.1043	88.0972	1.1098	88.1146	1.1153	88.1319	1.1208	88.1493	1.1263	88.1666
8310	1.1592	88.2707	1.1647	88.2881	1.1702	88.3054	1.1757	88.3227	1.1811	88.3401
8320	1.2140	88.4441	1.2195	88.4614	1.2250	88.4788	1.2305	88.4961	1.2360	88.5134
8330	1.2688	88.6174	1.2743	88.6347	1.2798	88.6520	1.2853	88.6694	1.2907	88.6867
8340	1.3236	88.7906	1.3291	88.8079	1.3346	88.8252	1.3400	88.8425	1.3455	88.8598
8350	1.3783	88.9637	1.3838	88.9810	1.3893	88.9983	1.3947	89.0156	1.4002	89.0329
8360	1.4330	89.1366	1.4385	89.1539	1.4440	89.1712	1.4494	89.1885	1.4550	89.2058
8370	1.4877	89.3095	1.4932	89.3268	1.4986	89.3441	1.5041	89.3614	1.5096	89.3786
8380	1.5423	89.4823	1.5478	89.4996	1.5533	89.5168	1.5587	89.5341	1.5642	89.5514
8390	1.5969	89.6550	1.6024	89.6722	1.6079	89.6895	1.6133	89.7067	1.6188	89.7240
8400	1.6515	89.8275	1.6570	89.8448	1.6624	89.8620	1.6679	89.8793	1.6733	89.8965
8410	1.7061	90.0000	1.7115	90.0172	1.7170	90.0345	1.7224	90.0517	1.7279	90.0690
8420	1.7606	90.1724	1.7660	90.1896	1.7715	90.2068	1.7769	90.2241	1.7824	90.2413
8430	1.8150	90.3446	1.8205	90.3618	1.8259	90.3791	1.8314	90.3963	1.8368	90.4135
8440	1.8695	90.5168	1.8749	90.5340	1.8804	90.5512	1.8858	90.5684	1.8912	90.5856
8450	1.9239	90.6888	1.9293	90.7060	1.9348	90.7232	1.9402	90.7404	1.9456	90.7576
8460	1.9783	90.8608	1.9837	90.8780	1.9891	90.8952	1.9946	90.9124	2.0000	90.9295
8470	2.0326	91.0326	2.0380	91.0498	2.0435	91.0670	2.0489	91.0842	2.0543	91.1014
8480	2.0869	91.2044	2.0923	91.2216	2.0978	91.2387	2.1032	91.2559	2.1086	91.2731
8490	2.1412	91.3760	2.1466	91.3932	2.1520	91.4104	2.1575	91.4275	2.1629	91.4447
8500	2.1954	91.5476	2.2009	91.5647	2.2063	91.5819	2.2117	91.5990	2.2171	91.6162
8510	2.2497	91.7190	2.2551	91.7362	2.2605	91.7533	2.2659	91.7705	2.2713	91.7876
8520	2.3038	91.8904	2.3093	91.9075	2.3147	91.9246	2.3201	91.9418	2.3255	91.9589
8530	2.3580	92.0616	2.3634	92.0788	2.3688	92.0959	2.3742	92.1130	2.3797	92.1301
8540	2.4121	92.2328	2.4175	92.2499	2.4229	92.2670	2.4284	92.2841	2.4338	92.3012
8550	2.4662	92.4038	2.4716	92.4209	2.4770	92.4380	2.4824	92.4551	2.4878	92.4722
8560	2.5203	92.5748	2.5257	92.5919	2.5311	92.6090	2.5365	92.6260	2.5419	92.6431
8570	2.5743	92.7456	2.5797	92.7627	2.5851	92.7798	2.5905	92.7969	2.5959	92.8139
8580	2.6283	92.9164	2.6337	92.9334	2.6391	92.9505	2.6445	92.9676	2.6499	92.9846
8590	2.6823	93.0870	2.6876	93.1041	2.6930	93.1211	2.6984	93.1382	2.7038	93.1552
8600	2.7362	93.2576	2.7416	93.2746	2.7470	93.2917	2.7524	93.3087	2.7577	93.3258
8610	2.7901	93.4280	2.7955	93.4451	2.8009	93.4621	2.8063	93.4791	2.8116	93.4962
8620	2.8440	93.5984	2.8493	93.6154	2.8547	93.6324	2.8601	93.6495	2.8655	93.6665
8630	2.8978	93.7686	2.9032	93.7856	2.9086	93.8027	2.9139	93.8197	2.9193	93.8367
8640	2.9516	93.9388	2.9570	93.9558	2.9624	93.9729	2.9677	93.9898	2.9731	94.0068
	90+	200+	90+	200+	90+	200+	90+	200+	90+	200+

Tables of Square Roots

5		6		7		8		9		N
√N	√10N	√N	√10N	√N	√10N	√N	√10N	√N	√10N	
90+	200+	90+	200+	90+	200+	90+	200+	90+	200+	
0.5815	86.4437	0.5870	86.4612	0.5925	86.4786	0.5980	86.4961	0.6035	86.5135	8200
0.6366	86.6182	0.6422	86.6357	0.6477	86.6531	0.6532	86.6705	0.6587	96.6880	8210
0.6918	86.7926	0.6973	86.8100	0.7028	86.8275	0.7083	86.8449	0.7138	86.8623	8220
0.7469	86.9669	0.7524	86.9843	0.7579	87.0017	0.7634	87.0192	0.7689	87.0366	8230
0.8020	87.1411	0.8075	87.1585	0.8130	87.1759	0.8185	87.1933	0.8240	87.2107	8240
0.8570	87.3152	0.8625	87.3326	0.8680	87.3500	0.8735	87.3674	0.8790	87.3848	8250
0.9120	87.4891	0.9175	87.5065	0.9230	87.5239	0.9285	87.5413	0.9340	87.5587	8260
0.9670	87.6630	0.9725	87.6804	0.9780	87.6978	0.9835	87.7151	0.9890	87.7325	8270
1.0220	87.8368	1.0275	87.8541	1.0330	87.8715	1.0385	87.8889	1.0439	87.9062	8280
1.0769	88.0104	1.0824	88.0278	1.0879	88.0451	1.0934	88.0625	1.0988	88.0799	8290
1.1318	88.1840	1.1373	88.2013	1.1427	88.2187	1.1482	88.2360	1.1537	88.2534	8300
1.1866	88.3574	1.1921	88.3748	1.1976	88.3921	1.2031	88.4094	1.2086	88.4268	8310
1.2414	88.5308	1.2469	88.5481	1.2524	88.5654	1.2579	88.5827	1.2634	88.6001	8320
1.2962	88.7040	1.3017	88.7213	1.3072	88.7386	1.3127	88.7560	1.3181	88.7733	8330
1.3510	88.8771	1.3564	88.8944	1.3619	88.9118	1.3674	88.9291	1.3729	88.9464	8340
1.4057	89.0502	1.4112	89.0675	1.4166	89.0848	1.4221	89.1021	1.4276	89.1194	8350
1.4604	89.2231	1.4658	89.2404	1.4713	89.2577	1.4768	89.2750	1.4822	89.2922	8360
1.5150	89.3959	1.5205	89.4132	1.5260	89.4305	1.5314	89.4478	1.5369	89.4650	8370
1.5696	89.5686	1.5751	89.5859	1.5806	89.6032	1.5860	89.6204	1.5915	89.6377	8380
1.6242	89.7413	1.6297	89.7585	1.6351	89.7758	1.6406	89.7930	1.6461	89.8103	8390
1.6788	89.9138	1.6842	89.9310	1.6897	89.9483	1.6951	89.9655	1.7006	89.9828	8400
1.7333	90.0862	1.7388	90.1034	1.7442	90.1207	1.7497	90.1379	1.7551	90.1551	8410
1.7878	90.2585	1.7932	90.2757	1.7987	90.2930	1.8041	90.3102	1.8096	90.3274	8420
1.8423	90.4307	1.8477	90.4479	1.8531	90.4651	1.8586	90.4824	1.8640	90.4996	8430
1.8967	90.6028	1.9021	90.6200	1.9076	90.6372	1.9130	90.6544	1.9184	90.6716	8440
1.9511	90.7748	1.9565	90.7920	1.9619	90.8092	1.9674	90.8264	1.9728	90.8436	8450
2.0054	90.9467	2.0109	90.9639	2.0163	90.9811	2.0217	90.9983	2.0272	91.0155	8460
2.0598	91.1185	2.0652	91.1357	2.0706	91.1529	2.0761	91.1701	2.0815	91.1872	8470
2.1141	91.2902	2.1195	91.3074	2.1249	91.3246	2.1303	91.3417	2.1358	91.3589	8480
2.1683	91.4618	2.1737	91.4790	2.1792	91.4961	2.1846	91.5133	2.1900	91.5304	8490
2.2226	91.6333	2.2280	91.6505	2.2334	91.6676	2.2388	91.6848	2.2442	91.7019	8500
2.2768	91.8047	2.2822	91.8219	2.2876	91.8390	2.2930	91.8561	2.2984	91.8733	8510
2.3309	91.9760	2.3363	91.9932	2.3418	92.0103	2.3472	92.0274	2.3526	92.0445	8520
2.3851	92.1472	2.3905	92.1643	2.3959	92.1815	2.4013	92.1986	2.4067	92.2157	8530
2.4392	92.3183	2.4446	92.3354	2.4500	92.3525	2.4554	92.3696	2.4608	92.3867	8540
2.4932	92.4893	2.4986	92.5064	2.5041	92.5235	2.5095	92.5406	2.5149	92.5577	8550
2.5473	92.6602	2.5527	92.6773	2.5581	92.6944	2.5635	92.7115	2.5689	92.7285	8560
2.6013	92.8310	2.6067	92.8481	2.6121	92.8652	2.6175	92.8822	2.6229	92.8993	8570
2.6553	93.0017	2.6607	93.0188	2.6661	93.0358	2.6715	93.0529	2.6769	93.0700	8580
2.7092	93.1723	2.7146	93.1894	2.7200	93.2064	2.7254	93.2235	2.7308	93.2405	8590
2.7631	93.3428	2.7685	93.3598	2.7739	93.3769	2.7793	93.3939	2.7847	93.4110	8600
2.8170	93.5132	2.8224	93.5302	2.8278	93.5473	2.8332	93.5643	2.8386	93.5813	8610
2.8709	93.6835	2.8763	93.7005	2.8816	93.7176	2.8870	93.7346	2.8924	93.7516	8620
2.9247	93.8537	2.9301	93.8707	2.9355	93.8877	2.9408	93.9047	2.9462	93.9218	8630
2.9785	94.0238	2.9839	94.0408	2.9892	94.0578	2.9946	94.0748	3.0000	94.0918	8640
90+	200+	90+	200+	90+	200+	90+	200+	90+	200+	

Tables of Square Roots

N	0 √N	√10N	1 √N	√10N	2 √N	√10N	3 √N	√10N	4 √N	√10N
	90+	200+	90+	200+	90+	200+	90+	200+	90+	200+
8650	3.0054	94.1088	3.0108	94.1258	3.0161	94.1428	3.0215	94.1598	3.0269	94.1768
8660	3.0591	94.2788	3.0645	94.2958	3.0699	94.3128	3.0752	94.3297	3.0806	94.3467
8670	3.1128	94.4486	3.1182	94.4656	3.1236	94.4826	3.1289	94.4996	3.1343	94.5166
8680	3.1665	94.6184	3.1719	94.6354	3.1773	94.6523	3.1826	94.6693	3.1880	94.6863
8690	3.2202	94.7881	3.2255	94.8050	3.2309	94.8220	3.2363	94.8389	3.2416	94.8559
8700	3.2738	94.9576	3.2792	94.9746	3.2845	94.9915	3.2899	95.0085	3.2952	95.0254
8710	3.3274	95.1271	3.3327	95.1440	3.3381	95.1610	3.3435	95.1779	3.3488	95.1949
8720	3.3809	95.2965	3.3863	95.3134	3.3916	95.3303	3.3970	95.3473	3.4024	95.3642
8730	3.4345	95.4657	3.4398	95.4827	3.4452	95.4996	3.4505	95.5165	3.4559	95.5334
8740	3.4880	95.6349	3.4933	95.6518	3.4987	95.6687	3.5040	95.6856	3.5094	95.7026
8750	3.5414	95.8040	3.5468	95.8209	3.5521	95.8378	3.5575	95.8547	3.5628	95.8716
8760	3.5949	95.9730	3.6002	95.9899	3.6056	96.0068	3.6109	96.0236	3.6162	96.0405
8770	3.6483	96.1419	3.6536	96.1587	3.6590	96.1756	3.6643	96.1925	3.6696	96.2094
8780	3.7017	96.3106	3.7070	96.3275	3.7123	96.3444	3.7177	96.3613	3.7230	96.3781
8790	3.7550	96.4793	3.7603	96.4962	3.7657	96.5131	3.7710	96.5300	3.7763	96.5468
8800	3.8083	96.6479	3.8136	96.6648	3.8190	96.6816	3.8243	96.6985	3.8296	96.7154
8810	3.8616	96.8164	3.8669	96.8333	3.8723	96.8501	3.8776	96.8670	3.8829	96.8838
8820	3.9149	96.9848	3.9202	97.0017	3.9255	97.0185	3.9308	97.0354	3.9361	97.0522
8830	3.9681	97.1532	3.9734	97.1700	3.9787	97.1868	3.9840	97.2036	3.9894	97.2205
8840	4.0213	97.3214	4.0266	97.3382	4.0319	97.3550	4.0372	97.3718	4.0425	97.3886
8850	4.0744	97.4895	4.0798	97.5063	4.0851	97.5231	4.0904	97.5399	4.0957	97.5567
8860	4.1276	97.6575	4.1329	97.6743	4.1382	97.6911	4.1435	97.7079	4.1488	97.7247
8870	4.1807	97.8255	4.1860	97.8422	4.1913	97.8590	4.1966	97.8758	4.2019	97.8926
8880	4.2338	97.9933	4.2391	98.0101	4.2444	98.0268	4.2497	98.0436	4.2550	98.0604
8890	4.2868	98.1610	4.2921	98.1778	4.2974	98.1946	4.3027	98.2113	4.3080	98.2281
8900	4.3398	98.3287	4.3451	98.3454	4.3504	98.3622	4.3557	98.3790	4.3610	98.3957
8910	4.3928	98.4962	4.3981	98.5130	4.4034	98.5297	4.4087	98.5465	4.4140	98.5632
8920	4.4458	98.6637	4.4510	98.6804	4.4563	98.6972	4.4616	98.7139	4.4669	98.7306
8930	4.4987	98.8311	4.5040	98.8478	4.5093	98.8645	4.5145	98.8812	4.5198	98.8980
8940	4.5516	98.9983	4.5569	99.0151	4.5621	99.0318	4.5674	99.0485	4.5727	99.0652
8950	4.6044	99.1655	4.6097	99.1822	4.6150	99.1989	4.6203	99.2156	4.6256	99.2324
8960	4.6573	99.3326	4.6626	99.3493	4.6678	99.3660	4.6731	99.3827	4.6784	99.3994
8970	4.7101	99.4996	4.7154	99.5163	4.7206	99.5330	4.7259	99.5497	4.7312	99.5664
8980	4.7629	99.6665	4.7681	99.6832	4.7734	99.6999	4.7787	99.7165	4.7840	99.7332
8990	4.8156	99.8333	4.8209	99.8500	4.8262	99.8666	4.8314	99.8833	4.8367	99.9000
9000	4.8683	*0.0000	4.8736	*0.0167	4.8789	*0.0333	4.8841	*0.0500	4.8894	*0.0667
9010	4.9210	0.1666	4.9263	0.1833	4.9316	0.1999	4.9368	0.2166	4.9421	0.2332
9020	4.9737	0.3331	4.9789	0.3498	4.9842	0.3664	4.9895	0.3831	4.9947	0.3997
9030	5.0263	0.4996	5.0316	0.5162	5.0368	0.5329	5.0421	0.5495	5.0474	0.5661
9040	5.0789	0.6659	5.0842	0.6826	5.0894	0.6992	5.0947	0.7158	5.0999	0.7324
9050	5.1315	0.8322	5.1367	0.8488	5.1420	0.8654	5.1473	0.8820	5.1525	0.8987
9060	5.1840	0.9983	5.1893	1.0150	5.1945	1.0316	5.1998	1.0482	5.2050	1.0648
9070	5.2365	1.1644	5.2418	1.1810	5.2470	1.1976	5.2523	1.2142	5.2575	1.2308
9080	5.2890	1.3304	5.2943	1.3470	5.2995	1.3636	5.3048	1.3802	5.3100	1.3967
9090	5.3415	1.4963	5.3467	1.5129	5.3520	1.5294	5.3572	1.5460	5.3625	1.5626
	90+	300+	90+	300+	90+	300+	90+	300+	90+	300+

5 √N	5 √10N	6 √N	6 √10N	7 √N	7 √10N	8 √N	8 √10N	9 √N	9 √10N	N
90+	200+	90+	200+	90+	200+	90+	200+	90+	200+	
3.0323	94.1938	3.0376	94.8108	3.0430	94.2278	3.0484	94.2448	3.0537	94.2618	8650
3.0860	94.3637	3.0914	94.3807	3.0967	94.3977	3.1021	94.4147	3.1075	94.4317	8660
3.1397	94.5335	3.1450	94.5505	3.1504	94.5675	3.1558	94.5845	3.1612	94.6014	8670
3.1933	94.7032	3.1987	94.7202	3.2041	94.7372	3.2094	94.7541	3.2148	94.7711	8680
3.2470	94.8729	3.2523	94.8898	3.2577	94.9068	3.2631	94.9237	3.2684	94.9407	8690
3.3006	95.0424	3.3059	95.0593	3.3113	95.0763	3.3167	95.0932	3.3220	95.1101	8700
3.3542	95.2118	3.3595	95.2287	3.3649	95.2457	3.3702	95.2626	3.3756	95.2795	8710
3.4077	95.3811	3.4131	95.3980	3.4184	95.4150	3.4238	95.4319	3.4291	95.4488	8720
3.4612	95.5503	3.4666	95.5673	3.4719	95.5842	3.4773	95.6011	3.4826	95.6180	8730
3.5147	95.7195	3.5201	95.7364	3.5254	95.7533	3.5307	95.7702	3.5361	95.7871	8740
3.5682	95.8885	3.5735	95.9054	3.5788	95.9223	3.5842	95.9392	3.5895	95.9561	8750
3.6216	96.0574	3.6269	96.0743	3.6323	96.0912	3.6376	96.1081	3.6429	96.1250	8760
3.6750	96.2263	3.6803	96.2431	3.6856	96.2600	3.6910	96.2769	3.6963	96.2938	8770
3.7283	96.3950	3.7337	96.4119	3.7390	96.4287	3.7443	96.4456	3.7497	96.4625	8780
3.7817	96.5637	3.7870	96.5805	3.7923	96.5974	3.7977	96.6142	3.8030	96.6311	8790
3.8350	96.7322	3.8403	96.7491	3.8456	96.7659	3.8509	96.7827	3.8563	96.7996	8800
3.8882	96.9007	3.8936	96.9175	3.8989	96.9343	3.9042	96.9512	3.9095	96.9680	8810
3.9415	97.0690	3.9468	97.0858	3.9521	97.1027	3.9574	97.1195	3.9628	97.1363	8820
3.9947	97.2373	4.0000	97.2541	4.0053	97.2709	4.0106	97.2877	4.0160	97.3046	8830
4.0479	97.4054	4.0532	97.4223	4.0585	97.4391	4.0638	97.4559	4.0691	97.4727	8840
4.1010	97.5735	4.1063	97.5903	4.1116	97.6071	4.1169	97.6239	4.1223	97.6407	8850
4.1541	97.7415	4.1594	97.7583	4.1647	97.7751	4.1701	97.7919	4.1754	97.8087	8860
4.2072	97.9094	4.2125	97.9262	4.2178	97.9429	4.2231	97.9597	4.2284	97.9765	8870
4.2603	98.0772	4.2656	98.0939	4.2709	98.1107	4.2762	98.1275	4.2815	98.1443	8880
4.3133	98.2449	4.3186	98.2616	4.3239	98.2784	4.3292	98.2952	4.3345	98.3119	8890
4.3663	98.4125	4.3716	98.4292	4.3769	98.4460	4.3822	98.4627	4.3875	98.4795	8900
4.4193	98.5900	4.4246	98.5967	4.4299	98.6135	4.4352	98.6302	4.4405	98.6469	8910
4.4722	98.7474	4.4775	98.7641	4.4828	98.7809	4.4881	98.7976	4.4934	98.8143	8920
4.5251	98.9147	4.5304	98.9314	4.5357	98.9482	4.5410	98.9649	4.5463	98.9816	8930
4.5780	99.0819	4.5833	99.0986	4.5886	99.1154	4.5939	99.1321	4.5992	99.1488	8940
4.6309	99.2491	4.6361	99.2658	4.6414	99.2825	4.6467	99.2992	4.6520	99.3159	8950
4.6837	99.4161	4.6890	99.4328	4.6942	99.4495	4.6995	99.4662	4.7048	99.4829	8960
4.7365	99.5830	4.7418	99.5997	4.7470	99.6164	4.7523	99.6331	4.7576	99.6498	8970
4.7892	99.7499	4.7945	99.7666	4.7998	99.7833	4.8051	99.7999	4.8103	99.8166	8980
4.8420	99.9167	4.8472	99.9333	4.8525	99.9500	4.8578	99.9667	4.8631	99.9833	8990
4.8947	*0.0833	4.8999	*0.1000	4.9052	*0.1166	4.9105	*0.1333	4.9158	*0.1500	9000
4.9474	0.2499	4.9526	0.2665	4.9579	0.2832	4.9632	0.2999	4.9684	0.3165	9010
5.0000	0.4164	5.0053	0.4330	5.0105	0.4497	5.0158	0.4663	5.0211	0.4829	9020
5.0526	0.5828	5.0579	0.5994	5.0631	0.6160	5.0684	0.6327	5.0737	0.6493	9030
5.1052	0.7491	5.1105	0.7657	5.1157	0.7823	5.1210	0.7989	5.1262	0.8156	9040
5.1578	0.9153	5.1630	0.9319	5.1683	0.9485	5.1735	0.9651	5.1788	0.9817	9050
5.2103	1.0814	5.2155	1.0980	5.2208	1.1146	5.2260	1.1312	5.2313	1.1478	9060
5.2628	1.2474	5.2680	1.2640	5.2733	1.2806	5.2795	1.2972	5.2838	1.3138	9070
5.3153	1.4133	5.3205	1.4299	5.3258	1.4465	5.3310	1.4631	5.3362	1.4797	9080
5.3677	1.5792	5.3730	1.5958	5.3782	1.6123	5.3834	1.6289	5.3887	1.6455	9090
90+	300+	90+	300+	90+	300+	90+	300+	90+	300+	

Tables of Square Roots

N	0 √N	0 √10N	1 √N	1 √10N	2 √N	2 √10N	3 √N	3 √10N	4 √N	4 √10N
	90+	300+	90+	300+	90+	300+	90+	300+	90+	300+
9100	5.3939	1.6621	5.3992	1.6786	5.4044	1.6952	5.4096	1.7118	5.4149	1.7284
9110	5.4463	1.8278	5.4516	1.8443	5.4568	1.8609	5.4620	1.8775	5.4673	1.8940
9120	5.4987	1.9934	5.5039	2.0099	5.5092	2.0265	5.5144	2.0430	5.5196	2.0596
9130	5.5510	2.1589	5.5563	2.1754	5.5615	2.1920	5.5667	2.2085	5.5720	2.2251
9140	5.6033	2.3243	5.6086	2.3409	5.6138	2.3574	5.6190	2.3739	5.6243	2.3905
9150	5.6556	2.4897	5.6609	2.5062	5.6661	2.5227	5.6713	2.5393	5.6765	2.5558
9160	5.7079	2.6549	5.7131	2.6714	5.7183	2.6880	5.7236	2.7045	5.7288	2.7210
9170	5.7601	2.8201	5.7653	2.8366	5.7706	2.8531	5.7758	2.8696	5.7810	2.8861
9180	5.8123	2.9851	5.8175	3.0017	5.8228	3.0182	5.8280	3.0347	5.8332	3.0512
9190	5.8645	3.1501	5.8697	3.1666	5.8749	3.1831	5.8801	3.1996	5.8853	3.2161
9200	5.9166	3.3150	5.9218	3.3315	5.9271	3.3480	5.9323	3.3645	5.9375	3.3809
9210	5.9687	3.4798	5.9740	3.4963	5.9792	3.5128	5.9844	3.5292	5.9896	3.5457
9220	6.0208	3.6445	6.0260	3.6610	6.0312	3.6775	6.0365	3.6939	6.0417	3.7104
9230	6.0729	3.8092	6.0781	3.8256	6.0833	3.8421	6.0885	3.8585	6.0937	3.8750
9240	6.1249	3.9737	6.1301	3.9901	6.1353	4.0066	6.1405	4.0230	6.1457	4.0395
9250	6.1769	4.1381	6.1821	4.1546	6.1873	4.1710	6.1925	4.1874	6.1977	4.2039
9260	6.2289	4.3025	6.2341	4.3189	6.2393	4.3353	6.2445	4.3518	6.2497	4.3682
9270	6.2808	4.4667	6.2860	4.4832	6.2912	4.4996	6.2964	4.5160	6.3016	4.5324
9280	6.3328	4.6309	6.3379	4.6473	6.3431	4.6637	6.3483	4.6802	6.3535	4.6966
9290	6.3846	4.7950	6.3898	4.8114	6.3950	4.8278	6.4002	4.8442	6.4054	4.8606
9300	6.4365	4.9590	6.4417	4.9754	6.4469	4.9918	6.4521	5.0082	6.4572	5.0246
9310	6.4883	5.1229	6.4935	5.1393	6.4987	5.1557	6.5039	5.1721	6.5091	5.1885
9320	6.5401	5.2868	6.5453	5.3031	6.5505	5.3195	6.5557	5.3359	6.5609	5.3523
9330	6.5919	5.4505	6.5971	5.4669	6.6023	5.4832	6.6075	5.4996	6.6126	5.5160
9340	6.6437	5.6141	6.6488	5.6305	6.6540	5.6469	6.6592	5.6632	6.6644	5.6796
9350	6.6954	5.7777	6.7006	5.7940	6.7057	5.8104	6.7109	5.8267	6.7161	5.8431
9360	6.7471	5.9412	6.7523	5.9575	6.7574	5.9739	6.7626	5.9902	6.7678	6.0065
9370	6.7988	6.1046	6.8039	6.1209	6.8091	6.1372	6.8143	6.1536	6.8194	6.1699
9380	6.8504	6.2679	6.8556	6.2842	6.8607	6.3005	6.8659	6.3168	6.8710	6.3332
9390	6.9020	6.4311	6.9072	6.4474	6.9123	6.4637	6.9175	6.4800	6.9227	6.4963
9400	6.9536	6.5942	6.9588	6.6105	6.9639	6.6268	6.9691	6.6431	6.9742	6.6594
9410	7.0052	6.7572	7.0103	6.7735	7.0155	6.7898	7.0206	6.8061	7.0258	6.8224
9420	7.0567	6.9202	7.0618	6.9365	7.0670	6.9528	7.0721	6.9691	7.0773	6.9853
9430	7.1082	7.0831	7.1133	7.0993	7.1185	7.1156	7.1236	7.1319	7.1288	7.1482
9440	7.1597	7.2458	7.1648	7.2621	7.1700	7.2784	7.1751	7.2946	7.1802	7.3109
9450	7.2111	7.4085	7.2163	7.4248	7.2214	7.4411	7.2265	7.4573	7.2317	7.4736
9460	7.2625	7.5711	7.2677	7.5874	7.2728	7.6036	7.2780	7.6199	7.2831	7.6361
9470	7.3139	7.7337	7.3191	7.7499	7.3242	7.7661	7.3293	7.7824	7.3345	7.7986
9480	7.3653	7.8961	7.3704	7.9123	7.3756	7.9286	7.3807	7.9448	7.3858	7.9610
9490	7.4166	8.0584	7.4218	8.0747	7.4269	8.0909	7.4320	8.1071	7.4372	8.1234
9500	7.4679	8.2207	7.4731	8.2369	7.4782	8.2531	7.4833	8.2694	7.4885	8.2856
9510	7.5192	8.3829	7.5244	8.3991	7.5295	8.4153	7.5346	8.4315	7.5397	8.4477
9520	7.5705	8.5450	7.5756	8.5612	7.5807	8.5774	7.5859	8.5936	7.5910	8.6098
9530	7.6217	8.7070	7.6268	8.7232	7.6320	8.7394	7.6371	8.7556	7.6422	8.7718
9540	7.6729	8.8689	7.6780	8.8851	7.6832	8.9013	7.6883	8.9175	7.6934	8.9337
	90+	300+	90+	300+	90+	300+	90+	300+	90+	300+

Tables of Square Roots

5 √N	5 √10N	6 √N	6 √10N	7 √N	7 √10N	8 √N	8 √10N	9 √N	9 √10N	N
90+	300+	90+	300+	90+	300+	90+	300+	90+	300+	
5.4201	1.7449	5.4254	1.7615	5.4306	1.7781	5.4358	1.7946	5.4411	1.8112	9100
5.4725	1.9106	5.4777	1.9271	5.4830	1.9437	5.4882	1.9603	5.4935	1.9768	9110
5.5249	2.0761	5.5301	2.0927	5.5353	2.1093	5.5406	2.1258	5.5458	2.1424	9120
5.5772	2.2416	5.5824	2.2582	5.5877	2.2747	5.5929	2.2913	5.5981	2.3078	9130
5.6295	2.4070	5.6347	2.4235	5.6400	2.4401	5.6452	2.4566	5.6504	2.4731	9140
5.6818	2.5723	5.6870	2.5888	5.6922	2.6054	5.6974	2.6219	5.7027	2.6384	9150
5.7340	2.7375	5.7392	2.7540	5.7445	2.7705	5.7497	2.7871	5.7549	2.8036	9160
5.7862	2.9026	5.7914	2.9191	5.7967	2.9356	5.8019	2.9521	5.8071	2.9686	9170
5.8384	3.0676	5.8436	3.0841	5.8488	3.1006	5.8541	3.1171	5.8593	3.1336	9180
5.8906	3.2326	5.8958	3.2491	5.9010	3.2656	5.9062	3.2820	5.9114	3.2985	9190
5.9427	3.3974	5.9479	3.4139	5.9531	3.4304	5.9583	3.4469	5.9635	3.4633	9200
5.9948	3.5622	6.0000	3.5787	6.0052	3.5951	6.0104	3.6116	6.0156	3.6281	9210
6.0469	3.7269	6.0521	3.7433	6.0573	3.7598	6.0625	3.7762	6.0677	3.7927	9220
6.0989	3.8914	6.1041	3.9079	6.1093	3.9243	6.1145	3.9408	6.1197	3.9572	9230
6.1509	4.0559	6.1561	4.0724	6.1613	4.0888	6.1665	4.1052	6.1717	4.1217	9240
6.2029	4.2203	6.2081	4.2368	6.2133	4.2532	6.2185	4.2696	6.2237	4.2861	9250
6.2549	4.3846	6.2601	4.4011	6.2653	4.4175	6.2705	4.4339	6.2756	4.4503	9260
6.3068	4.5488	6.3120	4.5653	6.3172	4.5817	6.3224	4.5981	6.3276	4.6145	9270
6.3587	4.7130	6.3639	4.7294	6.3691	4.7458	6.3743	4.7622	6.3795	4.7786	9280
6.4106	4.8770	6.4158	4.8934	6.4210	4.9098	6.4261	4.9262	6.4313	4.9426	9290
6.4624	5.0410	6.4676	5.0574	6.4728	5.0738	6.4780	5.0902	6.4832	5.1065	9300
6.5142	5.2048	6.5194	5.2212	6.5246	5.2376	6.5298	5.2540	6.5350	5.2704	9310
6.5660	5.3686	6.5712	5.3850	6.5764	5.4014	6.5816	5.4177	6.5867	5.4341	9320
6.6178	5.5323	6.6230	5.5487	6.6282	5.5651	6.6333	5.5814	6.6385	5.5978	9330
6.6695	5.6959	6.6747	5.7123	6.6799	5.7286	6.6851	5.7450	6.6902	5.7613	9340
6.7212	5.8594	6.7264	5.8758	6.7316	5.8921	6.7368	5.9085	6.7419	5.9248	9350
6.7729	6.0229	6.7781	6.0392	6.7833	6.0556	6.7884	6.0719	6.7936	6.0882	9360
6.8246	6.1862	6.8297	6.2025	6.8349	6.2189	6.8401	6.2352	6.8452	6.2515	9370
6.8762	6.3495	6.8814	6.3658	6.8865	6.3821	6.8917	6.3984	6.8969	6.4148	9380
6.9278	6.5126	6.9330	6.5290	6.9381	6.5453	6.9433	6.5616	6.9484	6.5779	9390
6.9794	6.6757	6.9845	6.6920	6.9897	6.7083	6.9948	6.7246	7.0000	6.7409	9400
7.0309	6.8387	7.0361	6.8550	7.0412	6.8713	7.0464	6.8876	7.0515	6.9039	9410
7.0824	7.0016	7.0876	7.0179	7.0927	7.0342	7.0979	7.0505	7.1030	7.0668	9420
7.1339	7.1645	7.1391	7.1807	7.1442	7.1970	7.1494	7.2133	7.1545	7.2296	9430
7.1854	7.3272	7.1905	7.3435	7.1957	7.3597	7.2008	7.3760	7.2060	7.3923	9440
7.2368	7.4898	7.2420	7.5061	7.2471	7.5224	7.2522	7.5386	7.2574	7.5549	9450
7.2882	7.6524	7.2934	7.6687	7.2985	7.6849	7.3036	7.7012	7.3088	7.7174	9460
7.3396	7.8149	7.3447	7.8311	7.3499	7.8474	7.3550	7.8636	7.3602	7.8798	9470
7.3910	7.9773	7.3961	7.9935	7.4012	8.0097	7.4064	8.0260	7.4115	8.0422	9480
7.4423	8.1396	7.4474	8.1558	7.4526	8.1720	7.4577	8.1883	7.4628	8.2045	9490
7.4936	8.3018	7.4987	8.3180	7.5038	8.3342	7.5090	8.3505	7.5141	8.3667	9500
7.5449	8.4639	7.5500	8.4801	7.5551	8.4964	7.5602	8.5126	7.5654	8.5288	9510
7.5961	8.6260	7.6012	8.6422	7.6064	8.6584	7.6115	8.6746	7.6166	8.6908	9520
7.6473	8.7880	7.6524	8.8041	7.6576	8.8203	7.6627	8.8365	7.6678	8.8527	9530
7.6985	8.9498	7.7036	8.9660	7.7088	8.9822	7.7139	8.9984	7.7190	9.0146	9540
90+	300+	90+	300+	90+	300+	90+	300+	90+	300+	

Tables of Square Roots

N	0 √N	0 √10N	1 √N	1 √10N	2 √N	2 √10N	3 √N	3 √10N	4 √N	4 √10N
	90+	300+	90+	300+	90+	300+	90+	300+	90+	300+
9550	7.7241	9.0307	7.7292	9.0469	7.7343	9.0631	7.7395	9.0793	7.7446	9.0955
9560	7.7753	9.1925	7.7804	9.2087	7.7855	9.2248	7.7906	9.2410	7.7957	9.2572
9570	7.8264	9.3542	7.8315	9.3703	7.8366	9.3865	7.8417	9.4027	7.8468	9.4188
9580	7.8775	9.5158	7.8826	9.5319	7.8877	9.5481	7.8928	9.5642	7.8979	9.5804
9590	7.9285	9.6773	7.9337	9.6934	7.9388	9.7095	7.9439	9.7257	7.9490	9.7418
9600	7.9796	9.8387	7.9847	9.8548	7.9898	9.8709	7.9949	9.8871	8.0000	9.9032
9610	8.0306	10.0000	8.0357	10.0161	8.0408	10.0323	8.0459	10.0484	8.0510	10.0645
9620	8.0816	10.1612	8.0867	10.1774	8.0918	10.1935	8.0969	10.2096	8.1020	10.2257
9630	8.1326	10.3224	8.1377	10.3385	8.1428	10.3546	8.1478	10.3707	8.1529	10.3869
9640	8.1835	10.4835	8.1886	10.4996	8.1937	10.5157	8.1988	10.5318	8.2039	10.5479
9650	8.2344	10.6445	8.2395	10.6606	8.2446	10.6767	8.2497	10.6928	8.2548	10.7089
9660	8.2853	10.8054	8.2904	10.8215	8.2955	10.8376	8.3006	10.8537	8.3056	10.8697
9670	8.3362	10.9662	8.3412	10.9823	8.3463	10.9984	8.3514	11.0145	8.3565	11.0305
9680	8.3870	11.1270	8.3921	11.1431	8.3972	11.1591	8.4022	11.1752	8.4073	11.1913
9690	8.4378	11.2876	8.4429	11.3037	8.4480	11.3198	8.4530	11.3358	8.4581	11.3519
9700	8.4886	11.4482	8.4937	11.4643	8.4987	11.4803	8.5038	11.4964	8.5089	11.5124
9710	8.5393	11.6087	8.5444	11.6248	8.5495	11.6408	8.5546	11.6569	8.5596	11.6729
9720	8.5901	11.7691	8.5951	11.7852	8.6002	11.8012	8.6053	11.8173	8.6103	11.8333
9730	8.6408	11.9295	8.6458	11.9455	8.6509	11.9615	8.6560	11.9776	8.6610	11.9936
9740	8.6914	12.0897	8.6965	12.1058	8.7016	12.1218	8.7066	12.1378	8.7117	12.1538
9750	8.7421	12.2499	8.7472	12.2659	8.7522	12.2819	8.7573	12.2979	8.7623	12.3139
9760	8.7927	12.4100	8.7978	12.4260	8.8028	12.4420	8.8079	12.4580	8.8130	12.4740
9770	8.8433	12.5700	8.8484	12.5860	8.8534	12.6020	8.8585	12.6180	8.8635	12.6340
9780	8.8939	12.7299	8.8989	12.7459	8.9040	12.7619	8.9090	12.7779	8.9141	12.7939
9790	8.9444	12.8898	8.9495	12.9057	8.9545	12.9217	8.9596	12.9377	8.9646	12.9537
9800	8.9949	13.0495	9.0000	13.0655	9.0051	13.0815	9.0101	13.0974	9.0152	13.1134
9810	9.0454	13.2092	9.0505	13.2252	9.0555	13.2411	9.0606	13.2571	9.0656	13.2730
9820	9.0959	13.3688	9.1010	13.3847	9.1060	13.4007	9.1110	13.4167	9.1161	13.4326
9830	9.1464	13.5283	9.1514	13.5443	9.1564	13.5602	9.1615	13.5761	9.1665	13.5921
9840	9.1968	13.6877	9.2018	13.7037	9.2069	13.7196	9.2119	13.7356	9.2169	13.7515
9850	9.2472	13.8471	9.2522	13.8630	9.2572	13.8790	9.2623	13.8949	9.2673	13.9108
9860	9.2975	14.0064	9.3026	14.0223	9.3076	14.0382	9.3126	14.0541	9.3177	14.0701
9870	9.3479	14.1656	9.3529	14.1815	9.3579	14.1974	9.3630	14.2133	9.3680	14.2292
9880	9.3982	14.3247	9.4032	14.3406	9.4082	14.3565	9.4133	14.3724	9.4183	14.3883
9890	9.4485	14.4837	9.4535	14.4996	9.4585	14.5155	9.4636	14.5314	9.4686	14.5473
9900	9.4987	14.6427	9.5038	14.6585	9.5088	14.6744	9.5138	14.6903	9.5188	14.7062
9910	9.5490	14.8015	9.5540	14.8174	9.5590	14.8333	9.5641	14.8492	9.5691	14.8651
9920	9.5992	14.9603	9.6042	14.9762	9.6092	14.9921	9.6143	15.0079	9.6193	15.0238
9930	9.6494	15.1190	9.6544	15.1349	9.6594	15.1508	9.6644	15.1666	9.6695	15.1825
9940	9.6995	15.2777	9.7046	15.2935	9.7096	15.3094	9.7146	15.3252	9.7196	15.3411
9950	9.7497	15.4362	9.7547	15.4521	9.7597	15.4679	9.7647	15.4838	9.7697	15.4996
9960	9.7998	15.5947	9.8048	15.6105	9.8098	15.6264	9.8148	15.6422	9.8198	15.6580
9970	9.8499	15.7531	9.8549	15.7689	9.8599	15.7847	9.8649	15.8006	9.8699	15.8164
9980	9.9000	15.9114	9.9050	15.9272	9.9100	15.9430	9.9150	15.9589	9.9200	15.9747
9990	9.9500	16.0696	9.9550	16.0854	9.9600	16.1013	9.9650	16.1171	9.9700	16.1329
	90+	300+	90+	300+	90+	300+	90+	300+	90+	300+

Tables of Square Roots

√N (5)	√10N (5)	√N (6)	√10N (6)	√N (7)	√10N (7)	√N (8)	√10N (8)	√N (9)	√10N (9)	N
90+	300+	90+	300+	90+	300+	90+	300+	90+	300+	
7.7497	9.1116	7.7548	9.1278	7.7599	9.1440	7.7650	9.1602	7.7701	9.1763	9550
7.8008	9.2735	7.8059	9.2895	7.8110	9.3057	7.8162	9.3218	7.8213	9.3380	9560
7.8519	9.4350	7.8570	9.4511	7.8621	9.4673	7.8673	9.4834	7.8724	9.4996	9570
7.9030	9.5965	7.9081	9.6127	7.9132	9.6288	7.9183	9.6450	7.9234	9.6611	9580
7.9541	9.7580	7.9592	9.7741	7.9643	9.7903	7.9694	9.8064	7.9745	9.8225	9590
8.0051	9.9193	8.0102	9.9355	8.0153	9.9516	8.0204	9.9677	8.0255	9.9839	9600
8.0561	10.0806	8.0612	10.0968	8.0663	10.1129	8.0714	10.1290	8.0765	10.1451	9610
8.1071	10.2416	8.1122	10.2580	8.1173	10.2741	8.1224	10.2902	8.1275	10.3063	9620
8.1580	10.4030	8.1631	10.4191	8.1682	10.4352	8.1733	10.4513	8.1784	10.4674	9630
8.2090	10.5640	8.2141	10.5801	8.2191	10.5962	8.2242	10.6123	8.2293	10.6284	9640
8.2599	10.7250	8.2649	10.7411	8.2700	10.7571	8.2751	10.7732	8.2802	10.7893	9650
8.3107	10.8858	8.3158	10.9019	8.3209	10.9180	8.3260	10.9341	8.3311	10.9502	9660
8.3616	11.0466	8.3667	11.0627	8.3717	11.0788	8.3768	11.0948	8.3819	11.1109	9670
8.4124	11.2073	8.4175	11.2234	8.4226	11.2395	8.4276	11.2555	8.4327	11.2716	9680
8.4632	11.3680	8.4683	11.3840	8.4733	11.4001	8.4784	11.4161	8.4835	11.4322	9690
8.5140	11.5285	8.5190	11.5445	8.5241	11.5606	8.5292	11.5766	8.5343	11.5927	9700
8.5647	11.6889	8.5698	11.7050	8.5748	11.7210	8.5799	11.7371	8.5850	11.7531	9710
8.6154	11.8493	8.6205	11.8654	8.6256	11.8814	8.6306	11.8974	8.6357	11.9135	9720
8.6661	12.0096	8.6712	12.0256	8.6762	12.0417	8.6813	12.0577	8.6864	12.0737	9730
8.7168	12.1698	8.7218	12.1858	8.7269	12.2019	8.7320	12.2179	8.7370	12.2339	9740
8.7674	12.3300	8.7725	12.3460	8.7775	12.3620	8.7826	12.3780	8.7877	12.3940	9750
8.8180	12.4900	8.8231	12.5060	8.8281	12.5220	8.8332	12.5380	8.8383	12.5540	9760
8.8686	12.6500	8.8737	12.6660	8.8787	12.6819	8.8838	12.6979	8.8888	12.7139	9770
8.9192	12.8098	8.9242	12.8258	8.9293	12.8418	8.9343	12.8578	8.9394	12.8738	9780
8.9697	12.9696	8.9747	12.9856	8.9798	13.0016	8.9848	13.0176	8.9899	13.0335	9790
9.0202	13.1294	9.0252	13.1453	9.0303	13.1613	9.0353	13.1773	9.0404	13.1932	9800
9.0707	13.2890	9.0757	13.3050	9.0808	13.3209	9.0858	13.3369	9.0909	13.3528	9810
9.1211	13.4486	9.1262	13.4645	9.1312	13.4805	9.1363	13.4964	9.1413	13.5124	9820
9.1716	13.6080	9.1766	13.6240	9.1817	13.6399	9.1867	13.6559	9.1917	13.6718	9830
9.2220	13.7674	9.2270	13.7834	9.2321	13.7993	9.2371	13.8152	9.2421	13.8312	9840
9.2724	13.9267	9.2774	13.9427	9.2824	13.9586	9.2875	13.9745	9.2925	13.9904	9850
9.3227	14.0860	9.3277	14.1019	9.3328	14.1178	9.3378	14.1337	9.3428	14.1496	9860
9.3730	14.2451	9.3781	14.2610	9.3831	14.2769	9.3881	14.2929	9.3932	14.3088	9870
9.4233	14.4042	9.4284	14.4201	9.4334	14.4360	9.4384	14.4519	9.4435	14.4678	9880
9.4736	14.5632	9.4786	14.5791	9.4837	14.5950	9.4887	14.6109	9.4937	14.6268	9890
9.5239	14.7221	9.5289	14.7380	9.5339	14.7539	9.5389	14.7698	9.5440	14.7856	9900
9.5741	14.8809	9.5791	14.8968	9.5841	14.9127	9.5892	14.9286	9.5942	14.9444	9910
9.6243	15.0397	9.6293	15.0556	9.6343	15.0714	9.6394	15.0873	9.6444	15.1032	9920
9.6745	15.1984	9.6795	15.2142	9.6845	15.2301	9.6895	15.2459	9.6945	15.2618	9930
9.7246	15.3569	9.7296	15.3728	9.7346	15.3886	9.7397	15.4045	9.7447	15.4204	9940
9.7747	15.5155	9.7798	15.5313	9.7848	15.5471	9.7898	15.5630	9.7948	15.5788	9950
9.8248	15.6739	9.8299	15.6897	9.8349	15.7056	9.8399	15.7214	9.8449	15.7372	9960
9.8749	15.8322	9.8799	15.8481	9.8849	15.8639	9.8899	15.8797	9.8949	15.8956	9970
9.9250	15.9905	9.9300	16.0063	9.9350	16.0222	9.9400	16.0380	9.9450	16.0538	9980
9.9750	16.1487	9.9800	16.1645	9.9850	16.1803	9.9900	16.1961	9.9950	16.2120	9990
90+	300+	90+	300+	90+	300+	90+	300+	90+	300+	

n	n^3	$\sqrt[3]{n}$	$\sqrt[3]{10n}$	$\sqrt[3]{100n}$	n	n^3	$\sqrt[3]{n}$	$\sqrt[3]{10n}$	$\sqrt[3]{100n}$
1	1	1.000 000	2.154 435	4.641 589					
2	8	1.259 921	2.714 418	5.848 035	50	125 000	3.684 031	7.937 005	17.09976
3	27	1.442 250	3.107 233	6.694 330	51	132 651	3.708 430	7.989 570	17.21301
4	64	1.587 401	3.419 952	7.368 063	52	140 608	3.732 511	8.041 452	17.32478
					53	148 877	3.756 286	8.092 672	17.43513
5	125	1.709 976	3.684 031	7.937 005	54	157 464	3.779 763	8.143 253	17.54411
6	216	1.817 121	3.914 868	8.434 327					
7	343	1.912 931	4.121 285	8.879 040	55	166 375	3.802 952	8.193 213	17.65174
8	512	2.000 000	4.308 869	9.283 178	56	175 616	3.825 862	8.242 571	17.75808
9	729	2.080 084	4.481 405	9.654 894	57	185 193	3.848 501	8.291 344	17.86316
					58	195 112	3.870 877	8.339 551	17.96702
10	1 000	2.154 435	4.641 589	10.00000	59	205 379	3.892 996	8.387 207	18.06969
11	1 331	2.223 980	4.791 420	10.32280					
12	1 728	2.289 428	4.932 424	10.62659	60	216 000	3.914 868	8.434 327	18.17121
13	2 197	2.351 335	5.065 797	10.91393	61	226 981	3.936 497	8.480 926	18.27160
14	2 744	2.410 142	5.192 494	11.18689	62	238 328	3.957 892	8.527 019	18.37091
					63	250 047	3.979 057	8.572 619	18.46915
15	3 375	2.466 212	5.313 293	11.44714	64	262 144	4.000 000	8.617 739	18.56636
16	4 096	2.519 842	5.428 835	11.69607					
17	4 913	2.571 282	5.539 658	11.93483	65	274 625	4.020 726	8.662 391	18.66256
18	5 832	2.620 741	5.646 216	12.16440	66	287 496	4.041 240	8.706 588	18.75777
19	6 859	2.668 402	5.748 897	12.38562	67	300 763	4.061 548	8.750 340	18.85204
					68	314 432	4.081 655	8.793 659	18.94536
20	8 000	2.714 418	5.848 035	12.59921	69	328 509	4.101 566	8.836 556	19.03778
21	9 261	2.758 924	5.943 922	12.80579					
22	10 648	2.802 039	6.036 811	13.00591	70	343 000	4.121 285	8.879 040	19.12931
23	12 167	2.843 867	6.126 926	13.20006	71	357 911	4.140 818	8.921 121	19.21997
24	13 824	2.884 499	6.214 465	13.38866	72	373 248	4.160 168	8.962 809	19.30979
					73	389 017	4.179 339	9.004 113	19.39877
25	15 625	2.924 018	6.299 605	13.57209	74	405 224	4.198 336	9.045 042	19.48695
26	17 576	2.962 496	6.382 504	13.75069					
27	19 683	3.000 000	6.463 304	13.92477	75	421 875	4.217 163	9.085 603	19.57434
28	21 952	3.036 589	6.542 133	14.09460	76	438 976	4.235 824	9.125 805	19.66095
29	24 389	3.072 317	6.619 106	14.26043	77	456 533	4.254 321	9.165 656	19.74681
					78	474 552	4.272 659	9.205 164	19.83192
30	27 000	3.107 233	6.694 330	14.42250	79	493 039	4.290 840	9.244 335	19.91632
31	29 791	3.141 381	6.767 899	14.58100					
32	32 768	3.174 802	6.839 904	14.73613	80	512 000	4.308 869	9.283 178	20.00000
33	35 937	3.207 534	6.910 423	14.88806	81	531 441	4.326 749	9.321 698	20.08299
34	39 304	3.239 612	6.979 532	15.03695	82	551 368	4.344 481	9.359 902	20.16550
					83	571 787	4.362 071	9.397 796	20.24694
35	42 875	3.271 066	7.047 299	15.18294	84	592 704	4.379 519	9.435 388	20.32793
36	46 656	3.301 927	7.113 787	15.32619					
37	50 653	3.332 222	7.179 054	15.46680	85	614 125	4.396 830	9.472 682	20.40828
38	54 872	3.361 975	7.243 156	15.60491	86	636 056	4.414 005	9.509 685	20.48800
39	59 319	3.391 211	7.306 144	15.74061	87	658 503	4.431 048	9.546 403	20.56710
					88	681 472	4.447 960	9.582 840	20.64560
40	64 000	3.419 952	7.368 063	15.87401	89	704 969	4.464 745	9.619 002	20.72351
41	68 921	3.448 217	7.428 959	16.00521					
42	74 088	3.476 027	7.488 872	16.13429	90	729 000	4.481 405	9.654 894	20.80084
43	79 507	3.503 398	7.547 842	16.26133	91	753 571	4.497 941	9.690 521	20.87759
44	85 184	3.530 348	7.605 905	16.38643	92	778 688	4.514 357	9.725 888	20.95379
					93	804 357	4.530 655	9.761 000	21.02944
45	91 125	3.556 893	7.663 094	16.50964	94	830 584	4.546 836	9.795 861	21.10454
46	97 336	3.583 048	7.719 443	16.63103					
47	103 823	3.608 826	7.774 980	16.75069	95	857 375	4.562 903	9.830 476	21.17912
48	110 592	3.634 241	7.829 735	16.86865	96	884 736	4.578 857	9.864 848	21.25317
49	117 649	3.659 306	7.883 735	16.98499	97	912 673	4.594 701	9.898 983	21.32671
					98	941 192	4.610 436	9.932 884	21.39975
50	125 000	3.684 031	7.937 005	17.09976	99	970 299	4.626 065	9.966 555	21.47229
					100	1 000 000	4.641 589	10.00000	21.54435

Tables of Cubes and Cube Roots

n	n^3	$\sqrt[3]{n}$	$\sqrt[3]{10n}$	$\sqrt[3]{100n}$
100	1 000 000	4.641 589	10.00000	21.54435
101	1 030 301	4.657 010	10.03322	21.61592
102	1 061 208	4.672 329	10.06623	21.68703
103	1 092 727	4.687 548	10.09902	21.75767
104	1 124 864	4.702 669	10.13159	21.82786
105	1 157 625	4.717 694	10.16396	21.89760
106	1 191 016	4.732 623	10.19613	21.96689
107	1 225 043	4.747 459	10.22809	22.03575
108	1 259 712	4.762 203	10.25986	22.10419
109	1 295 029	4.776 856	10.29142	22.17220
110	1 331 000	4.791 420	10.32280	22.23980
111	1 367 631	4.805 896	10.35399	22.30699
112	1 404 928	4.820 285	10.38499	22.37378
113	1 442 897	4.834 588	10.41580	22.44017
114	1 481 544	4.848 808	10.44644	22.50617
115	1 520 875	4.862 944	10.47690	22.57179
116	1 560 896	4.876 999	10.50718	22.63702
117	1 601 613	4.890 973	10.53728	22.70189
118	1 643 032	4.904 868	10.56722	22.76638
119	1 685 159	4.918 685	10.59699	22.83051
120	1 728 000	4.932 424	10.62659	22.89428
121	1 771 561	4.946 087	10.65602	22.95770
122	1 815 848	4.959 676	10.68530	23.02078
123	1 860 867	4.973 190	10.71441	23.08350
124	1 906 624	4.986 631	10.74337	23.14589
125	1 953 125	5.000 000	10.77217	23.20794
126	2 000 376	5.013 298	10.80082	23.26967
127	2 048 383	5.026 526	10.82932	23.33107
128	2 097 152	5.039 684	10.85767	23.39214
129	2 146 689	5.052 774	10.88587	23.45290
130	2 197 000	5.065 797	10.91393	23.51335
131	2 248 091	5.078 753	10.94184	23.57348
132	2 299 968	5.091 643	10.96961	23.63332
133	2 352 637	5.104 469	10.99724	23.69285
134	2 406 104	5.117 230	11.02474	23.75208
135	2 460 375	5.129 928	11.05209	23.81102
136	2 515 456	5.142 563	11.07932	23.86966
137	2 571 353	5.155 137	11.10641	23.92803
138	2 628 072	5.167 649	11.13336	23.98610
139	2 685 619	5.180 101	11.16019	24.04390
140	2 744 000	5.192 494	11.18689	24.10142
141	2 803 221	5.204 828	11.21346	24.15867
142	2 863 288	5.217 103	11.23991	24.21565
143	2 924 207	5.229 322	11.26623	24.27236
144	2 985 984	5.241 483	11.29243	24.32881
145	3 048 625	5.253 588	11.31851	24.38499
146	3 112 136	5.265 637	11.34447	24.44092
147	3 176 523	5.277 632	11.37031	24.49660
148	3 241 792	5.289 572	11.39604	24.55202
149	3 307 949	5.301 459	11.42165	24.60719
150	3 375 000	5.313 293	11.44714	24.66212

n	n^3	$\sqrt[3]{n}$	$\sqrt[3]{10n}$	$\sqrt[3]{100n}$
150	3 375 000	5.313 293	11.44714	24.66212
151	3 442 951	5.325 074	11.47252	24.71680
152	3 511 808	5.336 803	11.49779	24.77125
153	3 581 577	5.348 481	11.52295	24.82545
154	3 652 264	5.360 108	11.54800	24.87942
155	3 723 875	5.371 685	11 57295	24.93315
156	3 796 416	5.383 213	11.59778	24.98666
157	3 869 893	5.394 691	11.62251	25.03994
158	3 944 312	5.406 120	11.64713	25.09299
159	4 019 679	5.417 502	11.67165	25.14581
160	4 096 000	5.428 835	11.69607	25.19842
161	4 173 281	5.440 122	11.72039	25.25081
162	4 251 528	5.451 362	11.74460	25.30298
163	4 330 747	5.462 556	11.76872	25.35494
164	4 410 944	5.473 704	11.79274	25.40668
165	4 492 125	5.484 807	11.81666	25.45822
166	4 574 296	5.495 865	11.84048	25.50954
167	4 657 463	5.506 878	11.86421	25.56067
168	4 741 632	5.517 848	11.88784	25.61158
169	4 826 809	5.528 775	11.91138	25.66230
170	4 913 000	5.539 658	11.93483	25.71282
171	5 000 211	5.550 499	11.95819	25.76313
172	5 088 448	5.561 298	11.98145	25.81326
173	5 177 717	5.572 055	12.00463	25.86319
174	5 268 024	5.582 770	12.02771	25.91292
175	5 359 375	5.593 445	12.05071	25.96247
176	5 451 776	5.604 079	12.07362	26.01183
177	5 545 233	5.614 672	12.09645	26.06100
178	5 639 752	5.625 226	12.11918	26.10999
179	5 735 339	5.635 741	12.14184	26.15879
180	5 832 000	5.646 216	12.16440	26.20741
181	5 929 741	5.656 653	12.18689	26.25586
182	6 028 568	5.667 051	12.20929	26.30412
183	6 128 487	5.677 411	12.23161	26.35221
184	6 229 504	5.687 734	12.25385	26.40012
185	6 331 625	5.698 019	12.27601	26.44786
186	6 434 856	5.708 267	12.29809	26.49543
187	6 539 203	5.718 479	12.32009	26.54283
188	6 644 672	5.728 654	12.34201	26.59006
189	6 751 269	5.738 794	12.36386	26.63712
190	6 859 000	5.748 897	12.38562	26.68402
191	6 967 871	5.758 965	12.40731	26.73075
192	7 077 888	5.768 998	12.42893	26.77732
193	7 189 057	5.778 997	12.45047	26.82373
194	7 301 384	5.788 960	12.47194	26.86997
195	7 414 875	5.798 890	12.49333	26.91606
196	7 529 536	5.808 786	12.51465	26.96199
197	7 645 373	5.818 648	12.53590	27.00777
198	7 762 392	5.828 477	12.55707	27.05339
199	7 880 599	5.838 272	12.57818	27.09886
200	8 000 000	5.848 035	12.59921	27.14418

Tables of Cubes and Cube Roots

n	n^3	$\sqrt[3]{n}$	$\sqrt[3]{10n}$	$\sqrt[3]{100n}$	n	n^3	$\sqrt[3]{n}$	$\sqrt[3]{10n}$	$\sqrt[3]{100n}$
200	8 000 000	5.848 035	12.59921	27.14418	250	15 625 000	6.299 605	13.57209	29.24018
201	8 120 601	5.857 766	12.62017	27.18934	251	15 813 251	6.307 994	13.59016	29.27911
202	8 242 408	5.867 464	12.64107	27.23436	252	16 003 008	6.316 360	13.60818	29.31794
203	8 365 427	5.877 131	12.66189	27.27922	253	16 194 277	6.324 704	13.62616	29.35667
204	8 489 664	5.886 765	12.68265	27.32394	254	16 387 064	6.333 026	13.64409	29.39530
205	8 615 125	5.896 369	12.70334	27.36852	255	16 581 375	6.341 326	13.66197	29.43383
206	8 741 816	5.905 941	12.72396	27.41295	256	16 777 216	6.349 604	13.67981	29.47225
207	8 869 743	5.915 482	12.74452	27.45723	257	16 974 593	6.357 861	13.69760	29.51058
208	8 998 912	5.924 992	12.76501	27.50138	258	17 173 512	6.366 097	13.71534	29.54880
209	9 129 329	5.934 472	12.78543	27.54538	259	17 373 979	6.374 311	13.73304	29.58693
210	9 261 000	5.943 922	12.80579	27.58924	260	17 576 000	6.382 504	13.75069	29.62496
211	9 393 931	5.953 342	12.82609	27.63296	261	17 779 581	6.390 677	13.76830	29.66289
212	9 528 128	5.962 732	12.84632	27.67655	262	17 984 728	6.398 828	13.78586	29.70073
213	9 663 597	5.972 093	12.86648	27.72000	263	18 191 447	6.406 959	13.80337	29.73847
214	9 800 344	5.981 424	12.88659	27.76331	264	18 399 744	6.415 069	13.82085	29.77611
215	9 938 375	5.990 726	12.90663	27.80649	265	18 609 625	6.423 158	13.83828	29.81366
216	10 077 696	6.000 000	12.92661	27.84953	266	18 821 096	6.431 228	13.85566	29.85111
217	10 218 313	6.009 245	12.94653	27.89244	267	19 034 163	6.439 277	13.87300	29.88847
218	10 360 232	6.018 462	12.96638	27.93522	268	19 248 832	6.447 306	13.89030	29.92574
219	10 503 459	6.027 650	12.98618	27.97787	269	19 465 109	6.455 315	13.90755	29.96292
220	10 648 000	6.036 811	13.00591	28.02039	270	19 683 000	6.463 304	13.92477	30.00000
221	10 793 861	6.045 944	13.02559	28.06278	271	19 902 511	6.471 274	13.94194	30.03699
222	10 941 048	6.055 049	13.04521	28.10505	272	20 123 648	6.479 224	13.95906	30.07389
223	11 089 567	6.064 127	13.06477	28.14718	273	20 346 417	6.487 154	13.97615	30.11070
224	11 239 424	6.073 178	13.08427	28.18919	274	20 570 824	6.495 065	13.99319	30.14742
225	11 390 625	6.082 202	13.10371	28.23108	275	20 796 875	6.502 957	14.01020	30.18405
226	11 543 176	6.091 199	13.12309	28.27284	276	21 024 576	6.510 830	14.02716	30.22060
227	11 697 083	6.100 170	13.14242	28.31448	277	21 253 933	6.518 684	14.04408	30.25705
228	11 852 352	6.109 115	13.16169	28.35600	278	21 484 952	6.526 519	14.06096	30.29342
229	12 008 989	6.118 033	13.18090	28.39739	279	21 717 639	6.534 335	14.07780	30.32970
230	12 167 000	6.126 926	13.20006	28.43867	280	21 952 000	6.542 133	14.09460	30.36589
231	12 326 391	6.135 792	13.21916	28.47983	281	22 188 041	6.549 912	14.11136	30.40200
232	12 487 168	6.144 634	13.23821	28.52086	282	22 425 768	6.557 672	14.12808	30.43802
233	12 649 337	6.153 449	13.25721	28.56178	283	22 665 187	6.565 414	14.14476	30.47395
234	12 812 904	6.162 240	13.27614	28.60259	284	22 906 304	6.573 138	14.16140	30.50981
235	12 977 875	6.171 006	13.29503	28.64327	285	23 149 125	6.580 844	14.17800	30.54557
236	13 144 256	6.179 747	13.31386	28.68384	286	23 393 656	6.588 532	14.19456	30.58126
237	13 312 053	6.188 463	13.33264	28.72430	287	23 639 903	6.596 202	14.21109	30.61686
238	13 481 272	6.197 154	13.35136	28.76464	288	23 887 872	6.603 854	14.22757	30.65238
239	13 651 919	6.205 822	13.37004	28.80487	289	24 137 569	6.611 489	14.24402	30.68781
240	13 824 000	6.214 465	13.38866	28.84499	290	24 389 000	6.619 106	14.26043	30.72317
241	13 997 521	6.223 084	13.40723	28.88500	291	24 642 171	6.626 705	14.27680	30.75844
242	14 172 488	6.231 680	13.42575	28.92489	292	24 897 088	6.634 287	14.29314	30.79363
243	14 348 907	6.240 251	13.44421	28.96468	293	25 153 757	6.641 852	14.30944	30.82875
244	14 526 784	6.248 800	13.46263	29.00436	294	25 412 184	6.649 400	14.32570	30.86378
245	14 706 125	6.257 325	13.48100	29.04393	295	25 672 375	6.656 930	14.34192	30.89873
246	14 886 936	6.265 827	13.49931	29.08339	296	25 934 336	6.664 444	14.35811	30.93361
247	15 069 223	6.274 305	13.51758	29.12275	297	26 198 073	6.671 940	14.37426	30.96840
248	15 252 992	6.282 761	13.53580	29.16199	298	26 463 592	6.679 420	14.39037	31.00312
249	15 438 249	6.291 195	13.55397	29.20114	299	26 730 899	6.686 883	14.40645	31.03776
250	15 625 000	6.299 605	13.57209	29.24018	300	27 000 000	6.694 330	14.42250	31.07233

n	n^3	$\sqrt[3]{n}$	$\sqrt[3]{10n}$	$\sqrt[3]{100n}$	n	n^3	$\sqrt[3]{n}$	$\sqrt[3]{10n}$	$\sqrt[3]{100n}$
300	27 000 000	6.694 330	14.42250	31.07233	350	42 875 000	7.047 299	15.18294	32.71066
301	27 270 901	6.701 759	14.43850	31.10681	351	43 243 551	7.054 004	15.19739	32.74179
302	27 543 608	6.709 173	14.45447	31.14122	352	43 614 208	7.060 697	15.21181	32.77285
303	27 818 127	6.716 570	14.47041	31.17556	353	43 986 977	7.067 377	15.22620	32.80386
304	28 094 464	6.723 951	14.48631	31.20982	354	44 361 864	7.074 044	15.24057	32.83480
305	28 372 625	6.731 315	14.50218	31.24400	355	44 738 875	7.080 699	15.25490	32.86569
306	28 652 616	6.738 664	14.51801	31.27811	356	45 118 016	7.087 341	15.26921	32.89652
307	28 934 443	6.745 997	14.53381	31.31214	357	45 499 293	7.093 971	15.28350	32.92730
308	29 218 112	6.753 313	14.54957	31.34610	358	45 882 712	7.100 588	15.29775	32.95801
309	29 503 629	6.760 614	14.56530	31.37999	359	46 268 279	7.107 194	15.31198	32.98867
310	29 791 000	6.767 899	14.58100	31.41381	360	46 656 000	7.113 787	15.32619	33.01927
311	30 080 231	6.775 169	14.59666	31.44755	361	47 045 881	7.120 367	15.34037	33.04982
312	30 371 328	6.782 423	14.61229	31.48122	362	47 437 928	7.126 936	15.35452	33.08031
313	30 664 297	6.789 661	14.62788	31.51482	363	47 832 147	7.133 492	15.36864	33.11074
314	30 959 144	6.796 884	14.64344	31.54834	364	48 228 544	7.140 037	15.38274	33.14112
315	31 255 875	6.804 092	14.65897	31.58180	365	48 627 125	7.146 569	15.39682	33.17144
316	31 554 496	6.811 285	14.67447	31.61518	366	49 027 896	7.153 090	15.41087	33.20170
317	31 855 013	6.818 462	14.68993	31.64850	367	49 430 863	7.159 599	15.42489	33.23191
318	32 157 432	6.825 624	14.70536	31.68174	368	49 836 032	7.166 096	15.43889	33.26207
319	32 461 759	6.832 771	14.72076	31.71492	369	50 243 409	7.172 581	15.45286	33.29217
320	32 768 000	6.839 904	14.73613	31.74802	370	50 653 000	7.179 054	15.46680	33.32222
321	33 076 161	6.847 021	14.75146	31.78106	371	51 064 811	7.185 516	15.48073	33.35221
322	33 386 248	6.854 124	14.76676	31.81403	372	51 478 848	7.191 966	15.49462	33.38215
323	33 698 267	6.861 212	14.78203	31.84693	373	51 895 117	7.198 405	15.50849	33.41204
324	34 012 224	6.868 285	14.79727	31.87976	374	52 313 624	7.204 832	15.52234	33.44187
325	34 328 125	6.875 344	14.81248	31.91252	375	52 734 375	7.211 248	15.53616	33.47165
326	34 645 976	6.882 389	14.82766	31.94522	376	53 157 376	7.217 652	15.54996	33.50137
327	34 965 783	6.889 419	14.84280	31.97785	377	53 582 633	7.224 045	15.56373	33.53105
328	35 287 552	6.896 434	14.85792	32.01041	378	54 010 152	7.230 427	15.57748	33.56067
329	35 611 289	6.903 436	14.87300	32.04291	379	54 439 939	7.236 797	15.59121	33.59024
330	35 937 000	6.910 423	14.88806	32.07534	380	54 872 000	7.243 156	15.60491	33.61975
331	36 264 691	6.917 396	14.90308	32.10771	381	55 306 341	7.249 505	15.61858	33.64922
332	36 594 368	6.924 356	14.91807	32.14001	382	55 742 968	7.255 842	15.63224	33.67863
333	36 926 037	6.931 301	14.93303	32.17225	383	56 181 887	7.262 167	15.64587	33.70800
334	37 259 704	6.938 232	14.94797	32.20442	384	56 623 104	7.268 482	15.65947	33.73731
335	37 595 375	6.945 150	14.96287	32.23653	385	57 066 625	7.274 786	15.67305	33.76657
336	37 933 056	6.952 053	14.97774	32.26857	386	57 512 456	7.281 079	15.68661	33.79578
337	38 272 753	6.958 943	14.99259	32.30055	387	57 960 603	7.287 362	15.70014	33.82494
338	38 614 472	6.965 820	15.00740	32.33247	388	58 411 072	7.293 633	15.71366	33.85405
339	38 958 219	6.972 683	15.02219	32.36433	389	58 863 869	7.299 894	15.72714	33.88310
340	39 304 000	6.979 532	15.03695	32.39612	390	59 319 000	7.306 144	15.74061	33.91211
341	39 651 821	6.986 368	15.05167	32.42785	391	59 776 471	7.312 383	15.75405	33.94107
342	40 001 688	6.993 191	15.06637	32.45952	392	60 236 288	7.318 611	15.76747	33.96999
343	40 353 607	7.000 000	15.08104	32.49112	393	60 698 457	7.324 829	15.78087	33.99885
344	40 707 584	7.006 796	15.09568	32.52267	394	61 162 984	7.331 037	15.79424	34.02766
345	41 063 625	7.013 579	15.11030	32.55415	395	61 629 875	7.337 234	15.80759	34.05642
346	41 421 736	7.020 349	15.12488	32.58557	396	62 099 136	7.343 420	15.82092	34.08514
347	41 781 923	7.027 106	15.13944	32.61694	397	62 570 773	7.349 597	15.83423	34.11381
348	42 144 192	7.033 850	15.15397	32.64824	398	63 044 792	7.355 762	15.84751	34.14242
349	42 508 549	7.040 581	15.16847	32.67948	399	63 521 199	7.361 918	15.86077	34.17100
350	42 875 000	7.047 299	15.18294	32.71066	400	64 000 000	7.368 063	15.87401	34.19952

Tables of Cubes and Cube Roots

n	n^3	$\sqrt[3]{n}$	$\sqrt[3]{10n}$	$\sqrt[3]{100n}$	n	n^3	$\sqrt[3]{n}$	$\sqrt[3]{10n}$	$\sqrt[3]{100n}$
400	64 000 000	7.368 063	15.87401	34.19952	450	91 125 000	7.663 094	16.50964	35.56893
401	64 481 201	7.374 198	15.88723	34.22799	451	91 733 851	7.668 766	16.52186	35.59526
402	64 964 808	7.380 323	15.90042	34.25642	452	92 345 408	7.674 430	16.53406	35.62155
403	65 450 827	7.386 437	15.91360	34.28480	453	92 959 677	7.680 086	16.54624	35.64780
404	65 939 264	7.392 542	15.92675	34.31314	454	93 576 664	7.685 733	16.55841	35.67401
405	66 430 125	7.398 636	15.93988	34.34143	455	94 196 375	7.691 372	16.57056	35.70018
406	66 923 416	7.404 721	15.95299	34.36967	456	94 818 816	7.697 002	16.58269	35.72632
407	67 419 143	7.410 795	15.96607	34.39786	457	95 443 993	7.702 625	16.59480	35.75242
408	67 917 312	7.416 860	15.97914	34.42601	458	96 071 912	7.708 239	16.60690	35.77848
409	68 417 929	7.422 914	15.99218	34.45412	459	96 702 579	7.713 845	16.61897	35.80450
410	68 921 000	7.428 959	16.00521	34.48217	460	97 336 000	7.719 443	16.63103	35.83048
411	69 426 531	7.434 994	16.01821	34.51018	461	97 972 181	7.725 032	16.64308	35.85642
412	69 934 528	7.441 019	16.03119	34.53815	462	98 611 128	7.730 614	16.65510	35.88233
413	70 444 997	7.447 034	16.04415	34.56507	463	99 252 847	7.736 188	16.66711	35.90820
414	70 957 944	7.453 040	16.05709	34.59395	464	99 897 344	7.741 753	16.67910	35.93404
415	71 473 375	7.459 036	16.07001	34.62178	465	100 544 625	7.747 311	16.69108	35.95983
416	71 991 296	7.465 022	16.08290	34.64956	466	101 194 696	7.752 861	16.70303	35.98559
417	72 511 713	7.470 999	16.09578	34.67731	467	101 847 563	7.758 402	16.71497	36.01131
418	73 034 632	7.476 966	16.10864	34.70500	468	102 503 232	7.763 936	16.72689	36.03700
419	73 560 059	7.482 924	16.12147	34.73266	469	103 161 709	7.769 462	16.73880	36.06265
420	74 088 000	7.488 872	16.13429	34.76027	470	103 823 000	7.774 980	16.75069	36.08826
421	74 618 461	7.494 811	16.14708	34.78783	471	104 487 111	7.780 490	16.76256	36.11384
422	75 151 448	7.500 741	16.15986	34.81535	472	105 154 048	7.785 993	16.77441	36.13938
423	75 686 967	7.506 661	16.17261	34.84283	473	105 823 817	7.791 488	16.78625	36.16488
424	76 225 024	7.512 572	16.18534	34.87027	474	106 496 424	7.796 975	16.79807	36.19035
425	76 765 625	7.518 473	16.19806	34.89766	475	107 171 875	7.802 454	16.80988	36.21578
426	77 308 776	7.524 365	16.21075	34.92501	476	107 850 176	7.807 925	16.82167	36.24118
427	77 854 483	7.530 248	16.22343	34.95232	477	108 531 333	7.813 389	16.83344	36.26654
428	78 402 752	7.536 122	16.23608	34.97958	478	109 215 352	7.818 846	16.84519	36.29187
429	78 953 589	7.541 987	16.24872	35.00680	479	109 902 239	7.824 294	16.85693	36.31716
430	79 507 000	7.547 842	16.26133	35.03398	480	110 592 000	7.829 735	16.86865	36.34241
431	80 062 991	7.553 689	16.27393	35.06112	481	111 284 641	7.835 169	16.88036	36.36763
432	80 621 568	7.559 526	16.28651	35.08821	482	111 980 168	7.840 595	16.89205	36.39282
433	81 182 737	7.565 355	16.29906	35.11527	483	112 678 587	7.846 013	16.90372	36.41797
434	81 746 504	7.571 174	16.31160	35.14228	484	113 379 904	7.851 424	16.91538	36.44308
435	82 312 875	7.576 985	16.32412	35.16925	485	114 084 125	7.856 828	16.92702	36.46817
436	82 881 856	7.582 787	16.33662	35.19618	486	114 791 256	7.862 224	16.93865	36.49321
437	83 453 453	7.588 579	16.34910	35.22307	487	115 501 303	7.867 613	16 95026	36.51822
438	84 027 672	7.594 363	16.36156	35.24991	488	116 214 272	7.872 994	16.96185	36.54320
439	84 604 519	7.600 139	16.37400	35.27672	489	116 930 169	7.878 368	16.97343	36.56815
440	85 184 000	7.605 905	16.38643	35.30348	490	117 649 000	7.883 735	16.98499	36.59306
441	85 766 121	7.611 663	16.39883	35.33021	491	118 370 771	7.889 095	16.99654	36.61793
442	86 350 888	7.617 412	16.41122	35.35689	492	119 095 488	7.894 447	17.00807	36.64278
443	86 938 307	7.623 152	16.42358	35.38354	493	119 823 157	7.899 792	17.01959	36.66758
444	87 528 384	7.628 884	16.43593	35.41014	494	120 553 784	7.905 129	17.03108	36.69236
445	88 121 125	7.634 607	16.44826	35.43671	495	121 287 375	7.910 460	17.04257	36.71716
446	88 716 536	7.640 321	16.46057	35.46323	496	122 023 936	7.915 783	17.05404	36.74181
447	89 314 623	7.646 027	16.47287	35.48971	497	122 763 473	7.921 099	17.06549	36.76649
448	89 915 392	7.651 725	16.48514	35.51616	498	123 505 992	7.926 408	17.07693	36.79113
449	90 518 849	7.657 414	16.49740	35.54257	499	124 251 499	7.931 710	17.08835	36.81574
450	91 125 000	7.663 094	16.50964	35.56893	500	125 000 000	7.937 005	17.09976	36.84031

Tables of Cubes and Cube Roots

n	n³	∛n	∛10n	∛100n	n	n³	∛n	∛10n	∛100n
500	125 000 000	7.937 005	17.09976	36.84031	550	166 375 000	8.193 213	17.65174	38.02952
501	125 751 501	7.942 293	17.11115	36.86486	551	167 284 151	8.198 175	17.66243	38.05256
502	126 506 008	7.947 574	17.12253	36.88937	552	168 196 608	8.203 132	17.67311	38.07557
503	127 263 527	7.952 848	17.13389	36.91385	553	169 112 377	8.208 082	17.68378	38.09854
504	128 024 064	7.958 114	17.14524	36.93830	554	170 031 464	8.213 027	17.69443	38.12149
505	128 787 625	7.963 374	17.15657	36.96271	555	170 953 875	8.217 966	17.70507	38.14442
506	129 554 216	7.968 627	17.16789	36.98709	556	171 879 616	8.222 899	17.71570	38.16731
507	130 323 843	7.973 873	17.17919	37.01144	557	172 808 693	8.227 825	17.72631	38.19018
508	131 096 512	7.979 112	17.19048	37.03576	558	173 741 112	8.232 746	17.73691	38.21302
509	131 872 229	7.984 344	17.20175	37.06004	559	174 676 879	8.237 661	17.74750	38.23584
510	132 651 000	7.989 570	17.21301	37.08430	560	175 616 000	8.242 571	17.75808	38.25862
511	133 432 831	7.994 788	17.22425	37.10852	561	176 558 481	8.247 474	17.76864	38.28138
512	134 217 728	8.000 000	17.23548	37.13271	562	177 504 328	8.252 372	17.77920	38.30412
513	135 005 697	8.005 205	17.24669	37.15687	563	178 453 547	8.257 263	17.78973	38.32682
514	135 796 744	8.010 403	17.25789	37.18100	564	179 406 144	8.262 149	17.80026	38.34950
515	136 590 875	8.015 595	17.26908	37.20509	565	180 362 125	8.267 029	17.81077	38.37215
516	137 388 096	8.020 779	17.28025	37.22916	566	181 321 496	8.271 904	17.82128	38.39478
517	138 188 413	8.025 957	17.29140	37.25319	567	182 284 263	8.276 773	17.83177	38.41737
518	138 991 832	8.031 129	17.30254	37.27720	568	183 250 432	8.281 635	17.84224	38.43995
519	139 798 359	8.036 293	17.31367	37.30117	569	184 220 009	8.286 493	17.85271	38.46249
520	140 608 000	8.041 452	17.32478	37.32511	570	185 193 000	8.291 344	17.86316	38.48501
521	141 420 761	8.046 603	17.33588	37.34902	571	186 169 411	8.296 190	17.87360	38.50750
522	142 236 648	8.051 748	17.34696	37.37290	572	187 149 248	8.301 031	17.88403	38.52997
523	143 055 667	8.056 886	17.35804	37.39675	573	188 132 517	8.305 865	17.89444	38.55241
524	143 877 824	8.062 018	17.36909	37.42057	574	189 119 224	8.310 694	17.90485	38.57482
525	144 703 125	8.067 143	17.38013	37.44436	575	190 109 375	8.315 517	17.91524	38.59721
526	145 531 576	8.072 262	17.39116	37.46812	576	191 102 976	8.320 335	17.92562	38.61958
527	146 363 183	8.077 374	17.40218	37.49185	577	192 100 033	8.325 148	17.93599	38.64191
528	147 197 952	8.082 480	17.41318	37.51555	578	193 100 552	8.329 954	17.94634	38.66422
529	148 035 889	8.087 579	17.42416	37.53922	579	194 104 539	8.334 755	17.95669	38.68651
530	148 877 000	8.092 672	17.43513	37.56286	580	195 112 000	8.339 551	17.96702	38.70877
531	149 721 291	8.097 759	17.44609	37.58647	581	196 122 941	8.344 341	17.97734	38.73100
532	150 568 768	8.102 839	17.45704	37.61005	582	197 137 368	8.349 126	17.98765	38.75321
533	151 419 437	8.107 913	17.46797	37.63360	583	198 155 287	8.353 905	17.99794	38.77539
534	152 273 304	8.112 980	17.47889	37.65712	584	199 176 704	8.358 678	18.00823	38.79755
535	153 130 375	8.118 041	17.48979	37.68061	585	200 201 625	8.363 447	18.01850	38.81968
536	153 990 656	8.123 096	17.50068	37.70407	586	201 230 056	8.368 209	18.02876	38.84179
537	154 854 153	8.128 145	17.51156	37.72751	587	202 262 003	8.372 967	18.03901	38.86387
538	155 720 872	8.133 187	17.52242	37.75091	588	203 297 472	8.377 719	18.04925	38.88593
539	156 590 819	8.138 223	17.53327	37.77429	589	204 336 469	8.382 465	18.05947	38.90796
540	157 464 000	8.143 253	17.54411	37.79763	590	205 379 000	8.387 207	18.06969	38.92996
541	158 340 421	8.148 276	17.55493	37.82095	591	206 425 071	8.391 942	18.07989	38.95195
542	159 220 088	8.153 294	17.56574	37.84424	592	207 474 688	8.396 673	18.09008	38.97390
543	160 103 007	8.158 305	17.57654	37.86750	593	208 527 857	8.401 398	18.10026	38.99584
544	160 989 184	8.163 310	17.58732	37.89073	594	209 584 584	8.406 118	18.11043	39.01774
545	161 878 625	8.168 309	17.59809	37.91393	595	210 644 875	8.410 833	18.12059	39.03963
546	162 771 336	8.173 302	17.60885	37.93711	596	211 703 736	8.415 542	18.13074	39.06149
547	163 667 323	8.178 289	17.61959	37.96025	597	212 776 173	8.420 246	18.14087	39.08332
548	164 566 592	8.183 269	17.63032	37.98337	598	213 847 192	8.424 945	18.15099	39.10513
549	165 469 149	8.188 244	17.64104	38.00646	599	214 921 799	8.429 638	18 16111	39.12692
550	166 375 000	8.193 213	17.65174	38.02952	600	216 000 000	8.434 327	18.17121	39.14868

n	n^3	$\sqrt[3]{n}$	$\sqrt[3]{10n}$	$\sqrt[3]{100n}$	n	n^3	$\sqrt[3]{n}$	$\sqrt[3]{10n}$	$\sqrt[3]{100n}$
600	216 000 000	8.434 327	18.17121	39.14868	**650**	274 625 000	8.662 391	18.66256	40.20726
601	217 081 801	8.439 010	18.18130	39.17041	651	275 894 451	8.666 831	18.67212	40.22787
602	218 167 208	8.443 688	18.19137	39.19213	652	277 167 808	8.671 266	18.68168	40.24845
603	219 256 227	8.448 361	18.20144	39.21382	653	278 445 077	8.675 697	18.69122	40.26902
604	220 348 864	8.453 028	18.21150	39.23548	654	279 726 264	8.680 124	18.70076	40.28957
605	221 445 125	8.457 691	18.22154	39.25712	655	281 011 375	8.684 546	18.71029	40.31009
606	222 545 016	8.462 348	18.23158	39.27874	656	282 300 416	8.688 963	18.71980	40.33059
607	223 648 543	8.467 000	18.24160	39.30033	657	283 593 393	8.693 376	18.72931	40.35108
608	224 755 712	8.471 647	18.25161	39.32190	658	284 890 312	8.697 784	18.73881	40.37154
609	225 866 529	8.476 289	18.26161	39.34345	659	286 191 179	8.702 188	18.74830	40.39198
610	226 981 000	8.480 926	18.27160	39.36497	**660**	287 496 000	8.706 588	18.75777	40.41240
611	228 099 131	8.485 558	18.28158	39.38647	661	288 804 781	8.710 983	18.76724	40.43280
612	229 220 928	8.490 185	18.29155	39.40795	662	290 117 528	8.715 373	18.77670	40.45318
613	230 346 397	8.494 807	18.30151	39.42940	663	291 434 247	8.719 760	18.78615	40.47354
614	231 475 544	8.499 423	18.31145	39.45083	664	292 754 944	8.724 141	18.79559	40.49388
615	232 608 375	8.504 035	18.32139	39.47223	665	294 079 625	8.728 519	18.80502	40.51420
616	233 744 896	8.508 642	18.33131	39.49362	666	295 408 296	8.732 892	18.81444	40.53449
617	234 885 113	8.513 243	18.34123	39.51498	367	296 740 963	8.737 260	13.82386	40.55477
618	236 029 032	8.517 840	18.35113	39.53631	668	298 077 632	8.741 625	18.83326	40.57503
619	237 176 659	8.522 432	18.36102	39.55763	669	299 418 309	8.745 985	18.84265	40.59526
620	238 328 000	8.527 019	18.37091	39.57892	**670**	300 763 000	8.750 340	18.85204	40.61548
621	239 483 061	8.531 601	18.38078	39.60018	671	302 111 711	8.754 691	18.86141	40.63568
622	240 641 848	8.536 178	18.39064	39.62143	672	303 464 448	8.759 038	18.87078	40.65585
623	241 804 367	8.540 750	18.40049	39.64265	673	304 821 217	8.763 381	18.88013	40.67601
624	242 970 624	8.545 317	18.41033	39.66385	674	306 182 024	8.767 719	18.88948	40.69615
625	244 140 625	8.549 880	18.42016	39.68503	675	307 546 875	8.772 053	18.89882	40.71626
626	245 314 376	8.554 437	18.42998	39.70618	676	308 915 776	8.776 383	18.90814	40.73636
627	246 491 883	8.558 990	18.43978	39.72731	677	310 288 733	8.780 708	18.91746	40.75644
628	247 673 152	8.563 538	18.44958	39.74842	678	311 665 752	8.785 030	18.92677	40.77650
629	248 858 189	8.568 081	18.45937	39.76951	679	313 046 839	8.789 347	18.93607	40.79653
630	250 047 000	8.572 619	18.46915	39.79057	**680**	314 432 000	8.793 659	18.94536	40.81655
631	251 239 591	8.577 152	18.47891	39.81161	681	315 821 241	8.797 968	18.95465	40.83655
632	252 435 968	8.581 681	18.48867	39.83263	682	317 214 568	8.802 272	18.96392	40.85653
633	253 636 137	8.586 205	18.49842	39.85363	683	318 611 987	8.806 572	18.97318	40.87649
634	254 840 104	8.590 724	18.50815	39.87461	684	320 013 504	8.810 868	18.98244	40.89643
635	256 047 875	8.595 238	18.51788	39.89556	685	321 419 125	8.815 160	18.99169	40.91635
636	257 259 456	8.599 748	18.52759	39.91649	686	322 828 856	8.819 447	19.00092	40.93625
637	258 474 853	8.604 252	18.53730	39.93740	687	324 242 703	8.823 731	19.01015	40.95613
638	259 694 072	8.608 753	18.54700	39.95829	688	325 660 672	8.828 010	19.01937	40.97599
639	260 917 119	8.613 248	18.55668	39.97916	689	327 082 769	8.832 285	19.02858	40.99584
640	262 144 000	8.617 739	18.56636	40.00000	**690**	328 509 000	8.836 556	19.03778	41.01566
641	263 374 721	8.622 225	18.57602	40.02082	691	329 939 371	8.840 823	19.04698	41.03546
642	264 609 288	8.626 706	18.58568	40.04162	692	331 373 888	8.845 085	19.05616	41.05525
643	265 847 707	8.631 183	18.59532	40.06240	693	332 812 557	8.849 344	19.06533	41.07502
644	267 089 984	8.635 655	18.60495	40.08316	694	334 255 384	8.853 599	19.07450	41.09476
645	268 336 125	8.640 123	18.61458	40.10390	695	335 702 375	8.857 849	19.08366	41.11449
646	269 586 136	8.644 585	18.62419	40.12461	696	337 153 536	8.862 095	19.09281	41.13420
647	270 840 023	8.649 044	18.63380	40.14530	697	338 608 873	8.866 338	19.10195	41.15389
648	272 097 792	8.653 497	18.64340	40.16598	698	340 068 392	8.870 576	19.11108	41.17357
649	273 359 449	8.657 947	18.65298	40.18663	699	341 532 099	8.874 810	19.12020	41.19322
650	274 625 000	8.662 391	18.66256	40.20726	**700**	343 000 000	8.879 040	19.12931	41.21285

Tables of Cubes and Cube Roots

n	n³	$\sqrt[3]{n}$	$\sqrt[3]{10n}$	$\sqrt[3]{100n}$	n	n³	$\sqrt[3]{n}$	$\sqrt[3]{10n}$	$\sqrt[3]{100n}$
700	343 000 000	8.879 040	19.12931	41 21285	750	421 875 000	9.085 603	19.57434	42.17163
701	344 472 101	8.883 266	19.13842	41.23247	751	423 564 751	9.089 639	19.58303	42.19037
702	345 948 408	8.887 488	19.14751	41.25207	752	425 259 008	9.093 672	19.59172	42.20909
703	347 428 927	8.891 706	19.15660	41.27164	753	426 957 777	9.097 701	19.60040	42.22779
704	348 913 664	8.895 920	19.16568	41.29120	754	428 661 064	9.101 727	19.60908	42.24647
705	350 402 625	8.900 130	19.17475	41.31075	755	430 368 875	9.105 748	19.61774	42.26514
706	351 895 816	8.904 337	19.18381	41.33027	756	432 081 216	9.109 767	19.62640	42.28379
707	353 393 243	8.908 539	19.19286	41.34977	757	433 798 093	9.113 782	19.63505	42.30243
708	354 894 912	8.912 737	19.20191	41.36926	758	435 519 512	9.117 793	19.64369	42.32105
709	356 400 829	8.916 931	19.21095	41.38873	759	437 245 479	9.121 801	19.65232	42.33965
710	357 911 000	8.921 121	19.21997	41.40818	760	438 976 000	9.125 805	19.66095	42.35824
711	359 425 431	8.925 308	19.22899	41.42761	761	440 711 081	9.129 806	19.66957	42.37681
712	360 944 128	8.929 490	19.23800	41.44702	762	442 450 728	9.133 803	19.67818	42.39536
713	362 467 097	8.933 669	19.24701	41.46642	763	444 194 947	9.137 797	19.68679	42.41390
714	363 994 344	8.937 843	19.25600	41.48579	764	445 943 744	9.141 787	19.69538	42.43242
715	365 525 875	8.942 014	19.26499	41.50515	765	447 697 125	9.145 774	19.70397	42.45092
716	367 061 696	8.946 181	19.27396	41.52449	766	449 455 096	9.149 758	19.71256	42.46941
717	368 601 813	8.950 344	19.28293	41.54382	767	451 217 663	9.153 738	19.72113	42.48789
718	370 146 232	8.954 503	19.29189	41.56312	768	452 984 832	9.157 714	19.72970	42.50634
719	371 694 959	8.958 658	19.30084	41.58241	769	454 756 609	9.161 687	19.73826	42.52478
720	373 248 000	8.962 809	19.30979	41.60168	770	456 533 000	9.165 656	19.74681	42.54321
721	374 805 361	8.966 957	19.31872	41.62093	771	458 314 011	9.169 623	19.75535	42.56162
722	376 367 048	8.971 101	19.32765	41.64016	772	460 099 648	9.173 585	19.76389	42.58001
723	377 933 067	8.975 241	19.33657	41.65938	773	461 889 917	9.177 544	19.77242	42.59839
724	379 503 424	8.979 377	19.34548	41.67857	774	463 684 824	9.181 500	19.78094	42.61675
725	381 078 125	8.983 509	19.35438	41.69775	775	465 484 375	9.185 453	19.78946	42.63509
726	382 657 176	8.987 637	19.36328	41.71692	776	467 288 576	9.189 402	19.79797	42.65342
727	384 240 583	8.991 762	19.37216	41.73606	777	469 097 433	9.193 347	19.80647	42.67174
728	385 828 352	8.995 883	19.38104	41.75519	778	470 910 952	9.197 290	19.81496	42.69004
729	387 420 489	9.000 000	19.38991	41.77430	779	472 729 139	9.201 229	19.82345	42.70832
730	389 017 000	9.004 113	19.39877	41.79339	780	474 552 000	9.205 164	19.83192	42.72659
731	390 617 891	9.008 223	19.40763	41.81247	781	476 379 541	9.209 096	19.84040	42.74484
732	392 223 168	9.012 329	19.41647	41.83152	782	478 211 768	9.213 025	19.84886	42.76307
733	393 832 837	9.016 431	19.42531	41.85056	783	480 048 687	9.216 950	19.85732	42.78129
734	395 446 904	9.020 529	19.43414	41.86959	784	481 890 304	9.220 873	19.86577	42.79950
735	397 065 375	9.024 624	19.44296	41.88859	785	483 736 625	9.224 791	19.87421	42.81769
736	398 688 256	9.028 715	19.45178	41.90758	786	485 587 656	9.228 707	19.88265	42.83586
737	400 315 553	9.032 802	19.46058	41.92655	787	487 443 403	9.232 619	19.89107	42.85402
738	401 947 272	9.036 886	19.46938	41.94551	788	489 303 872	9.236 528	19.89950	42.87216
739	403 583 419	9.040 966	19.47817	41.96444	789	491 169 069	9.240 433	19.90791	42.89029
740	405 224 000	9.045 042	19.48695	41.98336	790	493 039 000	9.244 335	19.91632	42.90840
741	406 869 021	9.049 114	19.49573	42.00227	791	494 913 671	9.248 234	19.92472	42.92650
742	408 518 488	9.053 183	19.50449	42.02115	792	496 793 088	9.252 130	19.93311	42.94458
743	410 172 407	9.057 248	19.51325	42.04002	793	498 677 257	9.256 022	19.94150	42.96265
744	411 830 784	9.061 310	19.52200	42.05887	794	500 566 184	9.259 911	19.94987	42.98070
745	413 493 625	9.065 368	19.53074	42.07771	795	502 459 875	9.263 797	19.95825	42.99874
746	415 160 936	9.069 422	19.53948	42.09653	796	504 358 336	9.267 680	19.96661	43.01676
747	416 832 723	9.073 473	19.54820	42.11533	797	506 261 573	9.271 559	19.97497	43.03477
748	418 508 992	9.077 520	19.55692	42.13411	798	508 169 592	9.275 435	19.98332	43.05276
749	420 189 749	9.081 563	19.56563	42.15288	799	510 082 399	9.279 308	19.99166	43.07073
750	421 875 000	9.085 603	19.57434	42.17163	800	512 000 000	9.283 178	20.00000	43.08869

329

Tables of Cubes and Cube Roots

n	n^3	$\sqrt[3]{n}$	$\sqrt[3]{10n}$	$\sqrt[3]{100n}$	n	n^3	$\sqrt[3]{n}$	$\sqrt[3]{10n}$	$\sqrt[3]{100n}$
800	512 000 000	9.283 178	20.00000	43.08869	850	614 125 000	9.472 682	20.40828	43.96830
801	513 922 401	9.287 044	20.00833	43.10664	851	616 295 051	9.476 396	20.41628	43.98553
802	515 849 608	9.290 907	20.01665	43.12457	852	618 470 208	9.480 106	20.42427	44.00275
803	517 781 627	9.294 767	20.02497	43.14249	853	620 650 477	9.483 814	20.43226	44.01996
804	519 718 464	9.298 624	20.03328	43.16039	854	622 835 864	9.487 518	20.44024	44.03716
805	521 660 125	9.302 477	20.04158	43.17828	855	625 026 375	9.491 220	20.44821	44.05434
806	523 606 616	9.306 328	20.04988	43.19615	856	627 222 016	9.494 919	20.45618	44.07151
807	525 557 943	9.310 175	20.05816	43.21400	857	629 422 793	9.498 615	20.46415	44.08866
808	527 514 112	9.314 019	20.06645	43.23185	858	631 628 712	9.502 308	20.47210	44.10581
809	529 475 129	9.317 860	20.07472	43.24967	859	633 839 779	9.505 998	20.48005	44.12293
810	531 441 000	9.321 698	20.08299	43.26749	860	636 056 000	9.509 685	20.48800	44.14005
811	533 411 731	9.325 532	20.09125	43.28529	861	638 277 381	9.513 370	20.49593	44.15715
812	535 387 328	9.329 363	20.09950	43.30307	862	640 503 928	9.517 052	20.50387	44.17424
813	537 367 797	9.333 192	20.10775	43.32084	863	642 735 647	9.520 730	20.51179	44.19132
814	539 353 144	9.337 017	20.11599	43.33859	864	644 972 544	9.524 406	20.51971	44.20838
815	541 343 375	9.340 839	20.12423	43.35633	865	647 214 625	9.528 079	20.52762	44.22543
816	543 338 496	9.344 657	20.13245	43.37406	866	649 461 896	9.531 750	20.53553	44.24246
817	545 338 513	9.348 473	20.14067	43.39177	867	651 714 363	9.535 417	20.54343	44.25949
818	547 343 432	9.352 286	20.14889	43.40947	868	653 972 032	9.539 082	20.55133	44.27650
819	549 353 259	9.356 095	20.15710	43.42715	869	656 234 909	9.542 744	20.55922	44.29349
820	551 368 000	9.359 902	20.16530	43.44481	870	658 503 000	9.546 403	20.56710	44.31048
821	553 387 661	9.363 705	20.17349	43.46247	871	660 776 311	9.550 059	20.57498	44.32745
822	555 412 248	9.367 505	20.18168	43.48011	872	663 054 848	9.553 712	20.58285	44.34440
823	557 441 767	9.371 302	20.18986	43.49773	873	665 338 617	9.557 363	20.59071	44.36135
824	559 476 224	9.375 096	20.19803	43.51534	874	667 627 624	9.561 011	20.59857	44.37828
825	561 515 625	9.378 887	20.20620	43.53294	875	669 921 875	9.564 656	20.60643	44.39520
826	563 559 976	9.382 675	20.21436	43.55052	876	672 221 376	9.568 298	20.61427	44.41211
827	565 609 283	9.386 460	20.22252	43.56809	877	674 526 133	9.571 938	20.62211	44.42900
828	567 663 552	9.390 242	20.23066	43.58564	878	676 836 152	9.575 574	20.62995	44.44588
829	569 722 789	9.394 021	20.23880	43.60318	879	679 151 439	9.579 208	20.63778	44.46275
830	571 787 000	9.397 796	20.24694	43.62071	880	681 472 000	9.582 840	20.64560	44.47960
831	573 856 191	9.401 569	20.25507	43.63822	881	683 797 841	9.586 468	20.65342	44.49644
832	575 930 368	9.405 339	20.26319	43.65572	882	686 128 968	9.590 094	20.66123	44.51327
833	578 009 537	9.409 105	20.27130	43.67320	883	688 465 387	9.593 717	20.66904	44.53009
834	580 093 704	9.412 869	20.27941	43.69067	884	690 807 104	9.597 337	20.67684	44.54689
835	582 182 875	9.416 630	20.28751	43.70812	885	693 154 125	9.600 955	20.68463	44.56368
836	584 277 056	9.420 387	20.29561	43.72556	886	695 506 456	9.604 570	20.69242	44.58046
837	586 376 253	9.424 142	20.30370	43.74299	887	697 864 103	9.608 182	20.70020	44.59723
838	588 480 472	9.427 894	20.31178	43.76041	888	700 227 072	9.611 791	20.70798	44.61398
839	590 589 719	9.431 642	20.31986	43.77781	889	702 595 369	9.615 398	20.71575	44.63072
840	592 704 000	9.435 388	20.32793	43.79519	890	704 969 000	9.619 002	20.72351	44.64745
841	594 823 321	9.439 131	20.33599	43.81256	891	707 347 971	9.622 603	20.73127	44.66417
842	596 947 688	9.442 870	20.34405	43.82992	892	709 732 288	9.626 202	20.73902	44.68087
843	599 077 107	9.446 607	20.35210	43.84727	893	712 121 957	9.629 797	20.74677	44.69756
844	601 211 584	9.450 341	20.36014	43.86460	894	714 516 984	9.633 391	20.75451	44.71424
845	603 351 125	9.454 072	20.36818	43.88191	895	716 917 375	9.636 981	20.76225	44.73090
846	605 495 736	9.457 800	20.37621	43.89922	896	719 323 136	9.640 569	20.76998	44.74756
847	607 645 423	9.461 525	20.38424	43.91651	897	721 734 273	9.644 154	20.77770	44.76420
848	609 800 192	9.465 247	20.39226	43.93378	898	724 150 792	9.647 737	20.78542	44.78083
849	611 960 049	9.468 966	20.40027	43.95105	899	726 572 699	9.651 317	20.79313	44.79744
850	614 125 000	9.472 682	20.40828	43.96830	900	729 000 000	9.654 894	20.80084	44.81405

n	n^3	$\sqrt[3]{n}$	$\sqrt[3]{10n}$	$\sqrt[3]{100n}$	n	n^3	$\sqrt[3]{n}$	$\sqrt[3]{10n}$	$\sqrt[3]{100n}$
900	729 000 000	9.654 894	20.80084	44.81405	950	857 375 000	9.830 476	21.17912	45.62903
901	731 432 701	9.658 468	20.80854	44.83064	951	860 085 351	9.833 924	21.18655	45.64503
902	733 870 808	9.662 040	20.81623	44.84722	952	862 801 408	9.837 369	21.19397	45.66102
903	736 314 327	9.665 610	20.82392	44.86379	953	865 523 177	9.840 813	21.20139	45.67701
904	738 763 264	9.669 176	20.83161	44.88034	954	868 250 664	9.844 254	21.20880	45.69298
905	741 217 625	9.672 740	20.83929	44.89688	955	870 983 875	9.847 692	21.21621	45.70894
906	743 677 416	9.676 302	20.84696	44.91341	956	873 722 816	9.851 128	21.22361	45.72489
907	746 142 643	9.679 860	20.85463	44.92993	957	876 467 493	9.854 562	21.23101	45.74082
908	748 613 312	9.683 417	20.86229	44.94644	958	879 217 912	9.857 993	21.23840	45.75675
909	751 089 429	9.686 970	20.86994	44.96293	959	881 974 079	9.861 422	21.24579	45.77267
910	753 571 000	9.690 521	20.87759	44.97941	960	884 736 000	9.864 848	21.25317	45.78857
911	756 058 031	9.694 069	20.88524	44.99588	961	887 503 681	9.868 272	21.26055	45.80446
912	758 550 528	9.697 615	20.89288	45.01234	962	890 277 128	9.871 694	21.26792	45.82035
913	761 048 497	9.701 158	20.90051	45.02879	963	893 056 347	9.875 113	21.27529	45.83622
914	763 551 944	9.704 699	20.90814	45.04522	964	895 841 344	9.878 530	21.28265	45.85208
915	766 060 875	9.708 237	20.91576	45.06164	965	898 632 125	9.881 945	21.29001	45.86793
916	768 575 296	9.711 772	20.92338	45.07805	966	901 428 696	9.885 357	21.29736	45.88376
917	771 095 213	9.715 305	20.93099	45.09445	967	904 231 063	9.888 767	21.30470	45.89969
918	773 620 632	9.718 835	20.93860	45.11084	968	907 039 232	9.892 175	21.31204	45.91541
919	776 151 559	9.722 363	20.94620	45.12721	969	909 853 209	9.895 580	21.31938	45.93121
920	778 688 000	9.725 888	20.95379	45.14357	970	912 673 000	9.898 983	21.32671	45.94701
921	781 229 961	9.729 411	20.96138	45.15992	971	915 498 611	9.902 384	21.33404	45.96279
922	783 777 448	9.732 931	20.96896	45.17626	972	918 330 048	9.905 782	21.34136	45.97857
923	786 330 467	9.736 448	20.97654	45.19259	973	921 167 317	9.909 178	21.34868	45.99433
924	788 889 024	9.739 963	20.98411	45.20891	974	924 010 424	9.912 571	21.35599	46.01008
925	791 453 125	9.743 476	20.99168	45.22521	975	926 859 375	9.915 962	21.36329	46.02582
926	794 022 776	9.746 986	20.99924	45.24150	976	929 714 176	9.919 351	21.37059	46.04155
927	796 597 983	9.750 493	21.00680	45.25778	977	932 574 833	9.922 738	21.37789	46.05727
928	799 178 752	9.753 998	21.01435	45.27405	978	935 441 352	9.926 122	21.38518	46.07298
929	801 765 089	9.757 500	21.02190	45.29030	979	938 313 739	9.929 504	21.39247	46.08868
930	804 357 000	9.761 000	21.02944	45.30655	980	941 192 000	9.932 884	21.39975	46.10436
931	806 954 491	9.764 497	21.03697	45.32278	981	944 076 141	9.936 261	21.40703	46.12004
932	809 557 568	9.767 992	21.04450	45.33900	982	946 966 168	9.939 636	21.41430	46.13571
933	812 166 237	9.771 485	21.05203	45.35521	983	949 862 087	9.943 009	21.42156	46.15136
934	814 780 504	9.774 974	21.05954	45.37141	984	952 763 904	9.946 380	21.42883	46.16700
935	817 400 375	9.778 462	21.06706	45.38760	985	955 671 625	9.949 748	21.43608	46.18264
936	820 025 856	9.781 946	21.07456	45.40377	986	958 585 256	9.953 114	21.44333	46.19826
937	822 656 953	9.785 429	21.08207	45.41994	987	961 504 803	9.956 478	21.45058	46.21387
938	825 293 672	9.788 909	21.08956	45.43609	988	964 430 272	9.959 839	21.45782	46.22948
939	827 936 019	9.792 386	21.09706	45.45223	989	967 361 669	9.963 198	21.46506	46.24507
940	830 584 000	9.795 861	21.10454	45.46836	990	970 299 000	9.966 555	21.47229	46.26065
941	833 237 621	9.799 334	21.11202	45.48448	991	973 242 271	9.969 910	21.47952	46.27622
942	835 896 888	9.802 804	21.11950	45.50058	992	976 191 488	9.973 262	21.48674	46.29178
943	838 561 807	9.806 271	21.12697	45.51668	993	979 146 657	9.976 612	21.49396	46.30733
944	841 232 384	9.809 736	21.13444	45.53276	994	982 107 784	9.979 960	21.50117	46.32287
945	843 908 625	9.813 199	21.14190	45.54883	995	985 074 875	9.983 305	21.50838	46.33840
946	846 590 536	9.816 659	21.14935	45.56490	996	988 047 936	9.986 649	21.51558	46.35392
947	849 278 123	9.820 117	21.15680	45.58095	997	991 026 973	9.989 990	21.52278	46.36943
948	851 971 392	9.823 572	21.16424	45.59698	998	994 011 992	9.993 329	21.52997	46.38492
949	854 670 349	9.827 025	21.17168	45.61301	999	997 002 999	9.996 666	21.53716	46.40041
950	857 375 000	9.830 476	21.17912	45.62903	1000	1 000 000 000	10.000 000	21.54435	46.41589

	0	1	2	3	4	5	6	7	8	9
10	.00000	.00432	.00860	.01284	.01703	.02119	.02531	.02938	.03342	.03743
11	.04139	.04532	.04922	.05308	.05690	.06070	.06446	.06819	.07188	.07555
12	.07918	.08279	.08636	.08991	.09342	.09691	.10037	.10380	.10721	.11059
13	.11394	.11727	.12057	.12385	.12710	.13033	.13354	.13672	.13988	.14301
14	.14613	.14922	.15229	.15534	.15836	.16137	.16435	.16732	.17026	.17319
15	.17609	.17898	.18184	.18469	.18752	.19033	.19312	.19590	.19866	.20140
16	.20412	.20683	.20952	.21219	.21484	.21748	.22011	.22272	.22531	.22789
17	.23045	.23300	.23553	.23805	.24055	.24304	.24551	.24797	.25042	.25285
18	.25527	.25768	.26007	.26245	.26482	.26717	.26951	.27184	.27416	.27646
19	.27875	.28103	.28330	.28556	.28780	.29003	.29226	.29447	.29667	.29885
20	.30103	.30320	.30535	.30750	.30963	.31175	.31387	.31597	.31806	.32015
21	.32222	.32428	.32634	.32838	.33041	.33244	.33445	.33646	.33846	.34044
22	.34242	.34439	.34635	.34830	.35025	.35218	.35411	.35603	.35793	.35984
23	.36173	.36361	.36549	.36736	.36922	.37107	.37291	.37475	.37658	.37840
24	.38021	.38202	.38382	.38561	.38739	.38917	.39094	.39270	.39445	.39620
25	.39794	.39967	.40140	.40312	.40483	.40654	.40824	.40993	.41162	.41330
26	.41497	.41664	.41830	.41996	.42160	.42325	.42488	.42651	.42813	.42975
27	.43136	.43297	.43457	.43616	.43775	.43933	.44091	.44248	.44404	.44560
28	.44716	.44871	.45025	.45179	.45332	.45484	.45637	.45788	.45939	.46090
29	.46240	.46389	.46538	.46687	.46835	.46982	.47129	.47276	.47422	.47567
30	.47712	.47857	.48001	.48144	.48287	.48430	.48572	.48714	.48855	.48996
31	.49136	.49276	.49415	.49554	.49693	.49831	.49969	.50106	.50243	.50379
32	.50515	.50651	.50786	.50920	.51055	.51188	.51322	.51455	.51587	.51720
33	.51851	.51983	.52114	.52244	.52375	.52504	.52634	.52763	.52892	.53020
34	.53148	.53275	.53403	.53529	.53656	.53782	.53908	.54033	.54158	.54283
35	.54407	.54531	.54654	.54777	.54900	.55023	.55145	.55267	.55388	.55509
36	.55630	.55751	.55871	.55991	.56110	.56229	.56348	.56467	.56585	.56703
37	.56820	.56937	.57054	.57171	.57287	.57403	.57519	.57634	.57749	.57864
38	.57978	.58092	.58206	.58320	.58433	.58546	.58659	.58771	.58883	.58995
39	.59106	.59218	.59329	.59439	.59550	.59660	.59770	.59879	.59988	.60097
40	.60206	.60314	.60423	.60531	.60638	.60746	.60853	.60959	.61066	.61172
41	.61278	.61384	.61490	.61595	.61700	.61805	.61909	.62014	.62118	.62221
42	.62325	.62428	.62531	.62634	.62737	.62839	.62941	.63043	.63144	.63246
43	.63347	.63448	.63548	.63649	.63749	.63849	.63949	.64048	.64147	.64246
44	.64345	.64444	.64542	.64640	.64738	.64836	.64933	.65031	.65128	.65225
45	.65321	.65418	.65514	.65610	.65706	.65801	.65896	.65992	.66087	.66181
46	.66276	.66370	.66464	.66558	.66652	.66745	.66839	.66932	.67025	.67117
47	.67210	.67302	.67394	.67486	.67578	.67669	.67761	.67852	.67943	.68034
48	.68124	.68215	.68305	.68395	.68485	.68574	.68664	.68753	.68842	.68931
49	.69020	.69108	.69197	.69285	.69373	.69461	.69548	.69636	.69723	.69810
50	.69897	.69984	.70070	.70157	.70243	.70329	.70415	.70501	.70586	.70672
51	.70757	.70842	.70927	.71012	.71096	.71181	.71265	.71349	.71433	.71517
52	.71600	.71684	.71767	.71850	.71933	.72016	.72099	.72181	.72263	.72346
53	.72428	.72509	.72591	.72673	.72754	.72835	.72916	.72997	.73078	.73159
54	.73239	.73320	.73400	.73480	.73560	.73640	.73719	.73799	.73878	.73957
55	.74036	.74115	.74194	.74273	.74351	.74429	.74507	.74586	.74663	.74741
56	.74819	.74896	.74974	.75051	.75128	.75205	.75282	.75358	.75435	.75511
57	.75587	.75664	.75740	.75815	.75891	.75967	.76042	.76118	.76193	.76268
58	.76343	.76418	.76492	.76567	.76641	.76716	.76790	.76864	.76938	.77012
59	.77085	.77159	.77232	.77305	.77379	.77452	.77525	.77597	.77670	.77743

Table of Logarithms

	0	1	2	3	4	5	6	7	8	9
60	.77815	.77887	.77960	.78032	.78104	.78176	.78247	.78319	.78390	.78462
61	.78533	.78604	.78675	.78746	.78817	.78888	.78958	.79029	.79099	.79169
62	.79239	.79309	.79379	.79449	.79518	.79588	.79657	.79727	.79796	.79865
63	.79934	.80003	.80072	.80140	.80209	.80277	.80346	.80414	.80482	.80550
64	.80618	.80686	.80754	.80821	.80889	.80956	.81023	.81090	.81158	.81224
65	.81291	.81358	.81425	.81491	.81558	.81624	.81690	.81757	.81823	.81889
66	.81954	.82020	.82086	.82151	.82217	.82282	.82347	.82413	.82478	.82543
67	.82607	.82672	.82737	.82802	.82866	.82930	.82995	.83059	.83123	.83187
68	.83251	.83315	.83378	.83442	.83506	.83569	.83632	.83696	.83759	.83822
69	.83885	.83948	.84011	.84073	.84136	.84198	.84261	.84323	.84386	.84448
70	.84510	.84572	.84634	.84696	.84757	.84819	.84880	.84942	.85003	.85065
71	.85126	.85187	.85248	.85309	.85370	.85431	.85491	.85552	.85612	.85673
72	.85733	.85794	.85854	.85914	.85974	.86034	.86094	.86153	.86213	.86273
73	.86332	.86392	.86451	.86510	.86570	.86629	.86688	.86747	.86806	.86864
74	.86923	.86982	.87040	.87099	.87157	.87216	.87274	.87332	.87390	.87448
75	.87506	.87564	.87622	.87679	.87737	.87795	.87852	.87910	.87967	.88024
76	.88081	.88138	.88195	.88252	.88309	.88366	.88423	.88480	.88536	.88593
77	.88649	.88705	.88762	.88818	.88874	.88930	.88986	.89042	.89098	.89154
78	.89209	.89265	.89321	.89376	.89432	.89487	.89542	.89597	.89653	.89708
79	.89763	.89818	.89873	.89927	.89982	.90037	.90091	.90146	.90200	.90255
80	.90309	.90363	.90417	.90472	.90526	.90580	.90634	.90687	.90741	.90795
81	.90849	.90902	.90956	.91009	.91062	.91116	.91169	.91222	.91275	.91328
82	.91381	.91434	.91487	.91540	.91593	.91645	.91698	.91751	.91803	.91855
83	.91908	.91960	.92012	.92065	.92117	.92169	.92221	.92273	.92324	.92376
84	.92428	.92480	.92531	.92583	.92634	.92686	.92737	.92788	.92840	.92891
85	.92942	.92993	.93044	.93095	.93146	.93197	.93247	.93298	.93349	.93399
86	.93450	.93500	.93551	.93601	.93651	.93702	.93752	.93802	.93852	.93902
87	.93952	.94002	.94052	.94101	.94151	.94201	.94250	.94300	.94349	.94399
88	.94448	.94498	.94547	.94596	.94645	.94694	.94743	.94792	.94841	.94890
89	.94939	.94988	.95036	.95085	.95134	.95182	.95231	.95279	.95328	.95376
90	.95424	.95472	.95521	.95569	.95617	.95665	.95713	.95761	.95809	.95856
91	.95904	.95952	.95999	.96047	.96095	.96142	.96190	.96237	.96284	.96332
92	.96379	.96426	.96473	.96520	.96567	.96614	.96661	.96708	.96755	.96802
93	.96848	.96895	.96942	.96988	.97035	.97081	.97128	.97174	.97220	.97267
94	.97313	.97359	.97405	.97451	.97497	.97543	.97589	.97635	.97681	.97727
95	.97772	.97818	.97864	.97909	.97955	.98000	.98046	.98091	.98137	.98182
96	.98227	.98272	.98318	.98363	.98408	.98453	.98498	.98543	.98588	.98632
97	.98677	.98722	.98767	.98811	.98856	.98900	.98945	.98989	.99034	.99078
98	.99123	.99167	.99211	.99255	.99300	.99344	.99388	.99432	.99476	.99520
99	.99564	.99607	.99651	.99695	.99739	.99782	.99826	.99870	.99913	.99957

THE GREEK ALPHABET

A	α	alpha	N	ν	nu
B	β	beta	Ξ	ξ	xi
Γ	γ	gamma	O	o	omicron
Δ	δ	delta	Π	π	pi
E	ε	epsilon	P	ρ	rho
Z	ζ	zeta	Σ	σ	sigma
H	η	eta	T	τ	tau
Θ	θ	theta	Υ	υ	upsilon
I	ι	iota	Φ	φ	phi
K	κ	kappa	X	χ	chi
Λ	λ	lambda	Ψ	ψ	psi
M	μ	mu	Ω	ω	omega

Index

Index